ASHEVILLE-BUNCOMBE TECHNICAL INSTITUTE

NORTH CAROLINA
STATE BOARD OF EDUCATION
DEPT. OF COMMUNITY COLLEGES
LIBRARIES

D1684203

JUN 1 8 2025

E. Kowalski

Nuclear Electronics

With 337 Figures

Springer-Verlag New York Heidelberg Berlin 1970

Dr. Emil Kowalski

Lecturer, Institute of Applied Physics, University of Berne, Switzerland
Nucleonics Division, Landis & Gyr Ltd., Zug, Switzerland

The use of general descriptive names, trade names, trade marks etc. in this publication, even if the former are not especially identified, is not to be taken as a sign that such names, as understood by the Trade Marks and Merchandise Marks Act, may accordingly be used freely by anyone

This work is subject to copyright. All rights are reserved, whether the whole or part of the material is concerned, specifically those of translation, reprinting, re-use of illustrations, broadcasting, reproduction by photocopying machine or similar means, and storage in data banks. Under § 54 of the German Copyright Law where copies are made for other than private use, a fee is payable to the publisher, the amount of the fee to be determined by agreement with the publisher. © by Springer-Verlag Berlin · Heidelberg 1970. Library of Congress Catalog Card Number 72-98149. Printed in Germany

Title No. 1632

Preface

Electronics is the most important tool in nuclear radiation metrology. Without electronic instruments most of the problems concerned with measurement in pure or applied nuclear research, radiation protection or the use of radioactive isotopes in industrial process control would remain unsolved.

Conversely, the radiation metrology was one of the first areas, if not the first, outside communications in which electronic devices were successfully employed. The quantum nature of nuclear radiations determined the need to work with pulse-type signals and thus contributed substantially to the establishment of analog and digital pulse techniques in electronics. It was no coincidence that, as late as 1949, W. C. ELMORE and M. SANDS were able to call the first monograph on nuclear electronics quite simply "Electronics".

Despite these close interrelations between electronics and radiation measurement in nuclear physics, there is virtually no modern monograph dealing with the specialized electronic circuits and instruments used in measuring nuclear radiation. ELMORE and SANDS has long since become obsolete and, similarly, the excellent works covering special areas of nuclear electronics (e. g. A. B. GILLESPIE's "Signal, Noise and Resolution in Nuclear Counter Amplifiers", 1953; I. A. D. LEWIS and F. H. WELLS' "Millimicrosecond Pulse Techniques", 1959; or R. L. CHASE's "Nuclear Pulse Spectrometry", 1961) now lag well behind the latest advances in technology*). The reason for this state of affairs may well be that nuclear electronics, with about 300 original papers published per year, is enjoying such rapid growth that any summary is bound to be out of date within a year. There are also various works dealing with related fields, in particular digital techniques.

When in 1964 I began at the friendly suggestion of Professor K. P. MEYER to lecture on nuclear electronics, I soon felt the lack of any systematic review of the subject, and I started by writing down the script of my lectures. However, I soon found that it was desirable to supplement

---

* However, a most interesting book, published last year, "Instrumentation électronique en physique nucléaire", 1968, by J.-J. SAMUELI, J. PIGNERET and A. SARAZIN, at least plugs the gap with regard to the evaluation of the time and energy information of detector signals.

Preface

the text by a literature review. The present text arose out of a study of a thousand or so original papers; it could be called a monograph with the character of a textbook. The fact that the literature is covered up to the end of 1968 gives it the required up-to-dateness, and the attempt to create a systematic structur should ensure that the book retains its value as a reference work for some years to come.

The book is primarily addressed to the experimental physicist who, when designing an experiment, needs to be able to review the possibilities of the available instrumentation so as to instruct his technicians (electronic engineers) accordingly. It can also be of help in the training of electronic engineers and as a work of reference for the technicians whom almost every institute of physics, chemistry, biology, medicine etc. employs and who are expected to construct and maintain nuclear electronic apparatus, often without any special training.

To make the text as concise as possible, the general concepts of electronics are taken for granted and a certain basic knowledge is assumed. This should be acceptable to both the types of reader I have mentioned: the technicians have already acquired this knowledge and the physicists do not need it in order to understand the circuit principle. The text follows the order of construction of a piece of electronic apparatus: first the detectors and input circuits, then the analog portion, the analog-to-digital and time-to-digital converters and the digital analyzers, and finally the build-up of rather complex total systems are discussed.

I must here acknowledge the help and encouragement given me by Professor K. P. MEYER; I thank him most sincerely for his unflagging interest in my work. Dr. R. SIEGENTHALER performed valuable service in the early stages of the work, particularly in reviewing the literature, and I am much indebted to him. I am also grateful to Mr. J. MERTON for his kindness in helping to revise the English manuscript.

My thanks are due, too, to Springer-Verlag and especially to Dr. H. MAYER-KAUPP for accepting my work for publication despite its highly specialized field of interest. Last but not least, I have to thank my wife for her patience at the times when the manuscript received more of my attention than she did.

Cham, Switzerland, November 1969

E. K.

Contents

1.	**Introduction**	1
2.	**Radiation Detectors and Related Circuits**	4
2.1.	Ionization Chamber	4
2.1.1.	Energy Required for the Generation of One Charge Carrier Pair	5
2.1.2.	Mobility of the Charge Carriers	6
2.1.3.	The Pulse Shape	6
2.1.4.	Preamplifier Circuits	11
2.2.	Proportional Counters	14
2.2.1.	Detection Mechanism and Pulse Shape in the Proportional Counter	14
2.2.2.	Statistics of the Multiplication Process	16
2.2.3.	Preamplifier Circuits	17
2.3.	Geiger-Müller-Counters	20
2.3.1.	Detection Mechanism and Pulse Shape in the GM Counter	20
2.3.2.	Quenching Circuits	24
2.4.	Semiconductor Detectors	27
2.4.1.	Characteristic Properties of Semiconductor Detectors	27
2.4.2.	Energy Required to Form a Hole-Electron Pair	31
2.4.3.	The Pulse Shape in the *pn* and *pin* Detectors	34
2.4.4.	Preamplifiers and Related Circuits	38
2.5.	Scintillation and Čerenkov Counters	48
2.5.1.	Principle of a Scintillation Counter	48
2.5.2.	The Pulse Shape	51
2.5.3.	Photomultiplier Statistics and the Pulse Height	55
2.5.4.	Thermal Noise	56
2.5.5.	Signal Circuits Used in Scintillation Counters	60
2.5.6.	Auxiliary Circuits	66
2.5.7.	Scintillation Counter Stabilizer Circuits	70
2.5.8.	Čerenkov Counter	74

Contents

3.	**Analog Circuits**	75
3.1.	Linear Pulse Amplifiers	75
3.1.1.	General Considerations, Linearity	76
3.1.2.	The Transient Response of an Amplifier	84
3.1.3.	Pulse Shaping	90
3.1.4.	Sum Effects	106
3.1.5.	Overload Recovery	111
3.1.6.	Practical Design Criteria	114
3.1.7.	Amplifiers with Variable Gain	115
3.2.	Arithmetic Operations on Analog Signals	117
3.2.1.	Operational Amplifiers	117
3.2.2.	Arithmetic Operations on Pulse Amplitudes	120
3.2.3.	Practical Circuits	122
3.3.	Window Amplifiers	128
3.4.	Linear Gates	131
3.5.	Pulse Stretchers	137
3.6.	Fast Pulse Amplifiers	142
4.	**Analog-to-Digital Converters**	151
4.1.	Pulse Height Discriminators	152
4.1.1.	The Principle of a Multivibrator	152
4.1.2.	Integral Discriminators	158
4.1.3.	Differential Discriminators	169
4.1.4.	Multiple Arrays of Differential Discriminators	177
4.1.5.	Conservation of the Time Information in a Discriminator	179
4.1.6.	Fast Tunnel Diode Discriminators	183
4.2.	Digital Encoding of the Pulse Height	191
4.2.1.	Converters of the Wilkinson Type	192
4.2.2.	Other Converter Systems	201
4.3.	Pulse Shape Discriminators	205
5.	**Evaluation of the Time Information**	213
5.1.	General Considerations, Resolution	213
5.2.	Pulse Shapers for Coincidence Circuits and Time-to-Digital Converters	217
5.3.	Coincidence Circuits	227
5.3.1.	Ideal Coincidence Stage	227
5.3.2.	Practical Circuits	231

5.3.3.	The Chronotron Principle	240
5.4.	Digital Encoding of the Time Interval	243
5.4.1.	Direct Digital Encoding	244
5.4.2.	Principle of a Time-to-Pulse-Height Converter	246
5.4.3.	Start-Stop Converter	250
5.4.4.	Overlap Converter	256
5.4.5.	The Vernier Principle	261
5.5.	Auxiliary Circuits	264
6.	**Digital Circuits**	**268**
6.1.	Basic Digital Circuits	269
6.1.1.	Fundamentals of Boolean Algebra, Gates	269
6.1.2.	Circuitry of Different Logics	278
6.1.3.	The Flip-Flop	286
6.1.4.	Practical Flip-Flop Circuits	290
6.1.5.	Tunnel Diode Circuits	292
6.2.	Scalers and Registers	296
6.2.1.	Shift Registers	296
6.2.2.	Pulse Scalers	299
6.3.	Logical and Arithmetical Digital Circuits	308
6.4.	Memories	316
6.5.	Data Output	320
6.6.	Count Rate Meters	324
7.	**Data Processing**	**327**
7.1.	Simple Counting Systems	327
7.2.	Multiscaler Arrays	329
7.3.	Multichannel Analyzers	332
7.4.	Multiparameter Analyzers	336
7.5.	On-Line Computers	342
8.	**Appendix**	**347**
8.1.	Laplace Transform Calculus	347
8.1.1.	Networks	347
8.1.2.	Naive Operational Calculus	350
8.1.3.	Laplace Transformation	354
8.1.3.1.	Rules of the Laplace Transformation	355

	8.1.3.2.	Application of the Laplace Transformation in the Network Analysis.	357
	8.1.3.3.	Inverse Transformation of Rational Functions $F(p)$	359
	8.1.3.4.	Stability Considerations	361
	8.1.3.5.	Approximations	361
8.2.		Noise	364
	8.2.1.	General Considerations, Concept of Equivalent Noise Charge	364
	8.2.2.	Noise Sources	365
	8.2.3.	The Noise of an Amplifier with the Transfer Function $G(p)$	366
	8.2.4.	Noise in a Charge Sensitive Amplifier	368
	8.2.5.	Properties of Input Stages with Vacuum Tubes, Bipolar Transistors and FET	370
	8.2.6.	Noise and Resolution	373
9.		**References**	377
10.		**Subject Index**	399

1. Introduction

Electronics can be many things to many people [1.001]. In the most general sense, "electronics" is a method for solving a great variety of problems, usually non-electrical, by processing electrical signals chosen so as to represent the parameters under investigation.

Hence, an electronic apparatus consists essentially of three main parts: the input sensor, the signal processing part and the output equipment (Fig. 1.01). Different problems require different input sensors — microphones for acoustical, thermo-elements for thermal, position switches for mechanical or radiation detectors for nuclear applications. The output devices may indicate the results of the signal processing optically or acoustically, and can also initiate a control action (servo-mechanism), if desired.

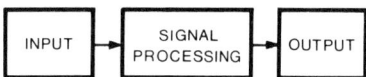

Fig. 1.01. Principle of a general electronic system

Since only measurable parameters can be dealt with, the processing of the electrical signals which represent these parameters is composed solely of arithmetical or logical operations. Therefore the same general "processing", but with different input sensors can be used to solve different problems, at least, theoretically. Although the actual equipment differs, for example, for communication, industrial and nuclear applications, similar circuits and design criteria will be found *in all cases. The object of this book is to lay emphasis on the part of signal processing which is characteristic of nuclear physics*, rather than to give a brief description of the well-known general principles of modern electronics, which have been expounded in a vast number of excellent textbooks. For the sake of clarity some important circuits common to other electronic techniques are briefly described. As a rule, however, some knowledge of basic electronics is assumed throughout the text.

Introduction

The only information available to physicists about nuclear reactions is obtained from the radiation emitted, its energy and momentum, the time and direction of its emission, etc. Nuclear metrology and radiation metrology are identical in this respect. The application of electrical methods to radiation measurements based on the ionizing properties of most nuclear radiations has been known for many years (viz. ionization chamber with electrometers). However, only pulse techniques can deal adequately with the quantum character of nuclear processes.

A current or voltage pulse is produced in the radiation detector, either directly by ionization of the active medium of the counter (e.g. proportional counter or semiconductor detector) or via complex intermediary processes, such as the scintillation counter. In both cases the detector pulses carry multiple information. To begin with, the pulse is correlated with the time instant of the nuclear process. The pulse manifests an emission of radiation inside the solid angle, under which the detector is seen from the source. The pulse height can often be used as a measure of the energy loss of the incident radiation in the detector. The pulse shapes vary for different types of radiation, or for different impact loci and angles at the detector.

The pulse carries most of its information in a continuous manner: its amplitude can in principle have any value and can appear at any time. The representation of data by a continuously variable parameter is known as the *analog* representation. The processing of analog signals is very difficult, since every small disturbance (e.g. of the voltage scale) affects the represented value adversely. Fortunately a very common problem is to register events whose parameters lie within a given range, irrespective of their actual value. In this case for every pulse a "yes-no" decision must be made, and consequently only the elementary "all or nothing" binary information has to be treated. This data representation is also known as *digital* representation.

The circuit or assembly which determines whether or not an event falls within the preselected range is known as the analog to digital converter. Specification of more than one selection range enables more than just the elementary "yes-no" information to be conserved. The digital representation in the latter case uses a code, which ordinates a given configuration of elementary signals to every parameter value (for more detailed information on this subject, see Chapter 6).

As can be seen, the processing of binary digital signals is very simple — quite a large disturbance is necessary to convert a "yes" signal into a "no" signal, and vice versa. Therefore, in designing electronic apparatus the general aim is to leave quickly the dangerous analog representation, convert the signal into a digital form (of course, conserving as much as possible of the initial information) and finally to process the digitized data. A common electronic apparatus for nuclear research consists of a detector, an analog part, an analog to digital converter ADC and a digital part for signal processing (Fig. 1.02).

Fig. 1.02. General electronic array for nuclear physics measurements

The framework of this booklet conforms to this simple scheme. Chapter 2 deals with various radiation detectors, preamplifiers and auxiliary circuits. In Chapter 3 the analog operation on the detector signals is discussed, especially the complex problem of pulse forming and amplification. Chapters 4 and 5 are devoted to the description of different types of analog to digital converters for the conversion of the pulse amplitude, the pulse shape, or the time interval between two pulses. The detectors and the analog part of the signal processing up to the analog to digital conversion are characteristic of nuclear electronics. The digital circuits, the fundamentals of which are treated in Chapter 6, can be found in data processing assemblies and computers, in industrial process measurement and control, and in nuclear measuring equipment. Finally, Chapter 7 reviews the internal organization of simple to complex scaler assemblies and analyzers for nuclear research.

In awareness of the importance of pulse techniques, this booklet is devoted only to the description of circuits for pulse signals. The discussion of electrometer amplifiers for integrating current ionization chambers is omitted. In addition, no review is made of circuits characteristic of reactor instrumentation, since this would need a detailed description of reactor control systems, which is well beyond the scope of this booklet.

2. Radiation Detectors and Related Circuits

Through the interaction of radioactive radiations with matter, atoms or molecules become ionized or at least excited. The operating principle of almost all radiation detectors is based on the detection of free electrons and ions (ionization chamber, proportional and GM tubes, spark chamber, semiconductor detector) or on the detection of light photons emitted by the excited atoms and molecules (scintillation counter). Čerenkov counters, based on the detection of Čerenkov radiation of particles faster than light in a radiator with an index of refraction $n>1$, represent an important exception.

2.1. Ionization Chamber

The principle of the ionization chamber is shown in Fig. 2.01. A charged particle loses an amount ΔE of its energy in the gas between the chamber electrodes, thus producing N pairs of charge carriers (electron + positive ion) with the total charge $Q = \pm N \cdot e$. Due to the electrostatic field between the electrodes the charge carriers move, inducing a current

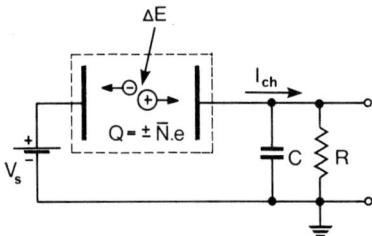

Fig. 2.01. Principle of the ionization chamber

I_{ch} in the chamber circuit, symbolized by a RC combination. The only condition concerning the chamber field strength in proportion to the voltage V_s is that it should be high enough to avoid recombination losses of charge carriers, and low enough to suppress charge multiplication effects, such as field emission or impact ionization. A detailed discussion of the physical properties of ionization chambers can be found in the review by FRANZEN and COCHRAN [2.001].

2.1.1. Energy Required for the Generation of One Charge Carrier Pair

The energy transfer to the orbital electrons of atoms is accomplished by a great number of heterogenous processes. Despite the complexity of the ionization process, the average energy amount W required for the generation of one charge-carrier pair remains constant and independent of the type and energy of the ionizing particle, or of the operating conditions of the detector [2.002]. Hence

$$\bar{N} = \frac{\Delta E}{W} \quad \text{and} \quad \bar{Q}_\pm = \pm \frac{\Delta E}{W} \cdot e. \tag{2.1}$$

The value of W is about 35 eV for air; more precise values for different gases can be found in [2.001] or in [2.003].

In the discussion of the statistical deviation of the charge carrier number N from the average value $\bar{N}$ given by (2.1), two different cases must be distinguished:

a) If ΔE is only a part of the total particle energy, i.e. if the linear dimensions of the so-called active volume of the chamber are smaller than the range of the particle of interest, the statistic of N obeys a theory of LANDAU [2.004], which has been reviewed in detail by BIRKHOFF [2.005]. The distribution of N is substantially wider than a corresponding Poisson distribution, due to the additional variation of the energy loss ΔE itself. The distribution is unsymmetrical, an excess of high values of N making the most probable number of charge carriers lower than the average value $\bar{N}$. Experimental work on this type of statistics is reviewed in [2.001].

The Landau variation of the energy loss in the entrance window of the detector or in an absorber may also influence the detector pulse height distribution [2.006].

b) If the whole particle energy is transferred to the chamber gas, the statistic deviates anew from a Poissonian, although the mean square deviation σ_N is smaller in this case. The problem has been dealt with by FANO [2.007]. We get

$$\sigma_N = \sqrt{f \cdot \bar{N}} \tag{2.2}$$

with the so-called Fano factor $f \leq 1$. Normally f amounts to some tenths. An undisturbed Poisson distribution (i.e. $f \approx 1$) can be assumed for the purpose of rough calculation.

These statistical effects are, of course, by no means characteristic only of the ionization chamber. The statistical variations of the charge carrier production or the excitation, e. g. in a proportional counter, semiconductor detector or in the scintillator of a scintillation counter, fit the same theory.

2.1.2. Mobility of the Charge Carriers

Besides the chaotic thermal movement, there is an additional movement of the charge carriers in the direction of the field strength, the velocity

$$w_{ion} = \mu_{ion} \frac{F}{p} \quad \text{and} \quad w_{el} = -\mu_{el} \frac{F}{p} \tag{2.3}$$

of which is proportional to the field strength F and to the reciprocal gas pressure p. The constants μ_{ion} and μ_{el} are known as the mobilities. They may themselves be functions of F/p.

The ion mobility μ_{ion} depends slightly on the actual chamber gas, the temperature and the ion mass. If the production of negative ions should occur, they have the same mobility as positive ones of the same mass. For rough calculations the value $\mu_{ion} \approx 1 \cdot 10^{-3} [\text{cm}^{-3} \cdot \text{mm Hg/V} \cdot \mu\text{s}]$ can be used. More precise values for a number of gases are given by WILKINSON [2.008].

The electron mobility μ_{el} is about 10^3 times higher than μ_{ion}. The situation is, however, complicated by the fact that w_{el} often lies within the range of thermic velocities, thus raising the kinetic energy of the electrons and affecting μ_{el}. The theoretical relationships and known experimental data on μ_{el} are discussed in [2.001].

2.1.3. The Pulse Shape

For the sake of simplicity we shall suppose that a point-like cloud of N positive ions and N electrons of the total charge $\pm Q = \pm N \cdot e$ is originated at the point x_0 of the chamber. The way of scaling x for three chamber geometries is shown in Fig. 2.02. The chamber current I_{ch} is composed of an ion and an electron component, so that

$$I_{ch} = I_{ion} + I_{el}. \tag{2.4}$$

As can easily be shown [2.001], the two components are

$$I_{ion} = Q \frac{F(x)}{V_s} w_{ion}(x) \quad \text{and} \quad I_{el} = -Q \frac{F(x)}{V_s} w_{el}(x). \tag{2.5}$$

Substituting (2.3) into (2.5) we get

$$I_{ion} = Q \frac{\mu_{ion} F^2(x)}{p \cdot V_s} \quad \text{and} \quad I_{el} = Q \frac{\mu_{el} F^2(x)}{p \cdot V_s} \tag{2.6}$$

where V_s is the voltage across the chamber. Because of the movement of the charge carriers through regions of different field strength $F(x)$, in general the chamber current is a function of the time t. The calculated pulse shapes for a parallel-plate chamber, a cylindrical chamber and a spherical chamber are shown in Fig. 2.02.

The calculation is briefly indicated below: The time t, at which the carrier with the velocity $w(x)$ reaches the point x, is

$$t(x) = \int_{x_0}^{x} \frac{dx}{w(x)}. \tag{2.7}$$

With $w(x)$ from (2.3) and $F(x)$ from Fig. 2.02, the relation (2.7) is used to determine the collection times T_{ion} and T_{el}, in which ions and electrons reach the positive (a) and the negative (b) electrode respectively. The inversion of (2.7) results in $x = x(t)$, which inserted in (2.6) immediately gives the current shapes.

Due to the substantially higher electron mobility $\mu_{el} \approx 10^3 \cdot \mu_{ion}$, the collection time T_{el} is roughly 10^3 times shorter than T_{ion}. The most common value for T_{el} is 1 µs.

Integration of the chamber current I_{ch} by a RC network yields voltage pulse shapes shown in Fig. 2.02 for the extreme case of $RC \to \infty$ (i.e. $RC \gg T_{ion}$). Despite the chamber geometry, the pulse always rises rapidly for $0 < t < T_{el}$, due to the fast electron movement. The final pulse height $V_{max} = \frac{Q}{C}$ is reached only after the very long time interval $t = T_{ion}$. In practical circuits the integrating time constant is chosen according to $T_{el} \ll RC \ll T_{ion}$ and hence only I_{el} is integrated, yielding a fast pulse with the height of about $V_{el}(T_{el})$. If this pulse height is used as a measure for ΔE, it must be considered that the relationship between V_{el} and Q (which is proportional to ΔE) depends on the origin x_0 of the primary ionization.

The pulse height spectrum $\eta(V)$ of the fast electron pulses V_{el} for monoenergetic events can easily be evaluated in the case of point-like ionization traces distributed uniformly throughout the whole chamber volume. This situation arises, for instance, if the chamber is filled by a gaseous α-emitting isotope. The calculated [2.001] spectra $\eta(V)$ are also summarized in Fig. 2.02.

In the parallel-plate chamber all pulses with amplitudes between zero and $V_{max} = Q/C = e \cdot \Delta E/C \cdot W$ have the same probability.

In the cylindrical chamber the volume differential is higher in the proximity of the tubular cathode and the field is concentrated in the neighbourhood of the anode wire. Therefore in the majority of events, the electron cloud must pass almost the whole potential difference V_s, thus favouring pulses having the maximum amplitude V_{max}. In the

Chamber geometry	Parallel plate chamber	Cylindrical chamber	Spherical chamber
	![parallel plate diagram]	![cylindrical diagram]	![spherical diagram]
Field strength $F(x)$	$F(x) = \dfrac{V_s}{b-a}$	$F(x) = \dfrac{V_s}{\log(b/a)} \cdot \dfrac{1}{x}$	$F(x) = V_s \dfrac{ab}{b-a} \cdot \dfrac{1}{x^2}$
Ion collection time T_{ion}	$T_{\mathrm{ion}} = \dfrac{p(b-a)}{V_s \cdot \mu_{\mathrm{ion}}}(b-x_0)$	$T_{\mathrm{ion}} = \dfrac{p \cdot \log(b/a)}{2 \cdot V_s \cdot \mu_{\mathrm{ion}}}(b^2 - x_0^2)$	$T_{\mathrm{ion}} = \dfrac{p(b-a)}{3ab \cdot V_s \cdot \mu_{\mathrm{ion}}}(b^3 - x_0^3)$
Electron collection time T_{el}	$T_{\mathrm{el}} = \dfrac{p(b-a)}{V_s \cdot \mu_{\mathrm{el}}}(x_0 - a)$	$T_{\mathrm{el}} = \dfrac{p \cdot \log(b/a)}{2 \cdot V_s \cdot \mu_{\mathrm{el}}}(x_0^2 - a^2)$	$T_{\mathrm{el}} = \dfrac{p(b-a)}{3ab \cdot V_s \cdot \mu_{\mathrm{el}}}(x_0^3 - a^3)$
Ion current $I_{\mathrm{ion}}(t)$ $0 < t < T_{\mathrm{ion}}$	$I_{\mathrm{ion}}(t) = \dfrac{Q \cdot V_s \cdot \mu_{\mathrm{ion}}}{p(b-a)^2}$	$I_{\mathrm{ion}}(t) = \dfrac{QV_s\mu_{\mathrm{ion}}}{p[\log(b/a)]^2}\left[x_0^2 + (b^2 - x_0^2)\dfrac{t}{T_{\mathrm{ion}}}\right]^{-1}$	$I_{\mathrm{ion}}(t) = \dfrac{QV_s\mu_{\mathrm{ion}}a^2 b^2}{p(b-a)^2}\left[x_0^3 + (b^3 - x_0^3)\dfrac{t}{T_{\mathrm{ion}}}\right]^{-\frac{4}{3}}$
Electron current $I_{\mathrm{el}}(t)$ $0 < t < T_{\mathrm{el}}$	$I_{\mathrm{el}}(t) = \dfrac{Q \cdot V_s \cdot \mu_{\mathrm{el}}}{p(b-a)^2}$	$I_{\mathrm{el}}(t) = \dfrac{QV_s\mu_{\mathrm{el}}}{p[\log(b/a)]^2}\left[x_0^2 - (x_0^2 - a^2)\dfrac{t}{T_{\mathrm{el}}}\right]^{-1}$	$I_{\mathrm{el}}(t) = \dfrac{QV_s\mu_{\mathrm{el}}a^2 b^2}{p(b-a)^2}\left[x_0^3 - (x_0^3 - a^3)\dfrac{t}{T_{\mathrm{el}}}\right]^{-\frac{4}{3}}$
Integrated voltage $V_{\mathrm{ion}}(t)$ $0 < t < T_{\mathrm{ion}}$	$V_{\mathrm{ion}}(t) = \dfrac{Q}{C}\dfrac{b - x_0}{b - a}\dfrac{t}{T_{\mathrm{ion}}}$	$V_{\mathrm{ion}}(t) = \dfrac{Q}{C} \cdot \dfrac{1}{2\log(b/a)}\left[\log\left\{x_0^2 + (b^2 - x_0^2)\dfrac{t}{T_{\mathrm{ion}}}\right\} - \log x_0^2\right]$	$V_{\mathrm{ion}}(t) = \dfrac{Q}{C}\dfrac{ab}{b-a}\left[\dfrac{1}{x_0} - \left\{x_0^3 + (b^3 - x_0^3)\dfrac{t}{T_{\mathrm{ion}}}\right\}^{-\frac{1}{3}}\right]$
Integrated voltage $V_{\mathrm{el}}(t)$ $0 < t < T_{\mathrm{el}}$	$V_{\mathrm{el}}(t) = \dfrac{Q}{C}\dfrac{x_0 - a}{b - a}\dfrac{t}{T_{\mathrm{el}}}$	$V_{\mathrm{el}}(t) = \dfrac{Q}{C}\dfrac{1}{2\log(b/a)}\left[\log x_0^2 - \log\left\{x_0^2 - (x_0^2 - a^2)\dfrac{t}{T_{\mathrm{el}}}\right\}\right]$	$V_{\mathrm{el}}(t) = \dfrac{Q}{C}\dfrac{ab}{b-a}\left[\left\{x_0^3 - (x_0^3 - a^3)\dfrac{t}{T_{\mathrm{el}}}\right\}^{-\frac{1}{3}} - \dfrac{1}{x_0}\right]$
Final voltage $V_{\mathrm{ion}}(T_{\mathrm{ion}})$	$\dfrac{Q}{C}\dfrac{b - x_0}{b - a}$	$\dfrac{Q}{C}\dfrac{\log(b/x_0)}{\log(b/a)}$	$\dfrac{Q}{C}\dfrac{a(b-x_0)}{x_0(b-a)}$

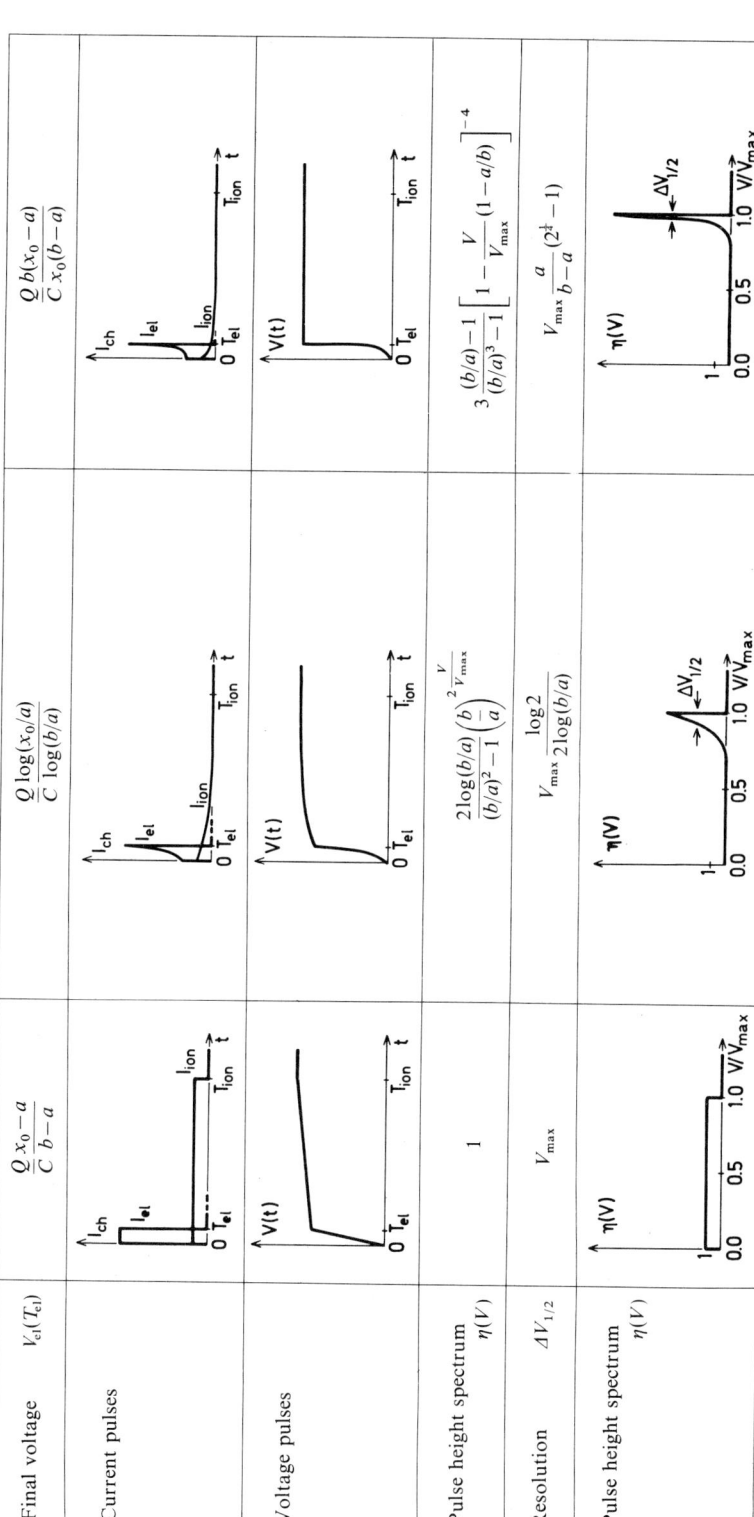

Fig. 2.02. The pulse shape in the ionization chamber

spherical chamber the situation is even more extreme. Despite the incomplete collection of the charge carriers during the "fast" operation ($RC \ll T_{ion}$), the ionization chambers with higher geometry enable an operation without the loss of energy information. The resolution of the chamber $\Delta V_{\frac{1}{2}}$(FWHM) amounts to a few per cent.

In cylindrical and spherical ionization chambers the rise time of the integrated voltage pulse $V_{el}(t)$ can be very short, but the pulse front is delayed in relation to the primary event by a time depending on the origin of ionization. The time information is thus conserved but incompletely. This effect must be taken into account in coincidence applications of these detectors.

According to FRISCH [2.010], the conservation of the energy information in a parallel-plate chamber can be achieved by separating the part of the chamber used for ionization from the part used for signal generation by means of a grid (Fig. 2.03). Independent of its origin, the electron cloud generated in an ionization event always passes the same potential difference V_{s1}. Consequently, the integral of the current $I_{ch} = I_{el}$ flowing in the grid-plate circuit is a direct measure of the total charge Q, which is, itself proportional to ΔE. A close grid shields the signal electrode very efficiently from the influence of the positive ion movement, although it captures many electrons. On the other hand, although a wide grid allows almost all electrons to pass, its shielding efficiency is too low. A detailed analysis of these effects is given by BUNEMAN, CRANSHAW and HARWEY [2.009]. Nearly all recent papers on ionization chambers ([2.011] to [2.014], [2.229]), concern the gridded chamber.

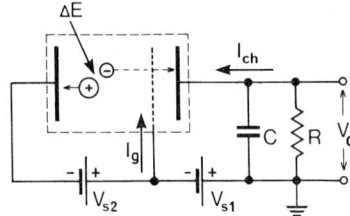

Fig. 2.03. The gridded ionization chamber

The grid current I_g (Fig. 2.03) can also be used as a signal when integrated by a RC network to voltage pulses. As BOCHAROV, VOROB'EV and KOMAR [2.015, 2.016] have shown, the grid signal carries information about the angle distribution of the ionization traces of particles emitted from the cathode.

The pulse shapes at the collecting electrode of a grid ionization chamber are shown in Fig. 2.04. The current pulse is delayed relative to the instant of ionization ($t = 0$) by the propagation time of the electron cloud from the origin up to the grid. The current pulse length T_r corresponds to the collection time of the electrons between the grid and the collection electrode.

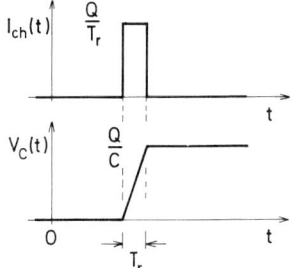

Fig. 2.04. The pulse shape in the gridded ionization chamber

2.1.4. Preamplifier Circuits

As has been shown in Chapter 2.1.1, an ionizing particle produces about 30000 electrons or about $5 \cdot 10^{-15}$ As charge per 1 MeV energy loss. With the integration capacity C in Fig. 2.01 or 2.04 being typically 20 pF, this amounts to a pulse height of only 0.25 mV/MeV. Hence the resulting pulse must be amplified. The noise of the amplifier system limits the possible energy resolution or the sensitivity of the assembly.

Normally the amplification is divided between a low-noise preamplifier and the main amplifier. The essential input circuit — the low-noise preamplifier — is disdussed in more detail in connection with semiconductor detectors (Chapter 2.4.4), so that we can restrict this chapter to the description of two examples.

Fig. 2.05 shows two basic preamplifier circuits: the voltage sensitive preamplifier and the charge sensitive preamplifier. The difference amplifier with the gain A has an inverting $(-)$ and a non-inverting $(+)$ input.

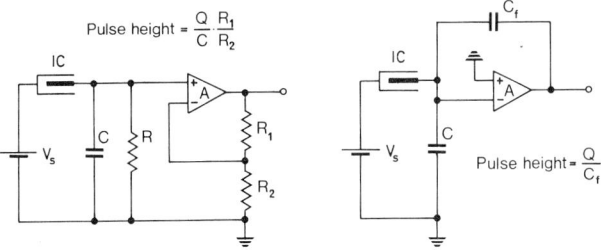

Fig. 2.05. Operating principle of the voltage amplifier (left side) and of the charge-sensitive amplifier (right side)

Both examples of practical circuits, which will be discussed briefly, belong to the voltage-sensitive type. Here the voltage pulse across the capacitor C is amplified, C representing the total capacity at the output of the ionization chamber (the sum of the chamber capacity, the capacity of cable between chamber and preamplifier input, and the input capacity of the preamplifier). The gain is stabilized by a feedback loop with the voltage divider R_1, R_2, to a value of approximately R_1/R_2, when $R_2 \ll R_1$ and $A \cdot R_2 \gg R_1$. The pulse shape is given on the one hand by the chamber geometry and on the other hand by the time constant RC of the input network. The input resistance of the preamplifier is included in R. Normally $RC \gg T_{el}$ is chosen (or $RC \gg T_r$ for a grid chamber) and the necessary differentiation takes place in the main amplifier first, since this gives a better signal-to-noise ratio, as will be shown later. In amplifier systems with double differentiation the input network may be used as the first differentiator, making the time

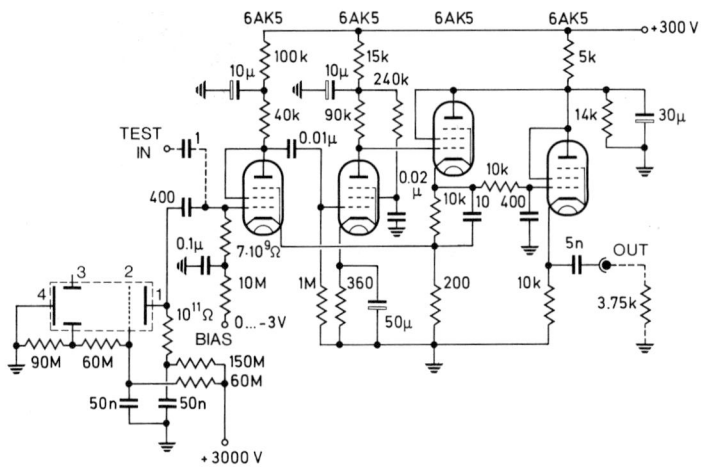

Fig. 2.06. A low noise preamplifier according to ENGELKEMEIR and MAGNUSSON [2.019]. Gain = 50. All electrolytic condensers paralleled with 10 nF ceramic condensers. Input capacity 23 pF. (1) = collecting electrode, (2) = grid, (3) = field shaping electrode, (4) cathode

constant RC of the order of T_{el}. The pulse height loss for different pulse shapes and various values of RC and other differentiator or integrator time constants is discussed by among others GILLESPIE [2.017], BALDINGER and FRANZEN [2.018], and especially for the case of a grid ionization chamber by TSUKUDA [2.012].

Preamplifier Circuits

In Fig. 2.06 the circuit diagram of a preamplifier for a grid ionization chamber according to ENGELKEMEIR and MAGNUSSON [2.019] is shown, with details of the chamber electrode connections. The triode-connected pentode 6AK5 has a small grid current ($6 \cdot 10^{-11}$ A) together with a relatively high transconductance (2.9 mA/V) which is important for low noise (Chapter 8.2). The gain of A = 50 is determined by R_1/R_2. The pulse shaping ($\tau_{int}=4\,\mu s$, $\tau_{diff}=19\,\mu s$) is done between the preamplifier and the main amplifier. The equivalent noise charge $Q_N = 330$ electrons corresponds to a mean square deviation $\sigma = 10$ keV, i.e. to a FWHM = 24 keV.

Fig. 2.07 shows a circuit by COTTINI, GATTI et al. [2.020]. The connection of the input stage as a cascode became common for these types of preamplifiers, because of inherent advantages such as high loop gain, low Miller capacity etc. The input tube operates with floating grid. Although the grid current in this case is about a factor of 2 higher than the minimum with more negative bias, this can be accepted, since it reduces the optimum time constant $\tau_0(8.220)$ for minimum noise. The amplifier has a gain of A = 100; the equivalent noise charge with 21 pF input capacity and $\tau_{diff}=\tau_{int}=1.5\,\mu s$ is $Q_N = 280$ electrons, so that $\sigma = 8.5$ keV and FWHM = 20 keV.

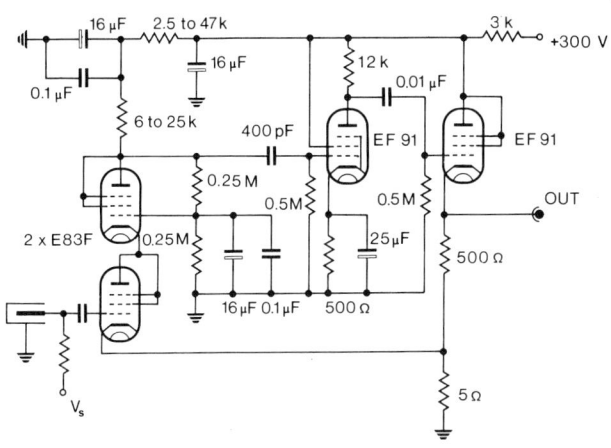

Fig. 2.07. Preamplifier using cascode circuit according to COTTINI et al. [2.020]

Due to the high loop gain, the voltage gain of these preamplifiers remains constant and independent of the age etc. of the components. However, since the energy information is carried primarily by the total pulse charge Q, whereas the voltage pulse height is given by Q/C, the stability of the system gain depends finally on the stability of the capacity C. Although in ionization chambers the "mechanical" capacity apart from possible shocks or vibrations remains constant, the grid cathode capacity, for example, is subject to changes with the dimensions of the space charge and thus with the filament voltage and the age of the tube [2.020]. This can be avoided with the aid of the circuit on the right of Fig. 2.05. By means of the capacitive feedback the preamplifier becomes charge sensitive. As will be seen in Chapter 2.4.4, the output pulse height of a charge-sensitive preamplifier with C_f as the feedback capacity, is Q/C_f, independent of C for a high loop gain $A \to \infty$. Charge-sensitive preamplifiers are especially necessary for semiconductor detectors, so this subject will be discussed in detail in Chapter 2.4.4 on semiconductor detector circuits. Of course, with appropriate modifications all these circuits can be used for ionization chambers. The performance of FET preamplifiers especially for gridded ionization chambers is discussed by MOON [2.230].

2.2. Proportional Counters

2.2.1. Detection Mechanism and Pulse Shape in the Proportional Counter

If in a part of the chamber volume the field strength rises to an extent such that electrons during their free passage between two collisions with the gas molecules can accumulate enough energy to ionize the impacted molecule, the number of charge carriers increases with every collision by a factor of 2. The chamber current is higher by the so-called multiplication factor $\overline{M} = 2^n$, n being the average number of collision of a primarily generated electron on its path to the anode.

As long as there is no interaction between the avalanches arising from different primary ionization events, M remains constant and the total charge Q is

$$Q = \overline{M} \frac{\Delta E}{W} e. \tag{2.8}$$

Q is proportional to the energy loss ΔE (hence *proportional* counter), although the proportionality constant is much higher than that in equation (2.1), valid for ionization chambers. According to HANNA, KIRKWOOD and PONTECORVO [2.021] the proportionality is conserved if Q does not exceed a critical charge Q_c, which is of the order of 10^6 to 10^7 electrons. For low-energy radiations liberating only a few primary electrons, M can be as high as 10^6 to 10^7. The physical and technological aspects of proportional counters have been reviewed by a number of authors. Besides the previously mentioned publications [2.001] and [2.003], only the recent review by CURRAN and WILSON [2.022] giving numerous bibliographic references is indicated.

The proportional counter usually has the well-known coaxial wire tube form, or any other geometrical form giving a non-homogeneous field (thin anode wire in a large cathode case). The multiplication takes place in the area of high field strength near the anode wire. Therefore, after the termination of the very fast multiplication process, the whole charge $\pm Q$ is situated in the direct proximity of the anode. The electrons must pass only a very small potential difference during their movement to the anode, so that the electron component of the signal current can be neglected[1]. The whole signal current is caused by the ion movement. The pulse shape can be seen from Fig. 2.02 ($I_{ion}(t)$ in cylindrical chamber) if $x_0 = a$ is introduced. The voltage pulse shape $V(t)$ at the integrating capacitor C is given in equation (2.9), assuming $b \gg a$ (cf. $V_{ion}(t)$ in Fig. 2.02)

$$V(t) = \frac{Q}{2 \cdot C \cdot \log(b/a)} \cdot \log\left(1 + \frac{b^2}{a^2} \cdot \frac{t}{T_{ion}}\right). \tag{2.9}$$

The voltage $V(t)$ rises at first very quickly and already reaches half of its maximum Q/C when $t_{\frac{1}{2}} = T_{ion} \cdot (a/b)$. Thereafter the voltage rises much more slowly. Since $a:b$ commonly approaches $1:10^3$, we get

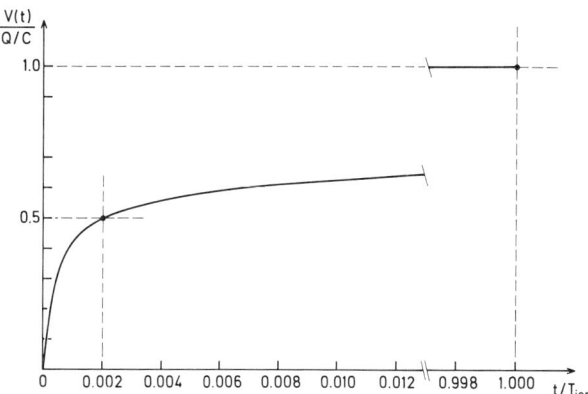

Fig. 2.08. The voltage pulse shape in a proportional counter ($a/b = 1/500$)

---

[1] STUCKENBERG (Thesis, Hamburg 1958, as quoted in [2.003]) however, in his experimental data on the electron component of the proportional counter signal, shows it as reaching about 10% of the total integrated current.

$t_\frac{1}{2} = 0.1 \ldots 1 \, \mu s$, even with the ion collection time T_{ion} in the millisecond range. The signal (2.9) can therefore be processed in medium-fast amplifiers with differentiating time constants of about $1 \, \mu s$, losing only one half of the pulse height. In Fig. 2.08 the voltage (2.9) is plotted as a function of time for $a : b = 1 : 500$.

If more than one primary electron is generated by an ionizing particle, several avalanches will be released, and the resulting signal consists of an overlap of several pulses $V(t)$ (2.9) with slightly different starting times. This does not adversely affect the overall rise time of the pulse.

However, in coincidence applications it must be considered that a primary electron of an avalanche can also be generated in a region of low field strength. In this case the signal is delayed by the time needed to move the electron towards a higher field where it can begin to release secondary charge carriers; this time is of the order of $0.1 \, \mu s$ (cf. RAMSEY [2.044]).

2.2.2. Statistics of the Multiplication Process

As has been shown in Chapter 2.1.1, the mean square deviation σ_N of the number N of primary charge carriers is $\sigma_N = \sqrt{f \cdot \overline{N}}$ (2.2). In the proportional counter the statistics of the multiplication process must also be considered. According to a simple theory of SNYDER [2.023], the distribution $P_1(M)$ of the total number M of avalanche electrons initiated by one primary electron is exponential

$$P_1(M) = \frac{1}{\overline{M}} e^{-M/\overline{M}}, \tag{2.10}$$

with $\overline{M}$ equal to the mean multiplication factor. The variance σ_M^2 of the gas multiplication M derived from equation (2.10), is

$$\sigma_M^2 = \overline{M}^2. \tag{2.11}$$

A general theory (CURRAN et al. [2.024], SAUTER [2.025], HOYT [2.026]) for the relative variance $(\sigma/\overline{N}\overline{M})^2$ of the number $N \cdot M$ of electrons reaching the anode wire gives the following relationship

$$\left(\frac{\sigma}{\overline{N}\overline{M}}\right)^2 = \left(\frac{\sigma_N}{\overline{N}}\right)^2 + \frac{1}{\overline{N}}\left(\frac{\sigma_M}{\overline{M}}\right)^2. \tag{2.12}$$

With a Fano factor $f = 1$ the relative variance of $N \cdot M$

$$\left(\frac{\sigma}{\overline{N}\overline{M}}\right)^2 = \frac{1}{\overline{N}} + \frac{1}{\overline{N}} \tag{2.13}$$

will be just doubled by the multiplication process, according to the simplified theory of SNYDER loc. cit.

This result is in contradiction to the experimental data published by CURRAN, COCKCROFT and ANGUS [2.024], indicating the amplitude distribution $P_1(M)$ of single electron pulses to be

$$P_1(M) \sim (M)^{\frac{1}{2}} \cdot \epsilon^{-M} \quad \text{with} \quad \sigma_M = 0.67 \cdot \overline{M}. \tag{2.14}$$

Contrary to (2.10), this distribution (2.14) exhibits a maximum at a most probable multiplication of $M_p \neq 0$. For the smaller variance σ_M^2, and since the Fano factor is $f < 1$, the variance σ remains substantially below the value given by (2.13). Experimental results of BISI and ZAPPA [2.027], HANNA et al. [2.021], WEST [2.028] and others can be described by the following empirical relationship

$$\frac{\sigma}{\overline{N} \cdot \overline{M}} = \frac{0.6}{\overline{N}^{0.4}} \tag{2.15}$$

as has been demonstrated by FRANZEN and COCHRAN [2.001].

The distribution (2.14) has been verified experimentally by other authors (SCHLUMBOHM [2.029]) and theoretically proved by BYRNE [2.030] and LAUSIART and MORUCCI [2.031]. PRESCOTT, [2.032] to [2.034], has examined the statistics of gas multiplication and the related problem of the photomultiplier statistics, pointing out that (2.10) as well as (2.14) can be understood as special cases of the more general POLYA distribution. The most recent review of the experimental material on single electron pulse spectra in proportional counters is given by CARVER and MITCHELL [2.231]. CHARLES and COOKE [2.232] recently surveyed the present state of the theory of the proportional counter resolution.

2.2.3. Preamplifier Circuits

Since the total charge of the proportional counter pulse must remain below the critical charge Q_c of 10^{-13} to 10^{-12} As (cf. Chapter 2.2.1), the maximum pulse height across a capacity of the order of 10 pF becomes 10 to 100 mV. Such small pulses must certainly be amplified; on the other hand, this pulse height is well over the noise level and normally no special precautions for low-noise operation need be taken. As with ionization chambers, the input capacity C (composed of the detector, cable and preamplifier input capacity) is stable enough to allow the use of voltage-sensitive preamplifiers. If even more stable operation, especially with low gas multiplication ($\overline{M} \sim 100$ and less), is desired, a charge-sensitive low-noise preamplifier (Chapter 2.4.4) may offer advantages.

Usually the preamplifier in proportional counter assemblies consists of a simple emitter-follower and functions only as an impedance converter for matching the high-resistance counter output to the low resistance (50...100 Ω) of the connection cable to the main amplifier. For low capacity C, the preamplifier is generally mounted directly on the counter housing.

Excessive thermal dissipation in the preamplifier must be avoided, since an axial temperature gradient adversely affects the energy resolution (MCCUTCHEN [2.035]). On the other hand, a radial temperature gradient increases the gas amplification and thus improves the resolution. NAGATANI and SAKAKI [2.036] report an improvement of the gas multiplication by an order of magnitude by heating the anode wire uniformly to 100 °C.

Fig. 2.09 shows a circuit diagram of a simple emitter-follower with two transistors in Darlington configuration. The signal is first differentiated with a time constant of 1 µs in the main amplifier, so that all differentiating time constants of the preamplifier must be substantially greater. For negative voltage pulses pnp transistors offer davantages. The silicon diodes D_1 and D_2 limit the pulse height and prevent the preamplifiers from being destroyed, e.g. during the switching-on of the high tension supply V_{HT} for the proportional counter. The 47 Ω potentiometer matches the output impedance to the cable.

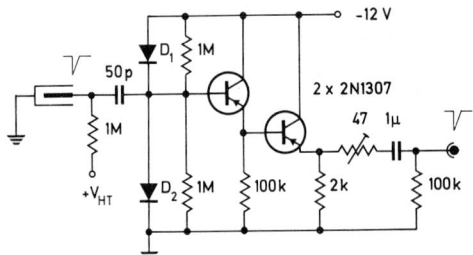

Fig. 2.09. Simple emitter follower for proportional counters

BENNETT [2.233] described a cascode preamplifier designed especially for use with proportional counters for proton recoil counting.

Strict design criteria must be applied if the preamplifier is to operate in an assembly with high-intensity ambient electromagnetic fields which induce noise in the interconnections. MAY and SEMTURS [2.037] have systematically analyzed this case. In a preamplifier—amplifier system (Fig. 2.10) the external fields induce high-frequency disturbing voltages V_1 to V_5 in the grounding loops, which must be prevented by means

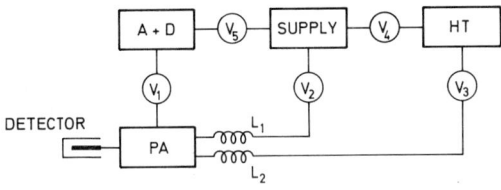

Fig. 2.10. The ground loops in a preamplifier-amplifier system, according to MAY and SEMTURS [2.037]

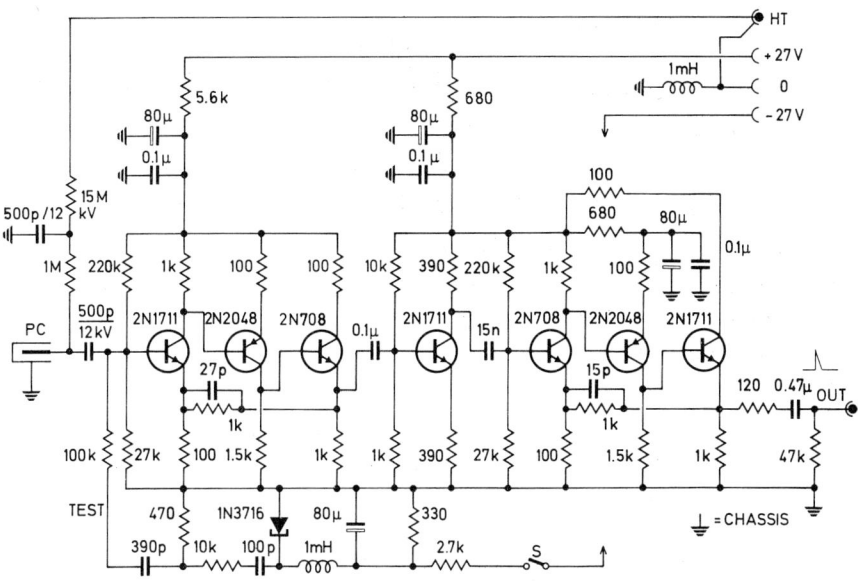

Fig. 2.11. Preamplifier according to MAY and SEMTURS [2.037]

of the chokes L_1 and L_2 from reaching the preamplifier input. Any interconnecting cable longer than 2 m must be double-shielded. The preamplifier must exhibit a high gain in order to diminish the influence of V_1, and the input stage of the amplifier must be mounted directly on the cable connector. Fig. 2.11 shows the detailed diagram of the preamplifier by MAY and SEMTURS. The detector and preamplifier housings are coupled mechanically and separated by a 1 mH choke from the common ground of the high tension and power supplies. The preamplifier has a gain of 50. An interesting detail is the built-in tunnel diode test-pulse generator which can be turned on by means of a switch S. Although the highest operating voltage of the counter is 4 kV, all high-voltage condensers are 12 kV types, located in a separate airtight box with P_2O_5 as the drying agent. This precaution avoids spurious counts due tu corona discharges or microscopic break-through events in the condenser dielectric. With the integral discriminator setting equivalent to 0.2 mV input voltage sensitivity, the preamplifier—amplifier system with a phantom counter produces less than 2 pulses per week at $V_{HT}=0$, and less than 2 pulses per day at $V_{HT}=4$ kV. The investigations of MAY and SEMTURS loc. cit., of course, also apply to preamplifiers for detectors other than proportional counters (cf. also Chapter 3.1.6).

2.3. Geiger-Müller-Counters

2.3.1. Detection Mechanism and Pulse Shape in the GM Counter

During the course of electron liberation in an avalanche in the proportional counter, short-wave light photons are also emitted. The emission probability increases rapidly with the counter operating voltage. By means of photoionization the photons may produce new electrons which serve as starting points for new avalanches. The probability per ion of the first avalanche of the emission of one photon initiating a second avalanche is denoted by ε. If the operating voltage is high, and the condition

$$\overline{M} \cdot \varepsilon \geqslant 1 \qquad (2.16)$$

is valid ($\overline{M}$ = gas multiplication factor), a single primary electron leads to the release of a vast number of avalanches, which finally surround the anode wire with a "hose" of positive ions and thus terminate the discharge.

In this case the total produced charge and therefore the amplitude of the detector voltage pulse remains constant (given by the counter dimensions and the field characteristics) and independent of the primary ionization. The pulse height may reach 100 V and more.

Counters operated in the voltage range defined by (2.16) are known as Geiger-Müller-Counters (GEIGER and MÜLLER [2.038]). If the operating voltage is raised much over the starting point $\overline{M}\varepsilon = 1$, stable Townsend discharge may take place and the counter may be destroyed. The admissible operating-voltage range is commonly called the counting plateau.

In the so-called non-self-quenching counters, which are filled with monoatomic or diatomic gases (particularly with noble gases), the axial propagation of the discharge results from the above mentioned-photons

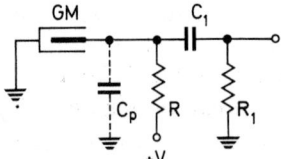

Fig. 2.12. Input circuit of a non-self-quenching GM counter

releasing electrons from the cathode material. The light positive ions on their path from anode to cathode gain sufficient energy to liberate new electrons. Once started, the discharge will therefore continue to pulsate infinitely and must be quenched by external means. For quenching, the operating voltage of the tube must be lowered below the starting value $\overline{M}\varepsilon = 1$ and held there until all the positive ions are neutralized. Since the voltage pulse of 100 V and more on the anode wire is high enough to quench the discharge, it suffices to make the time constant $R \cdot C_p$ (Fig. 2.12) of the input network longer than the ion collection time (which is of the order of few milliseconds). Since the parasitic capacity C_p is equal to 10 to 100 pF, this calls for very high resistances $R = 10^8 \ldots 10^9 \,\Omega$. The generation of quenching pulses with the aid of multivibrator circuits is discussed in the following Chapter 2.3.2.

In the so-called self-quenching counters, the quenching action is accomplished by the addition to the counting gas of heavy organic molecules, which have lower ionization energy than the light counting-gas molecules. During the propagation of the ion hose towards the cathode, the counting-gas molecules transfer their charge to the organic molecules. Thus only the slow, heavy organic ions reach the cathode, which cannot then liberate secondary electrons.

In addition, the added heavy molecules stop the photons emitted during the avalanche generation process with a mean free path of about 1 mm. The new avalanches do not therefore have their starting points on the cathode, but rather in the gas filling of the counter, in close proximity to the releasing avalanche. Because of this the ion hose spreads axially along the anode wire with a constant and not-too-high velocity. This velocity has been measured by numerous authors, and is commonly of the order of 10 cm/µs. The dependence of the propagation velocity on the counter operating voltage and on the partial pressure of the organic vapour is shown in Fig. 2.13 according to the experimental data published by ALDER et al. [2.039].

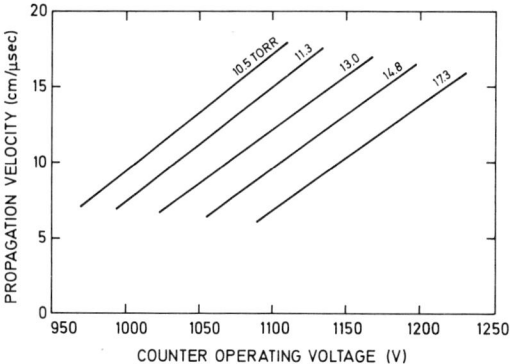

Fig. 2.13. Dependence of the ion hose propagation velocity on the counter operating voltage and on the partial pressure of the organic vapour according to ALDER et al. [2.039]. Filling gas = argon-ethylalcohol mixture, total pressure 80 Torr.

The halogen counters form a more-or-less separate category. Their counting gas is composed of noble gases with small ($\sim 1\%$) additions of halogen vapours (Cl_2, Br_2, I_2). The halogen molecules act — at least at low operating voltages — as a quenching agent. The exact mechanism of the halogen counter operation is not understood completely. Since halogen counters are very sensitive to capacitive load (cf. [2.040]), the signal is taken from the anode by means of a high-value resistor R (2 to 20 $M\Omega$, Fig. 2.14) in order to neutralize the influence of the connecting cable capacity or any other parasitic capacity C_p. For the rest, a high resistance R also extends the counting plateau, inasmuch as for higher operating voltage the halogen counter becomes non-self-quenching. The counting characteristics of halogen counters under different conditions are reviewed, e.g. by GEBAUER [2.041].

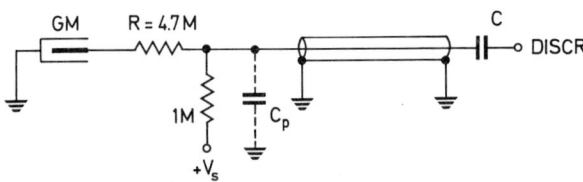

Fig. 2.14. Input circuit of a halogen-quenched GM counter

A comprehensive review of GM tubes can be found in the monograph by NEUERT [2.003].

As in the proportional counter, the signal pulse in the GM counter is formed by the ion component. Due to the final propagation velocity v_z of the ion hose along the anode wire, the current pulse consists of an overlap of delayed partial currents of the individual avalanches which have the shape $I_{ion}(t)$ for cylindrical chambers (Fig. 2.02). For simplicity the origin of primary ionization is assumed to be at one end ($z=0$) of the cylindrical counter of length l (the ordinate in the direction of the counter axis is denoted by z):

$$I_{GM, ion}(t) = \int_0^l I_{ion}\left(t - \frac{z}{v_z}\right) \frac{dz}{l} = \int_0^{T_l} I_{ion}(t-t') \frac{dt'}{T_l} \qquad (2.17)$$

with

$$T_l = \frac{l}{v_z}. \qquad (2.18)$$

T_l is the propagation time needed to spread the ion hose over the whole length l of the counter. In Fig. 2.15 a typical $I_{GM,ion}(t)$ from (2.17) is plotted as a function of time t. With the velocity v_z being about 10 cm/μs, and the length of a common GM tube being about 10 cm, the propagation time T_l is of the order of 1 μs. The characteristic time for an individual avalanche is $t_{\frac{1}{2}} = 0.1$ to 1 μs (cf. Chapter 2.2.1). The length of the current pulse in Fig. 2.15 and thus the rise time of the integrated voltage pulse for $RC \gg T_l$ is therefore given mainly by T_l.

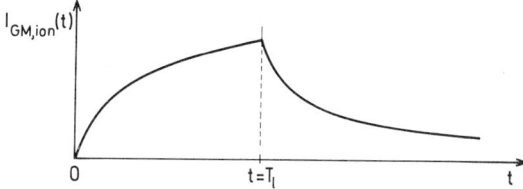

Fig. 2.15. The shape of the ion current pulse in a GM counter with primary ionization event at one end

If the primary ionization takes place not at the end, but at an arbitrary point $z_0 (0 < z_0 < l)$ of the anode wire, the ion current is composed of two components, one for each direction of ion hose propagation $z_0 \to 0$ and $z_0 \to l$, respectively. Therefore for $z_0 = l/2$ there is an effective decrease of T_l by a factor of 2. In the general case the shape of $I_{GM,ion}(t)$ has two maxima for $t = z_0/v_z$ and $t = (l - z_0)/v_z$. Hence, the rise time of the voltage pulse depends on the origin of ionization.

In addition, the primary electrons commonly liberated in the proximity of the cathode need a short time of the order of 10^{-7} s before they reach areas of high field strength and initiate the hose build-up (cf. for instance STEVENSON [2.042, 2.043], RAMSEY [2.044], PORTER and RAMSEY [2.045]). Even greater delays of more than 10 μs have been reported, which are caused by traces of electronegative gases in the counter filling. Such a molecule may capture the free electron and delay it substantially due to its much lower mobility. For this reason GM counters are used in coincidence assemblies only after special precautions have been taken (MANDEVILLE and SCHERB [2.046], MANDEVILLE [2.048]).

In the proportional counter the origin of the positive ions can be presumed to be in the nearest proximity of the anode wire (i.e. $x_0 = a$), so that the electron component of the current can be neglected. In a GM counter the high charge density affects the field surrounding the wire so that the ion hose has a finite diameter which, though very small, is still substantially greater than the wire diameter (i.e. $x_0 > a$). Hence the electron component may not be neglected in GM tubes.

The electron current pulse of a single avalanche (i.e. $I_{el}(t)$ from Fig. 2.02) is very short compared with T_l and can be regarded as a δ-function. The integration analogous to (2.17) hence yields a rectangular pulse of legth T_l for the electron component $I_{GM,el}(t)$, or the overlapping of two rectangular pulses, if the ionizing event does not occur at one end of the counter. The ratio of ion and electron components has been estimated theoretically by STUCKENBERG (as quoted in [2.003]) to be about 50%, which is in agreement with experiment. In Fig. 2.16 the current pulse shape $I_{GM}(t)$ is shown as measured by KELLEY, JORDAN and BELL [2.047]. The calculated part $I_{GM,ion}(t)$ has been subtracted from the experimental shape. The electron current pulse exhibits an approximately rectangular shape, and the areas under $I_{GM,ion}$ and $I_{GM,el}$ are roughly equal, as expected. The characteristic "ringing" of the electron current, which has been observed by other authors too, possibly indicates that the radius x_0 of the ion hose varies along the anode wire.

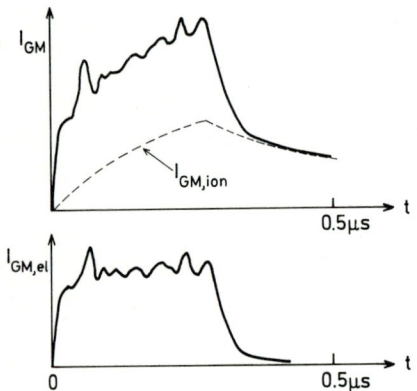

Fig. 2.16. The shape of the counter current pulses $I_{GM,el}$ and $I_{GM,ion}$ according to KELLEY et al. [2.047]

2.3.2. Quenching Circuits

The total charge of the ion hose lies in the order of 10^{-9} As, and the resulting pulse amplitudes are as high as 100 V. The connected electronics therefore need not amplify the signal but only form it and, if necessary, discriminate against noise, hum etc. Proper preamplifiers are not normally used. If a GM tube is to be connected with a distant electronic apparatus via a coaxial cable, the impedance transformation is accomplished by a small pulse transformer (cf. for instance [2.049]) rather than with active component circuits.

On the other hand there are numerous circuits for the external quenching of the counter discharge. These are used together with non-self-quenching counters and in order to diminish the dead time of the self-quenching ones. A simple circuit of this type is given by ELLIOT [2.050] (Fig. 2.17). The counter pulse triggers the monostable multivibrator with the double triode 6SN7, producing a negative pulse of

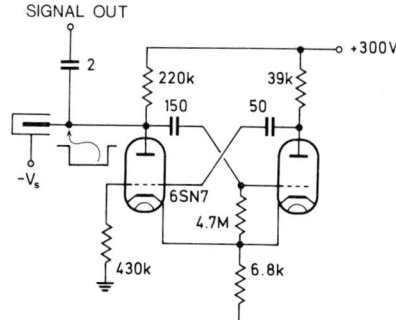

Fig. 2.17. Multivibrator type quenching circuit with a double triode according to ELLIOT [2.050]

about 200 V amplitude, which lowers the counter operating voltage below the starting value. With the component values given in the diagram the pulse length is approximately 1.5 ms. If during this time secondary electrons are released from the cathode, they do not give rise to new avalanches and the discharge breaks down. Other circuits ([2.051] to [2.055]) differ from that of Fig. 2.17 mainly in the manner in which the counter operating voltage V_s is connected, or by an unnecessary complexity.

CROWELL and LOW [2.056] describe another simple circuit, the principle of which is shown in Fig. 2.18. The triggering pulse for a

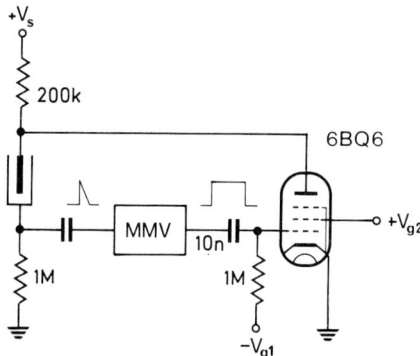

Fig. 2.18. Quenching circuit according to CROWELL and LOW [2.056] using a high voltage pentode

multivibrator is supplied by the cathode circuit of the counter. The multivibrator output pulse triggers a high-voltage vacuum tube (6BQ6) which is normally cut off, giving a negative pulse of about 800 V amplitude. The multivibrator has to supply a small amplitude signal only and can therefore be transistorized.

With self-quenching counters the integrating time constant RC or the length of a "quenching" pulse can be chosen smaller than the ion collection time T_{ion} (Fig. 2.02) as the discharge break-down occurs automatically. The maximum count rate is limited only by the intrinsic dead time of the counter. During the spread of the ion hose towards the cathode, the field is disturbed and the counter remains entirely insensitive for an instant (real dead time) and recovers slowly with the pulse height growing exponentially (recovery time). The situation is illustrated in Fig. 2.19. Externally the counter exhibits a dead time T_D

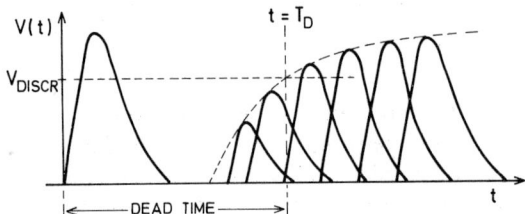

Fig. 2.19. The dead time situation in self-quenching GM counters

of about 200 µs depending on the discriminator level V_{DISCR}, as can easily be seen. Recently the dead time effects of GM tubes were investigated by KRAMERS [2.057] and GLAESER [2.234].

SIMPSON [2.058] attempted to reduce the dead time by reversing the polarity of the operating voltage for some 10 µs immediately after the counter pulse appeared. The principle of the circuit used is shown in Fig. 2.20. By the reverse field the positive ions are collected and neutralized on the temporarily negative anode wire, this action being extremely

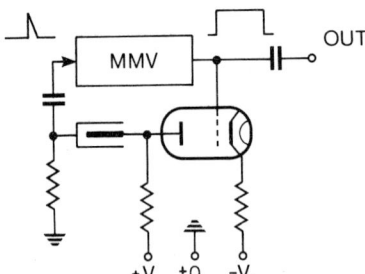

Fig. 2.20. Circuit for polarity reversal of the counter voltage according to SIMPSON [2.058]

fast because of the high field strength in the proximity of the wire and because of the small distance of the ions from the wire. Another simple circuit for the field reversal is published by HODSON [2.059]. SIMPSON loc. cit. succeeded in reaching a dead time of 20 µs, thus lowering T_D by a factor of 10.

The diminution of T_D is probably due not to the fast collection of ions on the anode wire but rather to the interruption of the ion hose built up by the fast voltage shut-down. Hence the field remains undisturbed along most of the wire and the counter remains ready to count. PORTER [2.060] describes a circuit (Fig. 2.21) which is very fast and which limits the ion hose to about 5 % of the wire length. The unsymmetrical monostable flips over as soon as the voltage pulse on the anode wire reaches 0.2 V and lowers the operating voltage by about 100 V in less than 100 ns, thus interrupting the hose build-up. The length of the inhibiting pulse is 750 ns, and the overall dead time of the assembly comes to 1.5 µs, due to recovery effects of the electronics. Of course, it must be considered that the counter remains insensitive in the neighbourhood of the point of primary particle impact so long as the hose fragment is not neutralized. For collimated particle beams the dead time of such a device can therefore be substantially longer.

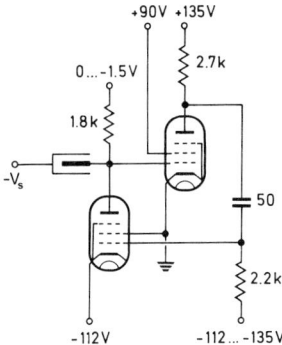

Fig. 2.21. Fast quenching circuit by PORTER [2.060]

2.4. Semiconductor Detectors

2.4.1. Characteristic Properties of Semiconductor Detectors

Analogous to the gas ionization in the ionization chamber, the generation of free charge carriers in solids can also be used for the detection of ionizing radiations[2]. Some important conditions must be fulfilled for

---

[2] GIBSON et al. [2.061] in this context quote an apt remark by McKAY: "Whenever a nuclear physicist observes a new effect caused by an atomic particle he tries to make a counter out of it."

this: The lifetime of charge carriers, i.e. the average time interval between their generation and their recombination or capture in traps, must be longer than the collection time. The short collection times required presuppose a high carrier mobility and a strong collection field. Despite the high field strength in the counter there should preferably be no background current, i.e. the counter medium should be insulating. Last but not least, a low energy requirement W per generation of one charge carrier pair is desirable for high signal amplitude and improved resolution. The reverse biased *pn* junction in semiconductors represents the only presently known solid state device with such characteristics.

A detailed treatment of the semiconductor properties and of the physical and technological aspects of the semiconductor detectors is outside the scope of this booklet. Readers interested in this subject will find the particular information in standard text books [2.062, 2.063], in review papers [2.003, 2.061, 2.064] and especially in the recent comprehensive review by GOULDING [2.065].

The situation in a reverse biased, extremely unsymmetrical, pn junction is shown in Fig. 2.22. The thin *n*-type semiconductor with very

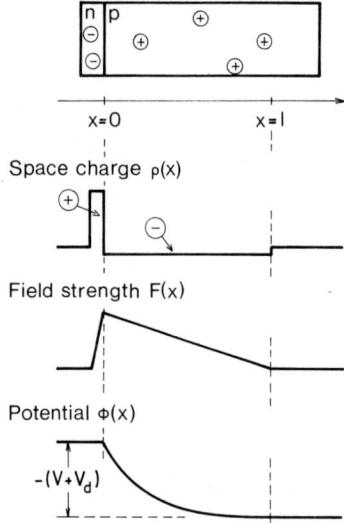

Fig. 2.22. Situation in a reverse biased *pn* junction

high donor concentration serves as the entrance window for the radiations, the *p*-type region being only moderately doped. The diffusion of the charge carriers in the depletion layer l causes a space charge distribution $\rho(x)$ with corresponding linear rising field strength $F(x)$ and parabolic potential $\Phi(x)$. The external reverse voltage V_s, increased by the so called diffusion voltage V_d (about 0.3 V in Ge, about 0.6 V in Si), appears as the potential difference between the *n*-type and the *p*-type region. Since no free holes can be delivered from the *n*-region there is actually no current in the depletion layer without external carrier generation. Hence, the charge carriers (electrons and holes) which are produced in the depletion layer by the ionizing particle, are collected by the field $F(x)$ and form the signal current in the same way as the ionization chamber.

Because of the parabolic shape of $\Phi(x)$ the width l of the depletion layer

$$l = \frac{\varepsilon}{2\pi \cdot e \cdot N_p} \sqrt{V_s + V_d} \tag{2.19}$$

is apparently proportional to the square root of $V_s + V_d$. In equation (2.19) ε denotes the dielectric constant of the semiconductor and N_p denotes the acceptor concentration in the *p*-type region. The capacity C_d of the depletion layer is inversely proportional to l:

$$C_d \approx \frac{e N_p}{2} S \frac{1}{\sqrt{V_s + V_d}}. \tag{2.20}$$

C_d thus depends on the reverse voltage V_s. S denotes the area of the junction. The maximum field strength $F(0)$ is

$$F(0) = 2 \frac{V_s + V_d}{l}. \tag{2.21}$$

Fig. 2.22 describes correctly the situation in semiconductor detectors of the so-called diffusion type (e.g. [2.066] to [2.068]) or in the surface-barrier detectors (e.g. [2.069] to [2.073]). The operating voltage V_s typically lies between 10 and 1000 V, and the depletion layer is as thick as 1 mm for the higher V_s. With a detector area of 2 cm² this gives a detector capacity C_d of the order of 10 to 100 pF.

Deeper depletion layers with approximately constant field strength can be achieved by introducing an intrinsic region between two highly doped *p*-type and *n*-type regions (Fig. 2.23). This structure is commonly called a *pin*-counter. The space charges are localized in the highly doped regions n^+ and p^+. The width l of the depletion layer is approximately independent of the bias voltage V_s, and the field strength in the *i*-region is almost constant and equal to

$$F(x) = \frac{V_s + V_d}{l}. \tag{2.22}$$

Detectors with a depletion layer as wide as 12 mm can be manufactured by using the ion drift method according to PELL [2.074, 2.075]. This involves compensating the acceptor concentration in an originally moderately p-doped region by lithium ions in the intermediate sites of the host lattice. Lithium drifted silicon and germanium detectors are now widely used in nuclear radiation spectroscopy (cf. e.g. [2.076] to [2.079]). Due to the great width l of the depletion layer, the capacity C_d of such detectors, even with larger areas (~ 5 cm^2), is only 10 pF.

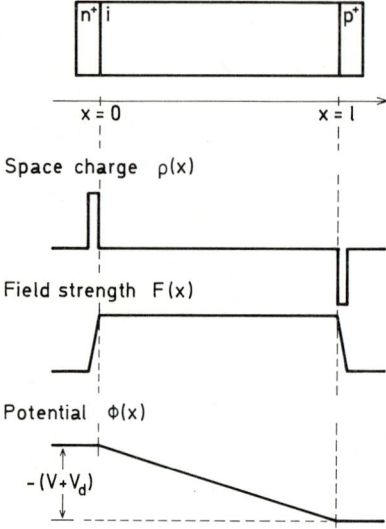

Fig. 2.23. Situation in a reverse biased *pin* counter

The leakage current I_d of a semiconductor detector, i.e. the quiet reverse current with no external ionization produced, consists basically of the diffusion current, the volume current and the surface current components. The diffusion current is caused by the minority charge carriers which are thermally generated less than their diffusion length away from the end of the depletion layer. Since the width l of the depletion layer is much greater than the diffusion length, the diffusion current can be neglected in comparison to the volume current I_{dv}.

By simplification, according to SAH, NOYCE and SHOCKLEY [2.080], I_{dv} can be represented as follows

$$I_{dv} = S \cdot l \cdot \frac{e \cdot n_i}{2\tau}, \qquad (2.23)$$

where S is the area of the junction, n_i the intrinsic charge carrier concentration and τ the recombination life time of the carriers. The temperature dependence of I_{dv} is essentially given by the term n_i which is proportional to

$$T^{3/2} \cdot \epsilon^{-E_g/2kT}, \tag{2.24}$$

with T in °K and E_g being the energy gap of the semiconductor. In slightly doped silicon (e.g. the i-region of a *pin*-counter) we have $\tau \approx 1$ ms, $n_i \approx 1.5 \cdot 10^{10}$ cm^{-3} for $T = 300$ °K, resulting in $I_{dv}/S \cdot l \approx 1$ µA/cm^3. A lithium drifted silicon detector with $S = 2$ cm^2 and $l = 5$ mm at room temperature therefore exhibits $I_{dv} \approx 1$ µA. Because of the different simplifications used in the derivation of (2.23) this relation does not correctly describe the leakage current, the experimental data being lower by a factor ranging between 2 to 20 [2.065].

If no special precautions are taken the surface leakage current can become very high. Since it depends on the little-understood surface conditions of the semiconductor, no mathematical treatment is possible. By means of improved counter manufacturing techniques, e.g. using the guard-ring principle (GOULDING and HANSEN [2.081]), and by means of surface protection by SiO$_2$, similar to the planar technology in the manufacture of transistors (e.g. MADDEN and GIBSON [2.082], HANSEN and GOULDING [2.083]), the surface component of the leakage current can be made negligible.

In all cases where the volume component I_{dv} dominates the leakage current, it can be reduced according to (2.24) by lowering the operating temperature.

2.4.2. Energy Required to Form a Hole-Electron Pair

The energy ΔE lost by an ionizing particle in a semiconductor, is divided between the excitation of the lattice vibrations and the production of free electrons and holes. The energy of the optical lattice vibrations is quantized, the quantum energy E_r (phonon energy) being given by the Raman frequency of the lattice which corresponds to about 50 meV. According to a theory of SHOCKLEY [2.082], the mean energy W required to form a hole-electron pair can be expressed as a function of the energy gap width E_g of the semiconductor

$$W = 2.2 E_g + n \cdot E_r. \tag{2.25}$$

Here n denotes the mean number of phonons produced per ionizing collision, which is of the order of 10 to 100. Since at least E_g energy is needed to form a charge-carrier pair, the efficiency η of the ionization process is

$$\eta = \frac{E_g}{W} = \frac{1}{2.2 + n(E_r/E_g)}. \tag{2.26}$$

Some characteristic properties of germanium and silicon are summarized in table Fig. 2.24 which is based essentially on the review by GOULDING [2.065]. In accordance with the theory (2.25), the experimental data ([2.065], [2.083] to [2.086]) yield W values of 3.6 eV and 2.9 eV for silicon and germanium, respectively. The values given in Fig. 2.24 represent a weighted average of the recent determinations. Apart from isolated exceptions [2.085] which can be explained in terms of parasitic effects [2.065], the same values of W have been measured for α and β particles, photons, heavy ions or fission fragments. Due to the light dependence of the energy gap E_g on the temperature T (cf. Fig. 2.24), the energy requirement W is also moderately temperature dependent [2.085, 2.087]. In addition a variation of E_g with the field strength in the pn junction has been reported [2.088].

		Silicon
Energy gap width E_g at 300 °K	(eV)	1.106
Energy gap width E_g at T °K	(eV)	$1.205 - 2.8 \cdot 10^{-4} T$
Intrinsic concentration n_i at 300 °K	(cm^{-3})	$1.5 \cdot 10^{10}$
Intrinsic concentration n_i at T °K	(cm^{-3})	$2.8 \cdot 10^{16} T^{3/2} \epsilon^{-6450/T}$
Electron mobility μ_n at 300 °K	(cm$^2 V^{-1} s^{-1}$)	1350
Hole mobility μ_p at 300 °K	(cm$^2 V^{-1} s^{-1}$)	480
Energy W per one hole-electron pair	(eV)	3.65 ± 0.05
		Germanium
Energy gap width E_g at 300 °K	(eV)	0.67
Energy gap width E_g at T °K	(eV)	$0.72 - 3.4 \cdot 10^{-4} T$
Intrinsic concentration n_i at 300 °K	(cm^{-3})	$2.4 \cdot 10^{13}$
Intrinsic concentration n_i at T °K	(cm^{-3})	$9.7 \cdot 10^{15} T^{3/2} \epsilon^{-4350/T}$
Electron mobility μ_n at 300 °K	(cm$^2 V^{-1} s^{-1}$)	3900
Hole mobility μ_p at 300 °K	(cm$^2 V^{-1} s^{-1}$)	1900
Energy W per one hole-electron pair	(eV)	2.95 ± 0.05

Fig. 2.24. Some characteristic properties of germanium and silicon

The statistical deviation σ_N of the number $N = \Delta E/W$ of electron and hole pairs produced by particles with the energy ΔE stopped completely in the semiconductor, is given by

$$\sigma_N = \sqrt{f \cdot \overline{N}} \qquad (2.27)$$

exactly as for the gas ionization. The Fano factor $f \leqslant 1$ [2.007] describes the diminution of the variation due to the correlation between the particular ionization events in consequence of the given total energy loss ΔE.

VAN ROOSBROECK [2.089] investigated theoretically the energy loss phenomena in semiconductors and calculated the efficiency η (2.26) and the Fano factor f as a function of the mean number n of phonons produced per ionizing collision. The calculations are valid for a special statistical model. In Fig. 2.25 the results of the calculations are shown and in Fig. 2.26 the Fano factor f is plotted as a function of the efficiency η. With the values from Fig. 2.24 we have $\eta_{Si} = 30\%$ and $\eta_{Ge} = 23\%$ thus giving $f_{Si} = 0.30$ and $f_{Ge} = 0.36$. GOULDING [2.065] reports measurements on germanium detectors yielding f-values of $f = 0.30 \pm 0.03$, which is in good agreement with the VAN ROOSBROECK's theory.

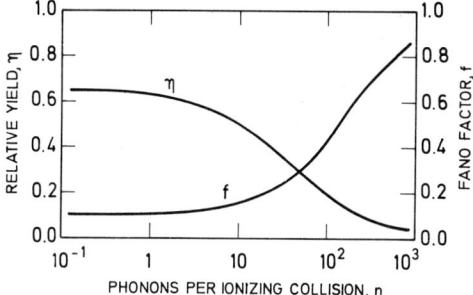

Fig. 2.25. The dependence of the efficiency η and of the Fano factor f on the phonon number n (according to VAN ROOSBROECK [2.089])

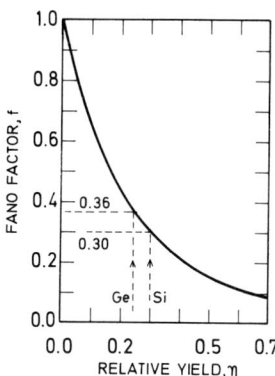

Fig. 2.26. Fano factor f as a function of η (from Fig. 2.25)

Other authors report f values between 0.15 and 0.20 (cf. BILGER [2.090]). Still smaller values ($f = 0.04$) are conceivable for some simplified models of ionization statistics [2.089]. The most recent reassessment of the Fano factor situation has been given by KLEIN [2.235]. Although a precise measurement of f is very difficult because of obstacles in determining the extent to which the total energy resolution of a system detector—amplifier is due to the statistical deviation σ_N, there is no doubt that σ_N is substantially smaller than would result from the non-disturbed Poisson distribution ($f=1$). Hence, the resolution for low particle energy is dominated by the detector and preamplifier noise.

2.4.3. The Pulse Shape in the *pn* and *pin* Detectors

Besides the chaotic thermal movement of the electrons and the holes, there is an additional movement of the charge carriers in the direction of the field strength F, the velocity of which

$$w_n = -\mu_n \cdot F \quad \text{and} \quad w_p = \mu_p \cdot F \tag{2.28}$$

is proportional to F, with μ_n and μ_p being the mobilities of the electrons and the holes, respectively. The relation (2.28) is similar to the relation

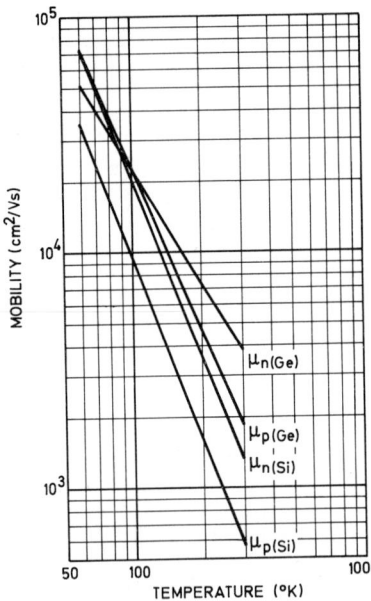

Fig. 2.27. The electron and hole mobilities in Ge and Si as a function of the temperature T (GOULDING [2.065])

(2.3). The mobilities μ_n and μ_p in the intrinsic germanium and silicon are plotted in Fig. 2.27 as a function of the temperature.

The situation is somewhat complicated by two effects. Firstly, the plots of Fig. 2.27 are valid only for pure intrinsic semiconductors. In

The Pulse Shape in the pn and pin Detectors

Chamber geometry	pin 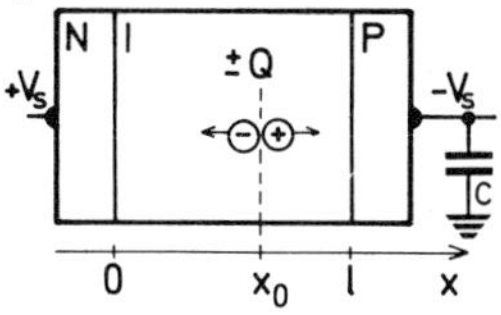	pn 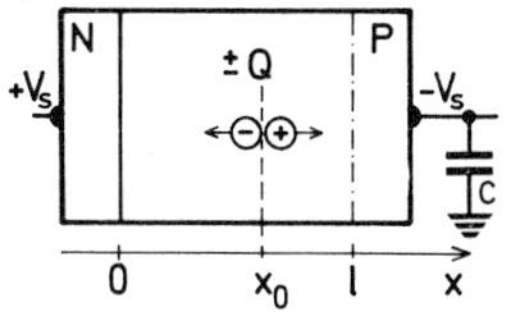
Field strength $F(x)$	$\dfrac{V_s}{l}$	$\dfrac{2V_s}{l}\left(1-\dfrac{x}{l}\right)$
Hole collection time T_p	$\dfrac{l}{\mu_p V_s}(l-x_0)$	∞
Electron collection time T_n	$\dfrac{l}{\mu_n V_s}x_0$	$\dfrac{l^2}{2\mu_n V_s}\log\dfrac{l}{l-x_0}$
Hole current $I_p(t)$ $0<t<T_p$	$Q\dfrac{\mu_p V_s}{l^2}$	$Q\dfrac{2\mu_p V_s}{l^2}\left(1-\dfrac{x_0}{l}\right)\epsilon^{-\tfrac{2\mu_p V_s}{l^2}t}$
Electron current $I_n(t)$ $0<t<T_n$	$Q\dfrac{\mu_n V_s}{l^2}$	$Q\dfrac{2\mu_n V_s}{l^2}\left(1-\dfrac{x_0}{l}\right)\epsilon^{+\tfrac{2\mu_n V_s}{l^2}t}$
Integrated voltage $V_p(t)$ $0<t<T_p$	$\dfrac{Q}{C}\left(1-\dfrac{x_0}{l}\right)\dfrac{t}{T_p}$	$\dfrac{Q}{C}\left(1-\dfrac{x_0}{l}\right)\left(1-\epsilon^{-\tfrac{2\mu_p V_s}{l^2}t}\right)$
Integrated voltage $V_n(t)$ $0<t<T_n$	$\dfrac{Q}{C}\dfrac{x_0}{l}\dfrac{t}{T_n}$	$\dfrac{Q}{C}\left(1-\dfrac{x_0}{l}\right)\left(\epsilon^{+\tfrac{2\mu_n V_s}{l^2}t}-1\right)$
Final voltage $V_p(T_p)$	$\dfrac{Q}{C}\left(1-\dfrac{x_0}{l}\right)$	$\dfrac{Q}{C}\left(1-\dfrac{x_0}{l}\right)$
Final voltage $V_n(T_n)$	$\dfrac{Q}{C}\dfrac{x_0}{l}$	$\dfrac{Q}{C}\dfrac{x_0}{l}$
Current pulses	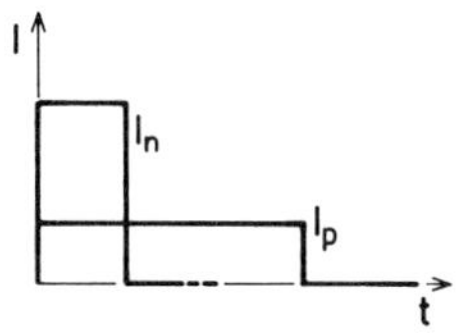	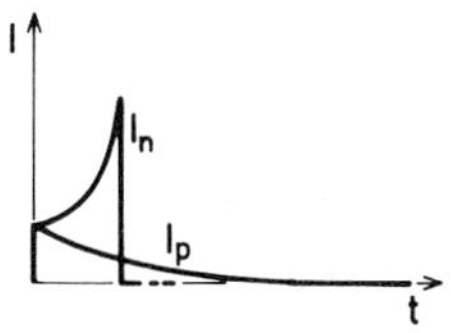
Voltage pulses	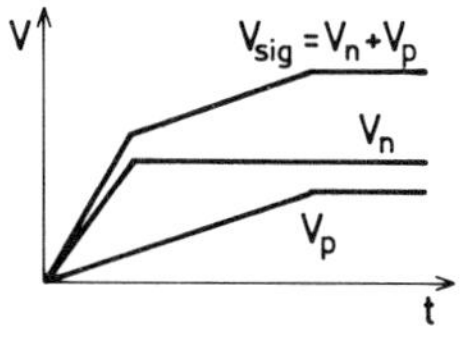	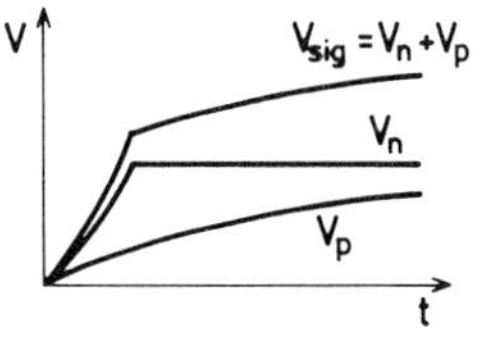

Fig. 2.28. The pulse shapes in the semiconductor detectors

heavily doped semiconductors the Coulomb interaction of the impurities with the charge carriers must be taken into account, resulting in lower mobilities. For higher doping concentrations the plots exhibit a maximum near $100\,°\mathrm{K}$. Below this point, lowering the temperature decreases the mobilities again. However, in moderately doped regions of semiconductor detectors, especially in the *pin* detector, Fig. 2.27 represents a good approximation.

Further deviations from Fig. 2.27 arise when the drift velocities w_n or w_p are in the order of thermal velocities ($\sim 10^7$ cm/sec) [2.091]. The mobilities decrease in this case with increasing field strength, resulting in a much slower increase in w as would correspond to (2.28), or even saturating after reaching a given value (e.g. $\sim 10^7$ cm/sec for electrons in Ge). Drift velocities of the order of 10^7 cm/sec in cooled Ge detectors with field strength of 10^3 V/cm are not unusual.

The shape of the signal current $I_{\text{sig}} = I_n + I_p$ consisting of an electron and a hole component can be calculated as in Chapter 2.1.3. In Fig. 2.28 the current pulse shapes $I_n(t)$, $I_p(t)$, the collection times T_n and T_p, and the shapes of integrated voltage pulses $V_n(t)$ and $V_p(t)$ of the electron and the hole component respectively, are summarized for the pn and pin detector. The calculated pulse shapes are based on the assumption of a point-like ionizing event at $x = x_0$ in $t = 0$, producing the total charge of $\pm Q$.

Taking into account the differences in the definition of the charge carrier mobilities in gas (2.3) and in semiconductors (2.28), the pulse shapes for the pin detector in Fig. 2.28 correspond entirely to the ones for the parallel-plate chamber in Fig. 2.02. Since it is $\mu_n \gtrsim \mu_p$ in semiconductor detectors, the hole component is also used for signal build-up. In cooled ($T \sim 100\,°\mathrm{K}$) germanium it is even $\mu_n = \mu_p$ and both charge carriers influence the pulse shape to the same extent. Because of the non-homogeneous field in the diffused *pn* junction or in the surface-barrier detector, the *pn* pulse resembles roughly the situation in a cylindrical ionization chamber. The hole component $I_p(t)$ of the detector current exhibits (in contradiction to the cylindrical ionization counters) an exponential decay with the time constant $l^2/2\mu_p V_s$, yielding an infinite hole collection time T_p. In pn detectors with p windows and with depletion layers in the moderately doped n region the given relations remain valid, accordingly interchanging the holes for the electrons and vice versa.

If the field-free p-type region in a *pn* detector is not negligibly thin it possesses a finite resistance R_d. R_d lies in the signal path and forms an integrating circuit with the detector capacity C_d and the (dynamical) preamplifier input capacity C_{in}. The influence of this integrator on the pulse shape can be easily calculated from the equivalent circuit in Fig.

2.29. The Laplace transform $\hat{V}_{sig}(p)$ of the voltage pulse $V_{sig}(t)$ is

$$\hat{V}_{sig}(p) = \hat{I}_{sig}(p) \frac{1}{p(C_d + C_{in})} \cdot \frac{1}{(1+p\tau)}, \qquad (2.29)$$

where

$$\tau = R_d \frac{C_d C_{in}}{C_d + C_{in}}.$$

The first term describes the current integration by the total capacity $C = C_d + C_{in}$. With a charge sensitive preamplifier it is $C_{in} \gg C_d$. Therefore, the time constant τ of the second term representing the deformation of the voltage pulse by an integrator circuit becomes equal to $\tau = R_d C_d$, independent of C_{in}. In very thin detectors it is commonly $\tau > T_n, T_p$ (or $\tau > l^2/2\mu_p V_s$ in a pn detector), the pulse shape being thus dominated by (2.29).

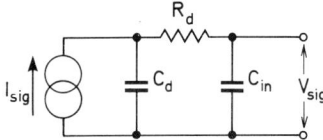

Fig. 2.29. Influence of the path resistance R on the pulse shape

Some remarks are necessary concerning the derivation of the pulse shape in pn detectors in Fig. 2.28. The signal current shape $I(t) = \frac{dq}{dt}$, i.e. the variation of the charge q at the preamplifier input, is commonly obtained from an analysis of the energy balance: the increase of the electrostatic energy $dE_{stat} = d(q^2/2C_d) = (q/C_d)dq = V_s \cdot dq$ of the detector capacity C_d must be equal to the work $Q \cdot d\phi = Q \cdot F(x) \cdot dx$ done by the field $F(x)$ during the movement of the charge Q over the distance dx or over the potential difference $d\phi$. From $V_s \cdot dq = Q \cdot F(x) \cdot dx$ it follows directly that $I = QFw/V_s$ (2.5). CAVALLERI, FABRI, GATTI and SVELTO [2.093] have pointed out that the relation $dE_{stat} = V_s \cdot dq$, not accounting for the changes in the space charge potential energy, does not apply because of the space charge in the pn junction. Hence the more strict theory by RAMO [2.094] (cf. also JEN [2.095, 2.096]) must be used. Instead of (2.5) this theory leads to the simple relation

$$I(t) = Q \frac{w}{l} \qquad (2.30)$$

where w is the drift velocity of the carriers. The pulse shapes given in Fig. 2.28 are based on this relation. The pulse shapes for different special cases have been calculated by different authors using the correct relation (2.30) ([2.097] to [2.100]). However many other authors use the incorrect relation (2.5) in theoretical considerations on pn detectors.

The pulse shapes according to Fig. 2.28 apply if short-range particles or low-energy gamma quanta are detected. Particles with ranges comparable to the depletion layer width l yields signals equal to the overlapp of similar pulses for every particular point of the ionization trace [2.097].

With maximum carrier drift velocities being about 10^7 cm/sec, the rise time of the voltage pulses in thin detectors ($l \sim 1$ mm) may amount to 1 nsec. In lithium drifted detectors ($l \sim 10$ mm) rise times of about 100 nsec can be expected. According to the ionization trace the rise time will be different, so that the semiconductor detector pulse is differently shaped for particles of different range. Thus using a pulse shape discriminator (cf. Chapter 4.3) different particle species can be distinguished.

The situation described is further complicated by the fact that some of the free charge carriers can be trapped on the lattice imperfections. On the one hand this gives rise to a pulse height defect ([2.101] to [2.103]), and on the other it leads to a delayed charge collection, i.e. to a slow (~ 1 µsec) component of the current shape [2.104].

Theoretical and experimental considerations concerning pulse shape in *pin* detectors are presented in papers [2.236] to [2.238].

2.4.4. Preamplifiers and Related Circuits

The semiconductor detector junction capacity C_d depends on the operating voltage V_s (2.20) and cannot be considered as a constant, even if V_s was stabilized. Therefore, for the integration of the current

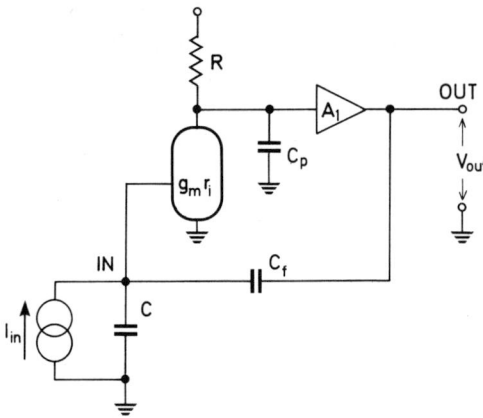

Fig. 2.30. Detailed equivalent circuit diagram of a charge-sensitive preamplifier

pulse a preamplifier of the charge sensitive type as shown on the right in Fig. 2.05 must be used. In Fig. 2.30 a detailed equivalent circuit of such a preamplifier is shown which will be used in the following analysis.

The input component of the preamplifier (vacuum tube, field effect or bipolar transistor, or combinations of these) will be described by its transconductance g_m and its internal resistance r_i. The dominant integration time constant τ_0 of the preamplifier is given by the output resistance $R/\!/r_i$ of the input stage and the parasitic capacity C_p

$$\tau_0 = (R/\!/r_i)C_p. \qquad (2.31)$$

If the behaviour of the amplifier should be aperiodic, all remaining time constants must be much shorter. Only a single-stage impedance convertor (cathode follower, emitter follower, Darlington emitter follower) can therefore be used as the amplifier A_1, the corresponding voltage gain A_1 being thus approximately equal to 1.

The voltage gain $-A(p)$ between the terminals IN and OUT without feedback comes to[3]

$$A(p) = \frac{g_m(R/\!/r_i)A_1}{1+p\tau_0} = \frac{A_0}{1+p\tau_0}. \qquad (2.32)$$

The capacitive attenuator C, C_f exhibits an attenuating factor of $b = C_f/(C+C_f)$. Since the gain of an amplifier with negative feedback is $-A(p)/(1+b\cdot A(p))$, the Laplace transform $\hat{V}_{out}$ of the output voltage is

$$\hat{V}_{out}(p) = -\hat{I}_{in} \cdot \frac{1}{p(C+C_f)} \cdot \frac{A(p)}{1+A(p)C_f/(C+C_f)}. \qquad (2.33)$$

By introduction of (2.32) we get

$$\hat{V}_{out}(p) = -\hat{I}_{in} \frac{1}{p[C_f+(C+C_f)/A_0]} \cdot \frac{1}{1+p\tau} \qquad (2.34)$$

with τ from (2.37). The second term in (2.34) describes the integration of the input current I_{in} by the feedback capacity C_f, slightly magnified by $(C+C_f)/A_0 \ll C_f$. Due to this additional term the pulse height of V_{out} depends somewhat on C which consists of the detector capacity C_d and different stray capacities; a simple calculation gives ($A_0 C_f \gg (C+C_f)$):

$$\frac{dV_{out}}{V_{out}} = -\frac{dC}{C} \cdot \frac{C}{A_0 \cdot C_f} \left(\frac{1}{1+(C+C_f)/A_0 C_f} \right) \approx -\frac{dC}{C} \cdot \frac{1}{A_{res}} \qquad (2.35)$$

where

$$A_{res} = A_0 \frac{C_f}{C} \qquad (2.36)$$

---

[3] The Laplace transform calculus will be used throughout this analysis.

is the so-called reserve gain, which is a measure of the insensitivity of the preamplifier to the variations in the input capacity C.

The third term in (2.34) represents an integrator with the time constant

$$\tau = \tau_0 \frac{C+C_f}{C+C_f+A_0 C_f} \approx \tau_0/A_{\text{res}}, \qquad (2.37)$$

which is independent of $R /\!/ r_i$ because of the relations (2.31), (2.32) and (2.36):

$$\tau = \frac{(R /\!/ r_i) C_p}{g_m (R /\!/ r_i) A_1 C_f/C} = \frac{C \cdot C_p}{g_m \cdot A_1 \cdot C_f}. \qquad (2.38)$$

The gain A_0 can thus be increased using a very high $R /\!/ r_i$ without at the same time making the amplifier slow. Fast preamplifiers need input components with high transconductance g_m. Since the feedback capacitor C_f must remain very small in order to obtain enough high signal voltage (commonly $C_f \sim 0.5 \cdots 5\,\text{pF}$, i.e. $C_f \ll C$) the only possibility to reach a high reserve gain A_{res} (2.36) is to increase the loop gain A_0.

Because of its dependence on C and g_m, the time constant τ depends on the operating conditions of the detector-preamplifier system and generally cannot be considered as constant. The integrator used for pulse shaping therefore must have a time constant $\tau_{\text{int}} \gg \tau$ and must be located outside the feedback loop. Hence the pulse shaping is commonly accomplished either between the preamplifier and the main amplifier, or first in one of the main amplifier stages (cf. Chapter 3.1.3).

The input component of the preamplifier Fig. 2.30 must exhibit low noise and high r_i and g_m at the same time. Where the noise of bipolar transistors can be accepted (e.g. together with detectors with very high C_d in high energy spectroscopy) or whenever field effect transistors are used, a high r_i will be obtained without difficulty, also with pentodes, though they are inferior to the low r_i triodes as far as noise is concerned, because of the additional partition noise (cf. e.g. [2.017]). However, as is well known, pentode amplification characteristics combined with the triode low noise characteristics, can be obtained in the cascode circuit Fig. 2.31 (GILLESPIE [2.017]). The internal resistance of the cascode T_1, T_2 is

$$r_{i,\text{casc}} = r_{i1} + r_{i2} + r_{i1} \cdot r_{i2} \cdot g_{m2} \gg r_{i1}, r_{i2}, \qquad (2.39)$$

where r_{i1}, g_{m1} and r_{i2}, g_{m2} are the parameters of the triode T_1 and T_2, respectively.

Preamplifiers and Related Circuits

The cathode follower T_3 in Fig. 2.31 raises the supply voltage of the cascode by means of the capacitor C_2 synchronously with the increase of the plate voltage of T_2 — this is known as bootstrap feedback.

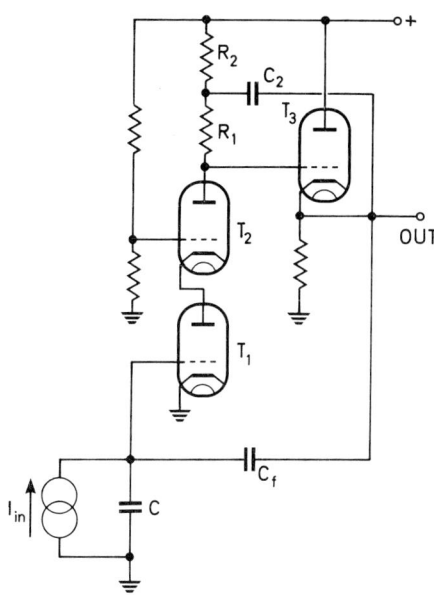

Fig. 2.31. Cascode preamplifier with bootstrap feedback

The load of T_2 is thus no longer R_1, but a dynamic resistance R_{dyn}. As can be shown easily by analysis of the circuit Fig. 2.31,

$$R_{dyn} = \frac{R_1}{1-A_3} \cdot \frac{p}{p+1/\tau_2} + (R_1+R_2)\frac{1/\tau_2}{p+1/\tau_2} \qquad (2.40)$$

with

$$\tau_2 = R_2 C_2 (1-A_3),$$

where $A_3 \approx 1$ denotes the voltage gain of the cathode follower T_3. If C_2 is great enough to make τ_2 longer than the maximum pulse length, which can be described symbolically by $\tau_2 \gg 1/p$ in the Laplace transform calculus, R_{dyn} is reduced to

$$R_{dyn} \approx \frac{R_1}{1-A_3} \gg R_1. \qquad (2.41)$$

In a cascode circuit with bootstrap feedback $r_{i,\text{casc}} // R_{\text{dyn}}$ is very high and the gain A_0 and the reserve gain A_{res} become sufficiently great.

A more detailed analysis of the cascode and of the bootstrap feedback can be found in various text books of electronics. The normal amplifier stage with bootstrapped load, the cascode with ohmic load and the cascode with bootstrapped load have been discussed recently by POENARU and VÎLCOV [2.105] with special emphasis on the applications in semiconductor detector preamplifiers.

Of course, the bootstrap technique is used to achieve a high load, not only in the previously described circuit, but also in circuits with FET or bipolar transistors as the input components. The impedance converter (i.e. the tube T_3 in Fig. 2.31) in the latter cases commonly consists of an emitter-follower with one to three transistors in Darlington configuration. Although the cascode is not necessary when using FET and bipolar transistors which already have high r_i, this circuit is used very often, especialy because of its high gain and low Miller capacity. Different hybrid cascode circuits can be used as well, preferably with vacuum triodes and field effect transistors instead of T_1, and bipolar transistors instead of T_2. Some examples of such circuits will be given in this chapter.

In what follows, some practical preamplifier circuits as described by different authors will be discussed, with special emphasis on the noise figures, the sensitivity and the mechanical lay-out. The input components which should be used under particular experimental situations can be seen from Fig. 8.25 in Appendix 8.2. A more detailed investigation of the advantages and disadvantages of vacuum tubes, bipolar transistors and FET has been carried out by BILGER [2.090] (cf. also [2.239]).

In Fig. 2.32 a preamplifier with the vacuum triode EC 1000 as input component according to GOULDING [2.065] is shown. The rest of the circuit is transistorized. Because of the different operating voltages of the two components involved, the hybrid cascode tube EC 1000—transistor 2N 3493 is coupled by a condenser 0.1 µF. An emitter-follower with another 2N 3493 serves as the impedance converter for the bootstrap condenser 0.047 µF. The feedback capacity $C_f = 1$ pF, shunted by a resistor 500 MΩ, is connected to the output via a condenser 0.01 µF which prevents the operating voltage of the triode grid from being affected by the output voltage level. The time constant of the integration network 500 MΩ × 1 pF is 500 µsec; $C_f = 1$ pF yields a sensitivity of 0.16 µV/hole-electron pair, or 44 mV/MeV in silicon. The rise time of the preamplifier amounts to about 15 nsec. The signal is amplified by two amplifier stages stabilized by the aid of feedback loops prior to being fed to the main amplifier.

Preamplifiers and Related Circuits

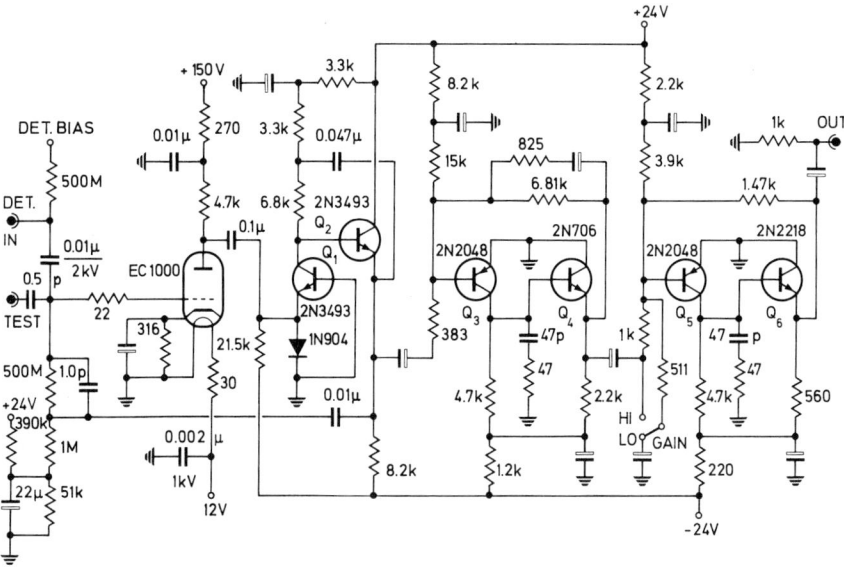

Fig. 2.32. A hybrid preamplifier using the triode EC1000 according to GOULDING [2.065]

The noise characteristics of this preamplifier are shown in Fig. 2.33. The pulse shaping circuit consists of one RC differentiator and two RC integrators with equal time constants. The total capacity $C = C_{ext} + C_{int}$ is composed of the preamplifier input capacity $C_{int} = 6\,\text{pF}$ and the

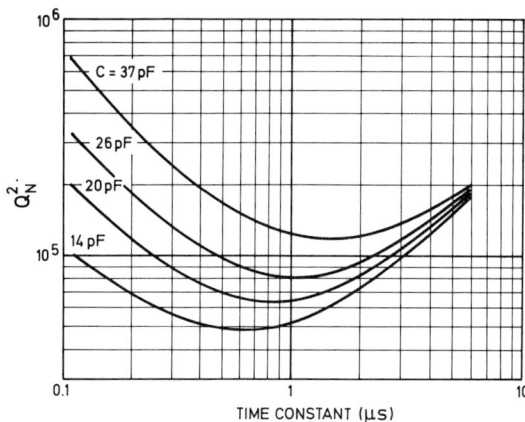

Fig. 2.33. The noise characteristics of the preamplifier Fig. 2.32

external capacity C_{ext} (detector, connections etc.). The grid current of EC 1000 under the specified operating conditions is about 1 nA. For semiconductor detectors with very small leakage current ($I_d \ll 1$ nA) affecting the total noise only by their capacity C_d, the equivalent noise charge Q_N can be read off directly from Fig. 2.33. For instance, with a germanium detector having $C_{ext} = 14$ pF (i.e. $C = 20$ pF), the Q_N is minimum at a time constant 0.8 μsec, giving a resolution FWHM $= 1.8$ keV. The selection of vacuum tubes for low-noise preamplifiers has been discussed, e.g. by COTTINI et al. [2.020], DUBRAU [2.106], BLANKENSHIP [2.107], BLANKENSHIP and PINASCO [2.108], LEVENTHAL [2.109] and BILGER [2.090]. Since the tube preamplifiers are used preferably with thin non-cooled detectors exhibiting a high leakage current and high capacity C_d, the transconductance of the tubes is a more severe criterion for their selection, as the grid current, according to the Appendix 8.2. Hence preference is given to triodes or triode-connected pentodes ECC88, E88CC, E180F, E280F, E810F (Europe), 6922, 7788 (USA), 6Ж9П (USSR), 6R-H2, 6EJ7 (Japan).

The amplifier A_1 in Fig. 2.30 must be noninverting and preferably also a single stage one, in order that the whole amplifier remains stable. The common solution of this dilemma — which has been discussed already — consists in renouncing $A_1 \gg 1$ and in increasing the cascode gain by the bootstrap. Besides the common bootstrap, which is a positive voltage feedback known from many tube preamplifier circuits [2.110, 2.111], a high load R_{dyn} can also be realised by means of a positive current feedback according to HAHN and MEYER [2.112]. The current feedback circuit has been investigated in hybrid tube-transistor preamplifiers [2.113] as well as in pure tube circuits [2.114]. KANDIAH [2.115] reported the use of a constant current transistor as the load of a vacuum tube cascode.

Sometimes a long-tailed pair is used as the noninverting amplifier A_1. Although it consists of two components it exhibits the properties of a single-stage amplifier due to the cathode or emitter coupling. Another advantage is the low Miller capacity. According to CHASE et al. [2.110] the long-tailed pair yields stable preamplifiers even if they are made very fast. TAKEDA [2.116] suggested the use of the so-called transitron for the amplifier A_1, the principle of which is shown in Fig. 2.34. In a conducting pentode the current division between the screen grid and the anode is controlled by the potential of the suppressor grid g_3. A positive voltage pulse on g_3 causes a decrease in screen grid current and thus a positive voltage pulse on g_2.

GOLDSWORTHY [2.240] increased the gain A_0 of the amplifier virtually to infinity by means of a small amount of positive feedback.

WAHL [2.117] describes a fast-slow preamplifier with E280F.

Bipolar transistors are suitable for preamplifiers (cf. Fig. 8.25 in Appendix "Noise") only if semiconductor detectors with high junction capacity C_d are used or if a high energy resolution is not required. For low parallel noise, the base current of the transistor should be low. On

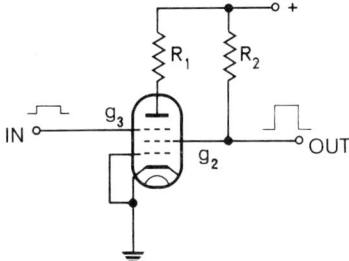

Fig. 2.34. The transitron circuit

the other hand, the transconductance $q_m \approx (e/kT) \cdot I_c$ depends on the collector current and thus on the base current. For small series noise therefore a high base current is required. Hence there is an optimum base current yielding a minimum total noise. The optimum base current amounts to 1...1000 µA, yet depends on the total input capacity C [2.118].

EMMER [2.118] investigated the properties of bipolar transistor preamplifiers and produced a circuit with 2N 1279 in the common bootstrapped cascode configuration. Fig. 2.35 shows another circuit by EMMER [2.119], where the lower transistor of the cascode is preceded by an emitter follower. The operating point of the input transistor is adjusted with the aid of the potentiometer POT. With an input current of 50 µA the amplifier exhibits a loop gain of about 1500, the noise (FWHM) for $C_{ext}=0$ being 25 keV(Si), for $C_{ext}=50$ pF 35 keV(Si). The preamplifier is completed by a voltage amplifier with $A=25$ and an impedance converter, which is of the usual layout and not shown in Fig. 2.35. SPLICHAL [2.120] reports a preamplifier with cascode and input emitter follower with 2N 697, with the emitter follower and the lower transistor of the cascode being connected to a Darlington stage. This makes the adjusting of the operating points more easy.

JONASSON [2.121] reports experiments on a preamplifier with a tunnel diode as the input component, the noise of which for $C_{ext}=0$ was about 12 keV (FWHM, Si). Due to difficulties connected with the use of negative resistance dipoles in amplifiers, tunnel diodes are not normally used in charge sensitive preamplifiers.

CHASE and RADEKA [2.122, 2.123] investigated low noise preamplifiers with parametric diodes. Despite a very complicated circuit the resulting noise is not much better than the noise of substantially simpler preamplifiers with cooled field effect transistors.

The use of FET in preamplifiers for semiconductor detectors is discussed by RADEKA [2.124]. He constructed a preamplifier with two

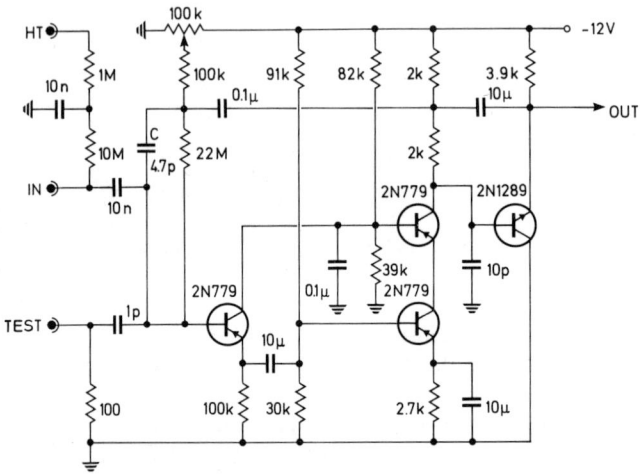

Fig. 2.35. Transistorized preamplifier according to EMMER [2.119]

n-channel field effect transistors FSP 401 in cascode configuration with bootstrapped load, the diagram of which is shown in Fig. 2.36. The type of FET used is not at all critical; RADEKA reports 2N 930, 2N 2586 or 2N 2252 as working equally well. The FSP 401 exhibits a small transconductance ($g_m \approx 0.2$ mA/V), and is advantageous preferably for detectors with very small capacity. With a total input capacity $C = 4$ pF the preamplifier noise is about 2 keV (FWHM, Si, room temperature), the dependence of the noise on the input capacity being about 0.5 keV/pF, with $\tau_{\mathrm{diff}} = \tau_{\mathrm{int}} = 1$ μsec.

Instead of the upper FET of the cascode a bipolar transistor can be used, though a low-noise type must be selected [2.124]. BLALOCK [2.125] described such a circuit. The input element is a p-channel FET 2N 2500 with $g_m = 1.5$ mA/V, combined with the bipolar transistor 2N 835 to form a cascode. The output stage consists of a White emitter-follower, which feeds the bootstrap loop. BLALOCK loc. cit. reports the noise of this preamplifier to be a minimum at a temperature of about 125 °K, which amounts to 2 keV (FWHM, Si) for $C = 25$ pF, $dQ_N/dC = 0.08$ keV/pF, with $\tau_{\mathrm{diff}} = \tau_{\mathrm{int}} = 1$ μsec. Preamplifiers with cooled FET are discussed in a number of papers ([2.126] to [2.130] and [2.241] to [2.244]). The minimum noise is commonly found at about $T=110\,°K$ (cf. also RADEKA [2.131]). Of course, only the input FET is cooled and is usually mounted together with the semiconductor detector in the same cryostat. The stray capacities can be reduced by appropriate arrangement of the FET close to the detector. Constructional details of such arrangements are given by SMITH and CLINE [2.127] and NYBAKEN and VALI [2.129].

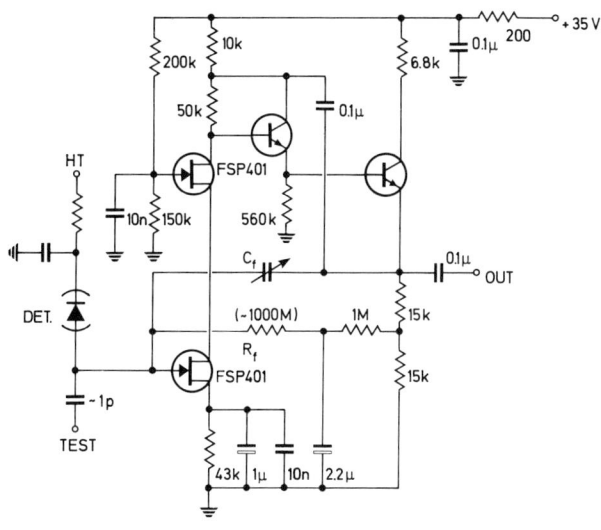

Fig. 2.36. Preamplifier circuit employing two FET according to RADEKA [2.124]

N-channel FET's such as 2N 3823, 2N 3819 usually exhibit lower noise than the p-channel types. With cooled detectors and FET preamplifiers, overall energy resolution of 0.7 keV has been reached (e.g. ELAD [2.128, 2.241]) which must be only partly attributed to the preamplifier noise. HARRIS and SHULER [2.243] reported 500 eV resolution with 2N 3823. Using germanium junction field effect transistors (JFET) the resolution can be improved to 0.28 keV FWHM (Ge) with $dQ_N/dC = 0.018\,\text{keV/pF}$ (ELAD and NAKAMURA [2.245] to [2.247]).

SMITH and CLINE [2.127] recommend the use of several parallel-connected FET's as the input element of preamplifiers for detectors having high capacity C_d. The preamplifier described by NYBAKEN and VALI [2.129] is realized in hybrid techniques using vacuum tubes (nuvistors) throughout the circuit with the exception of the input FET.

Most of the preamplifiers, including those in Fig. 2.32, 2.35 and 2.36, use a test signal input with a small coupling capacity of 0.5 to 1 pF. By means of this input, current pulses of constant and well-defined total charge can be applied to the preamplifier for calibration purposes. Since the test pulse height does not vary, the variations of the corresponding output pulse heights are caused solely by the preamplifier noise and the width of the test signal line corresponds directly to the contribution of the preamplifier to the total energy resolution (FAIRSTEIN [2.132, 2.133]). The mean noise voltage $\langle V_{out}^2 \rangle$ can also of course be measured directly by means of a voltmeter with square characteristic.

If high pulse rates are to be dealt with, a separation of the detector from the charge-sensitive preamplifier (or from the integration capacitor) with the aid of a bipolar transistor in common base configuration offers advantages according to ALBERIGI-QUARANTA et al. [2.134] (Fig. 2.37). The prototype of the circuit worked satisfactorily up to rates of 5 Mcps with a noise level of 35 keV (FWHM) at $C_d = 1$ pF, $dQ_N/dC = 1$ keV/pF.

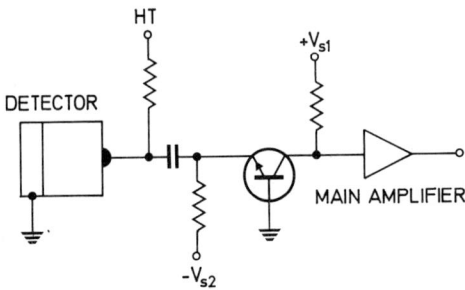

Fig. 2.37. Semiconductor detector input circuit for very high counting rates [2.134]

GOLDSWORTHY [2.248] proposed the reduction of the amplifier integrating time constant by connecting the shunt resistor to the output of a later stage instead of connecting it in parallel to the feedback capacitor C_f.

2.5. Scintillation and Čerenkov Counters

2.5.1. Principle of a Scintillation Counter

Apart from the complete ionization, some atoms and molecules are excited to higher energy levels during the interaction of ionizing radiations with matter, and consequently emit light during deexcitation. The scintillation counter is based on the detection of this light from an optically active medium (scintillator) by means of a photomultiplier.

There is a vast number of monographs and reviews on scintillation counters (e.g. [2.003], [2.135] to [2.138]) among which at least the comprehensive review by BIRKS [2.139] should be mentioned by name. Thus, it is hardly necessary to quote the original papers for common questions discussed in this chapter.

The structure of a scintillation counter is shown schematicaly in Fig. 2.38. Inorganic and organic monocrystalls, polycristalline layers, solutions of fluorescing organic compounds in organic solvents and plastics, or fluorescing glasses and gases are used as scintillators. The fluorescence mechanism is different for different scintillator types, and the amount of light produced per unit of energy ΔE absorbed in the scintillator depends generally on the particle species and the particle energy (cf. Chapter 6 and 11 in BIRKS [2.139]). The fluorescence photons release photo electrons from the photo cathode of the photomultiplier which are collected by the first dynode and fed into the multiplier system.

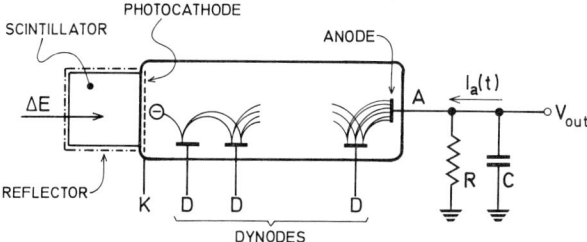

Fig. 2.38. Operating principle of a scintillation counter

The light loss between the scintillator and the photo cathode is minimized by using suitable reflectors and taking precautions to achieve an effective optical contact between the scintillator and the phototube window. At the anode the initial charge pulse appears magnified by the secondary emission of the n dynodes. In the approximation of an integration time constant RC being much greater than the pulse duration, the output voltage pulse height V_0 is

$$V_0 = \frac{\bar{A}}{C} e \alpha \Delta E p_{max} \int_0^\infty \sigma(\lambda)\pi(\lambda)d\lambda = \frac{\bar{A}}{C} e \bar{N}. \tag{2.42}$$

The particular symbols are as follows: e = electron charge; α = total photon yield of the scintillator, measured in photons per unit energy ΔE; $\sigma(\lambda)$ = emission spectrum of the scintillator, $\alpha \cdot \sigma(\lambda)$ exhibits the dimension photons per unit wavelength interval, $\sigma(\lambda)$ is normalized by $\int_0^\infty \sigma(\lambda)d\lambda = 1$;

$p_{max}\cdot\pi(\lambda)$ = probability of the release of one photo cathode electron reaching the multiplier system for every photon emitted from the scintillator, this probability has a maximum p_{max} for $\lambda = \lambda_{max}$, thus $\pi(\lambda_{max})=1$;
$\bar{A}$ = mean multiplier gain.

The term $p_{max}\cdot\pi(\lambda)$ is mainly given by the quantum efficiency of the photo cathode of the multiplier and by its spectrum sensitivity. It allows for light losses between the scintillator and the photo cathode, for the possible dependence of the reflectivity of the reflector on the light wavelength, as well as for the collection efficiency cathode—first dynode. The integral in (2.42) describes the spectral response matching of the system scintillator—photomultiplier.

> The emission spectrum $\sigma(\lambda)$ often depends also on the particle species [2.140]. If wavelength filters are used between the scintillator and the photo cathode, this can be used for discrimination between different particles [2.141]. Within classes of similar scintillators the shape of $\sigma(\lambda)$ is approximately constant and the specification of the mean wavelength of emission $\bar{\lambda}$ sufficiently characterises the properties of the scintillator concerned. Similarly, for the description of $\pi(\lambda)$ a single parameter often suffices. In this case the overlap integral (2.42) can be represented as a simple function of this parameter and $\bar{\lambda}$ (cf. Kowalski et al. [2.142] for liquid scintillators).

In the right hand side of the relation (2.42), the corresponding quantities are summarized in $\bar{N}$, the number of electrons reaching the multiplier system. The energy $W = \Delta E/\bar{N}$ required to form one photo cathode electron (in analogy to (2.1) and (2.25)) in the scintillation counter is relatively high because of the multistage conversion process. In NaI(Tl) W comes to 300 to 1000 eV for gamma quants. The corresponding values are about three times higher in organic scintillators, and about twenty times higher in glass scintillators.

The multiplier gain $\bar{A}$ is equal to the product of the secondary emission factors $\bar{\delta}_i$ of the particular dynodes. The actual value of $\bar{\delta}_i$ also allows for losses due to incomplete secondary electron collection. If all $\bar{\delta}_i$ are the same, $\bar{A}$ is

$$\bar{A} = \bar{\delta}^n, \qquad (2.43)$$

where n is the number of dynodes. Since $\bar{\delta}$ depends approximately linearly on the dynode voltage, the gain $\bar{A}$ is proportional to a high power of the overall photomultiplier voltage V_s, which must be therefore particularly stable (cf. Chapter 2.5.5). Besides this the gain $\bar{A}$ depends to some extent on external magnetic fields. Typical values of $\bar{A}$ are $10^6 \ldots 10^8$.

Besides the conventional scintillation counters consisting of a scintillator and a photomultiplier tube, any other photosensitive element (e.g. semiconductor photocell) can be used for the detection of the light pulses from the scintillator [2.249, 2.250].

2.5.2. The Pulse Shape

The scintillator atoms or molecules excited at the time $t=0$ by a primary event have a certain lifetime, so that the intensity L of the emitted light will decrease exponentially with time

$$L(t) = L_0 \, \epsilon^{-\frac{t}{\tau_{fl}}} H(t),^4 \tag{2.44}$$

where τ_{fl} is the fluorescence decay time of the scintillator. Since there is no delay between the photon impact and the photo electron release, the rate of the electron emission from the photo cathode exhibits the same time dependence

$$\frac{d\overline{N}}{dt} = \frac{\overline{N}}{\tau_{fl}} \epsilon^{-\frac{t}{\tau_{fl}}} H(t). \tag{2.45}$$

To be exact, $(1/\tau_{fl}) \cdot \epsilon^{-t/\tau_{fl}} \cdot dt$ denotes the probability of the emission of one photo electron in the time interval $(t; t+dt)$, and (2.45) therefore describes merely the mean emission rate, the relative statistical deviations of which increase with decreasing total number $\overline{N}$ of electrons released per primary event. Hence, the first electron of the pulse is delayed against the instant of primary event by a time which varies in a statistical manner. This situation must be regarded e. g. in coincidence devices [2.143, 2.144].

Due to the variations in the times of flight of the particular secondary electrons in the multiplier system, the emission of even one single photo cathode electron leads to a cloud of secondary electrons with a finite space spread, thus resulting in an anode pulse $i_a(t)$ of finite duration. Of course, $i_a(t)$ is delayed against the primary event by the total propagation time cathode-anode, which is constant for given operating conditions and must not be considered here. The particular shape of the single electron pulse $i_a(t)$ depends certainly on the actual dynode geometry, however, according to LEWIS and WELLS [2.145], it can be represented by a Gaussian

$$i_a(t) = \frac{\overline{A}e}{t_p \sqrt{\pi}} \epsilon^{-\left(\frac{t}{t_p}\right)^2} \tag{2.46}$$

with a sufficiently small error. The constant t_p describes the variations of the electron propagation time. Of course $i_a(t)$ is normalized: $\int_{-\infty}^{+\infty} i_a(t)\,dt = \overline{A}e$. Another approximation (a harmonic one) has been proposed by TANASESCU [2.146].

---

[4] $H(t)$ denotes the Heaviside step function (cf. Appendix 8.1):

$$H(t) = \begin{cases} 1 & \text{for } t > 0, \\ 0 & \text{for } t < 0. \end{cases}$$

Radiation Detectors and Related Circuits

The anode current pulse $I_a(t)$ for several released photo cathode electrons is composed of the overlap of several single electron pulses $i_a(t-t_i)$ timed correspondingly to the particular emission instants t_i. In the approximation of a large total number $\bar{N}$ of photo cathode electrons, $I_a(t)$ is given by the convolution of (2.45) and (2.46)

$$I_a(t) = \frac{\bar{A}e\bar{N}}{\tau_{fl}t_p} \cdot \frac{1}{\sqrt{\pi}} \int_{-\infty}^{+\infty} e^{-\left(\frac{t-t'}{t_p}\right)^2} \cdot e^{-\frac{t'}{\tau_{fl}}} \cdot H(t')dt'. \qquad (2.47)$$

The relation (2.47) can easily be transformed into a Gaussian error function [2.145] and evaluated. The anode current $I_a(t)$ as a function of t/τ_{fl} and t/t_p is plotted in Fig. 2.39 and 2.40. Again, to be exact, (2.47) has merely the significance of a probability distribution and applies exactly only if $\bar{N} \to \infty$.

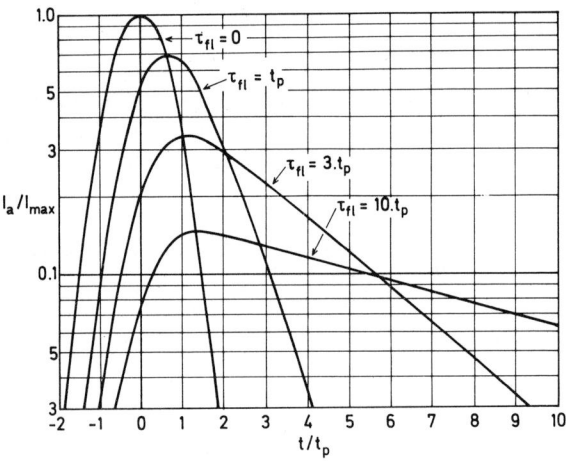

Fig. 2.39. Photomultiplier anode current I_a as a function of t/t_p

Plots similar to Fig. 2.39 are obtained in the course of measurements of the fluorescence decay times τ_{fl} of various scintillators. The plot for $\tau_{fl}=0$ corresponds to the resolution of the used photomultiplier assembly. The resolution can be measured by illuminating the photo cathode by very short light pulses of the duration of t_l ($t_l \ll t_p$). In fast phototubes it is $t_p \approx 1$ nsec. Light pulse generators with $t_l \lesssim 1$ nsec are described by various authors ([2.147] to [2.151] and [2.251] to [2.253]). However, the shortest light pulses with durations of a few psec are presumably produced in the phase-locked gas lasers ([2.152] to [2.155]). The pulses commonly lie in the near infrared and must be converted by means of non-linear optics [2.156] into the wavelength range where the spectral response of the photomultiplier is high.

MATHÉ [2.254] developed a technique for simulation of scintillation pulses by illuminating the photomultiplier cathode by a constant weak light source and controlling its grid by fast exponential pulses.

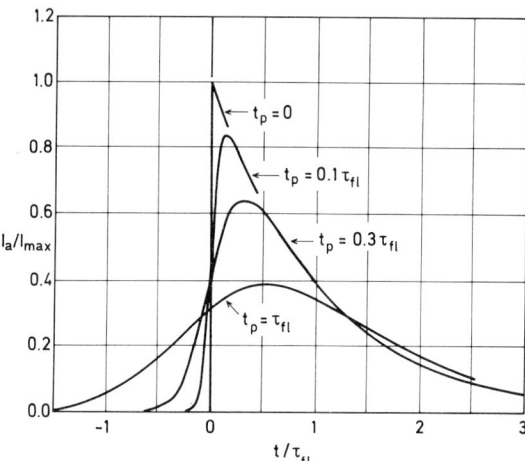

Fig. 2.40. Photomultiplier anode current I_a as a function of t/τ_{fl}

If the $I_a(t)$ corresponding to $\tau_{fl}=0$ is known, the corresponding value of τ_{fl} for any empirical plot $I_a(t)$ can be estimated easily. For this purpose the analysis according to NEWTON [2.157] and BAY et al. [2.158, 2.159] can be used. Although this analysis is conducted for another experimental situation (delayed coincidence measurements) it applies also in the case concerned.

Fluorescense decay time measurements are reported by numerous authors (apart from the general reviews [2.135] to [2.139] cf. the recent papers [2.160] to [2.162]). Inorganic scintillators are commonly slower (τ_{fl} of NaI(Tl) = 0.25 μsec; τ_{fl} of CsI(Tl) for gamma quants = 1 μsec) than organic scintillators (τ_{fl} of anthracene crystal = 30 nsec; τ_{fl} of liquid scintillators between 1 to 5 nsec). However, the simplified description of the fluorescence decay by $L(t)$ (2.44) proves to be insufficient, since in addition to the dominant fast fluorescence component, most of the scintillators exhibit other, slower, components. Thus $L(t)$ must be represented by the sum of two or more exponential terms with different time constants. The relative weight of the particular components is generally a function of the particle species. In Fig. 2.41 for instance the shapes of current pulses $I_a(t)$ in a scintillation counter with stilbene are shown, the exciting radiations being alpha particles, protons and gamma quants, respectively (BOLLINGER and THOMAS [2.228]). The difference in the pulse shapes can be used in pulse shape discriminators (Chapter 4.3) for distinguishing between different particles.

The anode current $I_a(t)$ (2.47) is integrated in a RC network at the output of the photomultiplier (Fig. 2.38) in order to obtain a voltage pulse with amplitude proportional to ΔE. Dependent on the time constant $\tau = RC$, more or less of the total $\overline{N}$ photocathode electrons are released until the voltage pulse reaches its maximum (2.45). For small statistical deviation of the pulse height, $\tau \gg \tau_{fl}$ and thus $\tau \gg t_p$

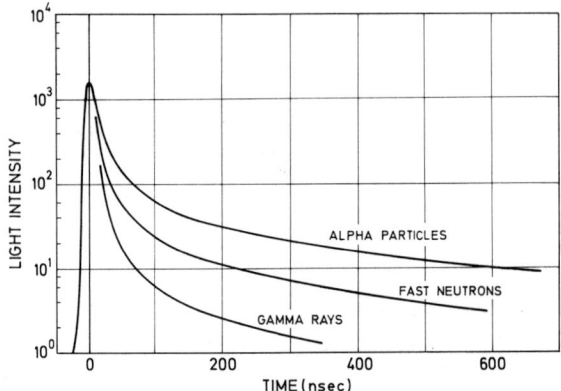

Fig. 2.41. Current pulse shapes in scintillation counter with stilbene scintillator for alpha particles, protons and gamma quants according to BOLLINGER and THOMAS [2.228]

must be realized. In the case of t_p being negligible (i.e. $t_p=0$), $i_a(t)$ (2.46) becomes a δ-function, $I_a(t)$ remains pure exponential and the output voltage $V_{out}(t)$ in Fig. 2.3.8 is

$$V_{out}(t) = -V_0 \cdot \frac{\tau}{\tau_{fl} - \tau} \left(\epsilon^{-\frac{t}{\tau_{fl}}} - \epsilon^{-\frac{t}{\tau}} \right) H(t), \qquad (2.48)$$

with $V_0 = \overline{A} e \overline{N}/C$ from (2.42). In Fig. 2.42 the pulse shape $V_{out}(t)/V_0$ for different ratios τ/τ_{fl} is plotted [2.163]. Obviously $\tau > \tau_{fl}$, or better $\tau > 10 \cdot \tau_{fl}$, must be chosen so as to avoid a pulse height loss and a reduction in the number of photocathode electrons contributing to the signal. If any differentiating time constant other than τ, and later in the amplifier, limits the pulse length, this condition remains valid accordingly. For slow scintillators, such as e.g. NaI(Tl) with $\tau_{fl} = 0.25$ μsec, the appropriate combination is to use $\tau \approx 100$ μsec and to limit the pulse length, for instance by means of a delay line differentiator in the main amplifier, to 1 μsec.

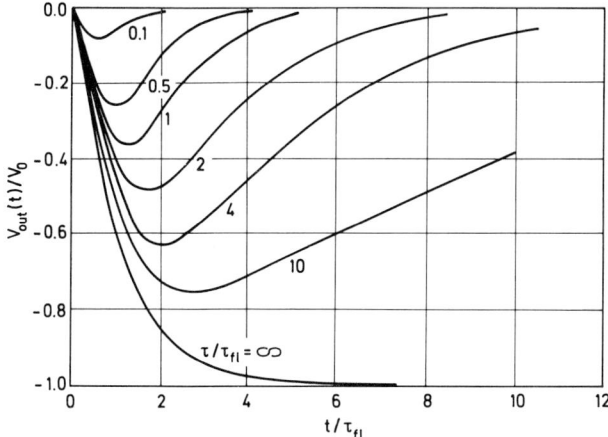

Fig. 2.42. Photomultiplier integrated voltage output pulse for various τ/τ_{fl}

2.5.3. Photomultiplier Statistics and the Pulse Height

With infinite or sufficiently long time constants τ, the voltage pulse height is given by $V_0 = \bar{A} e \bar{N}/C$ (2.48). Its deviation σ_V — similar to the proportional counter — depends apart from the statistical deviation σ_N of the number N of photo cathode electrons (2.42), on the deviation σ_A of the multiplier gain. If all particular processes taking part in the release of photo cathode electrons possess a normal distribution, the relation

$$\left(\frac{\sigma_V}{V_0}\right)^2 = \left(\frac{\sigma_N}{\bar{N}}\right)^2 + \frac{1}{\bar{N}}\left(\frac{\sigma_A}{\bar{A}}\right)^2 \tag{2.49}$$

can be derived in analogy to (2.12). Since $\sigma_N = \sqrt{\bar{N}}$, (2.49) simplifies to

$$\left(\frac{\sigma_V}{V_0}\right)^2 = \frac{1}{\bar{N}}\left[1 + \left(\frac{\sigma_A}{\bar{A}}\right)^2\right] \tag{2.50}$$

(for more details cf. e.g. [2.139] p. 148 and BREITENBERGER [2.164]). If $\tau \gg \tau_{fl}$ cannot be presumed, then instead of $\bar{N}$, only that part of the total photo cathode electron number which in fact influences the pulse height must be introduced in (2.50).

The variance $(\sigma_A/\bar{A})^2$ of the multiplier gain can be calculated from

$$\left(\frac{\sigma_A}{\bar{A}}\right)^2 = \frac{1}{\bar{\delta}-1}, \tag{2.51}$$

if all dynodes exhibit the same secondary emission factor $\bar{\delta}$, or from

$$\left(\frac{\sigma_A}{\bar{A}}\right)^2 = \frac{\bar{\delta}}{\bar{\delta}_1} \cdot \frac{1}{\bar{\delta}-1}, \qquad (2.52)$$

if all except the first dynodes have $\bar{\delta}$, and the first dynodes have $\bar{\delta}_1$ as the mean stage gain. In (2.51) and (2.52) a Poisson distribution is assumed for δ.

This assumption is a useful though rough approximation of the actual situation. PRESCOTT [2.033, 2.034] investigated the contribution of the multiplier system to the total statistical deviation σ_V under less limiting conditions. Papers concerning this subject (up to 1962) are reviewed in a comprehensive paper by PRESCOTT and TAKHAR [2.032].

Using (2.51) or (2.52), (2.50) can be written

$$\left(\frac{\sigma_V}{V_0}\right)^2 = \frac{1}{N} \frac{\bar{\delta}}{\bar{\delta}-1} \quad \text{or} \quad \left(\frac{\sigma_V}{V_0}\right)^2 = \frac{1}{N}\left(1 + \frac{\bar{\delta}}{\bar{\delta}_1(\bar{\delta}-1)}\right). \qquad (2.53)$$

The resolution (FWHM) amounts to 2.35 (σ_V/V_0). From (2.53) it can easily be seen, that $\bar{\delta}$ and especially $\bar{\delta}_1$, i.e. the photomultiplier operating voltage, must be high if the optimum possible resolutions of $1/\sqrt{N}$ shall not be the worse for the multiplier statistics.

In counting practice the energy resolution of the scintillation counter as a rule is lower than in (2.53). The following effects are responsible for this situation: The photon yield of the scintillator is dependent on the origin of light emission. The radiation can lose its energy in different ways in the scintillator, for instance a gamma quantum by one photo effect or by several gradual Compton effects. In the first case one electron of high energy is produced, in the second case the same energy is divided between several low-energy electrons. Due to the energy dependence of the light yield, the total amount of light produced in both cases is different. Also δ-rays formed during the stopping of a beta ray can contribute to the amplitude deviation. Further, the light coupling losses and the photo cathode sensitivity exhibit local variations. Experimental data concerning the energy resolution are reviewed in [2.138, 2.139].

Direct measurements of the statistical gain fluctuations occuring in the electron multiplication process are reportet by EBERHARDT [2.255].

2.5.4. Thermal Noise

Even with no external light source a "dark current" $\bar{I}_{ao}$ flows to the photomultiplier anode. The actual value of $\bar{I}_{ao}$ depends on the particular photo tube, a typical value being 0.1 µA. An obvious assumption is to attribute $\bar{I}_{ao}$ to the thermal emission of $\bar{n}_0$ electrons per second from the photo cathode. The cathode dark current $\bar{I}_{ko} = \bar{n}_0 \cdot e$ exhibits statistical variations with $d\langle I_{ko}^2\rangle/df = 2e\bar{I}_{ko}$ (8.205). Introducing $\bar{I}_{ao} = \bar{A}\cdot\bar{I}_{ko}$, we get

$$\frac{d\langle I_{ao}^2\rangle}{df} = 2e\bar{I}_{ao}\bar{A}\frac{\bar{\delta}}{\bar{\delta}-1}. \qquad (2.54)$$

The term with $\bar{\delta}$ describes the magnification of the noise due to the multiplier statistics. According to (8.214) and (8.207), the equivalent noise charge Q_N for pulse shaping by means of one integrator and one differentiator with equal time constants τ amounts to

$$Q_N = \sqrt{2\tau\frac{1}{e}\frac{\bar{I}_{ao}}{\bar{A}}\frac{\bar{\delta}}{\bar{\delta}-1}}. \qquad (2.55)$$

Here Q_N is expressed in photo cathode electrons and must be compared to $\bar{N}$. For instance, a dark current $\bar{I}_{ao}=0.1$ µA, $\bar{A}=10^6$, $\tau=1$ µsec and $\bar{\delta}=3$ results in $Q_N=1.4$ photo cathode electrons.

This approximative calculation is so far incorrect, since we have used two very simplified assumptions. Since the anode dark current also has other origins besides the thermal cathode emission of single electrons, it is $\bar{I}_{ko}<\bar{I}_{ao}/\bar{A}$. Thus in calculating Q_N from (2.55), the term $\bar{I}_{ao}/\bar{A}$ must be replaced by the much smaller $\bar{I}_{ko}$, yielding a lower Q_N. The other origins of anode dark current mentioned may be thermal emission from dynodes, field emission, scintillations in the glass envelope of the phototube (e.g. electroluminescence, K^{40}-content of the glass), optical or ion feedback anode to cathode, isolation currents, etc. On the other hand, in deriving (2.54), the validity of Poisson distribution has been assumed; this is not true, at least for δ, making Q_N somewhat higher.

Nevertheless, as can be seen from (2.55), the photomultiplier noise as a rule can be neglected in comparison with the statistical broadening of the pulse height resolution (2.53).

Hence the only inconvenience caused by the thermal emission of the photo cathode is the increase in background with discriminator settings equivalent to 1 to 10 photo cathode electrons or less. The variations of the multiplier gain A cause a disproportional amplification of some of the single-electron pulses, thus making their pulse heights equal to the mean amplitudes of multielectron pulses. The pulse height spectrum of the thermal noise background therefore corresponds to the distribution of multiplier gain A. This spectrum usually has an exponential shape monotonously decreasing from the smallest to the higher amplitudes, although some authors report a flat maximum at the pulse height equivalent to the mean amplitude of single electron pulses (cf. [2.034]). For pulse heights of more than about 10 electron equivalent, the spectrum shape becomes flat due to a multielectron component of the noise, the precise origin of which is not known at present (BAICKER [2.165]).

Several photomultiplier types have been developed with considerably reduced thermal noise (e.g. the S types by EMI, 8575 by RCA, and many others), whose emission rate is about 10 electrons per second per square centimeter of the cathode area. The noise generally can be reduced by cooling the phototube ([2.166] to [2.168], [2.171]), by connecting two phototubes viewing the same scintillator in coincidence (proposed by MORTON and ROBINSON [2.169]), or by combining of both precautions. The second method became widely used in liquid scintillation counters (PACKARD [2.170]).

From the electronic point of view, the discrimination between noise and signal pulses of a photomultiplier on account of their different shape is very interesting. The general aspects of the pulse shape discrimination are treated in Chapter 4.3, while here only those problems directly related to the scintillation counter are discussed. With the exception of the very rare multielectron component, the noise pulses are disproportionally amplified single-electron pulses. Restricting our considerations to two or more electron signal pulses (single-electron signal pulses are indistinguishable from the noise anyhow), the signal pulse is characterized by the correlated emission of several electrons from the photo cathode within a time interval corresponding approximately to the fluorescence decay time τ_{fl} of the scintillator. The principle of the discrimination is as follows: the phototube output pulse is differentiated with a time constant $\tau_{diff} \ll \tau_{fl}$ thus resolving the particular single-electron contributions. The first single-electron pulse opens a gate for e.g. $3 \cdot \tau_{fl}$ without producing a count pulse. First the surplus charge, corresponding to the second, third, etc. photo cathode electron, which can pass the open gate, forms a pulse at the output of this signal noise discriminator. In this way only the signal pulses with but 1 electron in $3 \cdot \tau_{fl}$ are lost, on the other hand all the noise pulses are eliminated (with the sole exception of chance coincidences within $3 \cdot \tau_{fl}$). This signal noise discrimination has been discussed by SWANK [2.172]. FORTE and ANZANI [2.173] describe a practical circuit of this type for slow inorganic scintillators (ZnS(Ag) with $\tau_{fl} = 5$ μsec).

If τ_{fl} is not too long and if the signal pulses consist of substantially more that 2 to 3 electrons, the device described by FORTE and ANZANI loc. cit. cannot be used without difficulties. The signal pulse in this case differs from the noise by a longer mean rise time only: the rise time of single-electron pulses is given solely by t_p (2.46) of the phototube whereas the rise time of a multielectron signal pulse consists of both t_p and the decay time τ_{fl} of the scintillator. Correspondingly, the zero crossing of a double differentiated signal pulse occurs later than the of a single electron noise pulse. LANDIS and GOULDING [2.174, 2.175] describe a noise-suppressing system for NaI(Tl) scintillation counters using this effect. The output pulse of the phototube is divided between two channels. In the fast channel the "zero" time point is determined. The zero crossing point of the double differentiated slow channel pulse is delayed against this "zero" point by about 300 nsec for signal pulses and actually not delayed for noise pulses. The delay is measured by a coincidence circuit actuating a linear gate for signal events. Thus only signal pulses appear on the linear gate output. The influence of the noise suppression on the low energy part of the pulse height spectrum is demonstrated in Fig. 2.43.

A more flexible system for noise elimination is proposed by DAMERELL [2.176]. The principle can be seen from Fig. 2.44. The current pulse at the last dynode D_n is integrated on C_1 and differentiated by $R_1 C_1 < \tau_{fl}$, the pulse height being thus proportional to the amplitude of the first single electron sub-pulse. The differentiating time constant of the anode circuit is $R_2 C_2 \gg \tau_{fl}$, and the resulting pulse height is proportional to the total charge of the current pulse. The duration of the stretched anode and dynode pulses is equalized and the gain of the amplifier in the dynode channel is adjusted to yield exactly

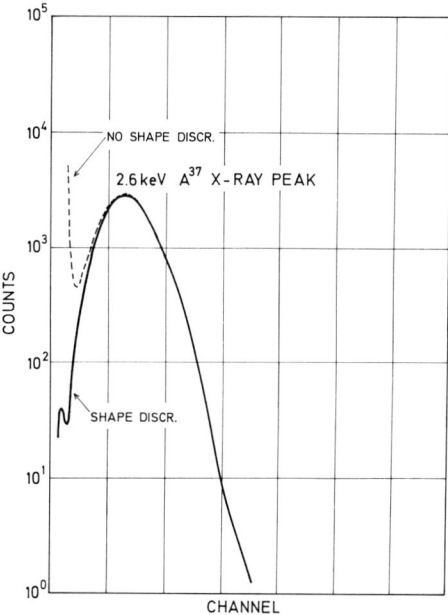

Fig. 2.43. Influence of photomultiplier noise suppression on the low energy part of the pulse height spectrum according to LANDIS and GOULDING [2.175]

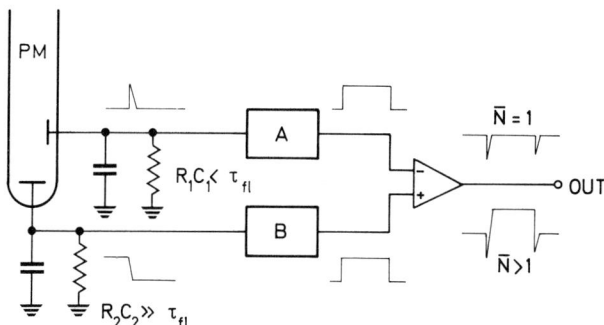

Fig. 2.44. Principle of the noise suppressing circuit by DAMERELL [2.167]. A = pulse stretcher and amplifier, B = inverting amplifier with delay line clipping and additional signal delay

the same dynode and anode pulse height in the case of single electron pulses. For multi-electron pulses obviously the anode pulse is higher. Therefore at the output of the difference amplifier only signal pulses appear. As can easily be seen, the pulse height spectrum remains conserved at the difference amplifier output, though the pulse height is a function of $N-1$ (and not N).

The circuit described by DAMERELL [2.176] worked satisfactorily for scintillators with $\tau_{fl} \geqslant 200$ nsec. Apparently the photomultiplier must be faster than the scintillator, i.e. $t_p < \tau_{fl}$, if the rise times of the signal and noise pulses shall differ substantially. Since very fast phototubes have recently been developed [2,177, 2.178, 2.256], there exists a possibility of using DAMERELL's principle also for noise discrimination in liquid scintillation counters with $\tau_{fl} \approx 1$ nsec.

CHEVALIER [2.250] solved the noise suppression problem by the construction of a new type of photomultiplier tube, in which the photo cathode electrons are first accelerated by about 50 kV and then directed to a high resolution semiconductor detector. Hence simple electron pulses are easily resolved.

2.5.5. Signal Circuits Used in Scintillation Counters

In scintillation counters the signal pulse amplitude usually is very high as compared with the noise level of common preamplifiers. The preamplifier need therefore not be of a low noise type and often its sole function is to match the photomultiplier output impedance to the characteristic impedance of the connecting shielded coaxial cable. The situation at the photomultiplier output is shown in Fig. 2.45 in some detail. The dynode voltages are supplied by a resistor chain (cf. Chapter 2.5.6). If the cathode lies on the ground potential, the whole high tension of about 1 kV is on the anode and the preamplifier must be connected via a high tension coupling condensor C_1. C_a and C_b denote the parasitic capacities of the photomultiplier output and preamplifier input respectively and $C_p = C_a + C_b$ denotes the total parasitic capacity.

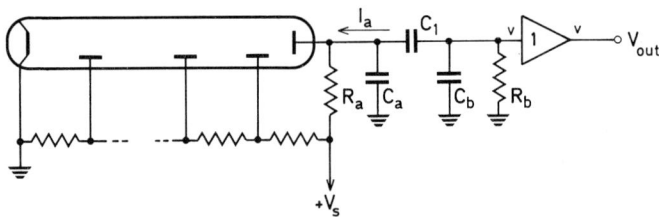

Fig. 2.45. Signal output of a photomultiplier

Mostly $C_a \approx C_b \approx 10$ pF. If the anode circuit is earthed and the high tension $-V_s$ is applied to the cathode, a special high voltage condenser C_1 is superfluous. R_a denotes the anode resistor, R_b is the preamplifier input resistance.

In the case of R_a, R_b approaching infinity, the voltage $\hat{V}_{out}$ becomes

$$\hat{V}_{out} = -\frac{1}{p(C_a+C_b)} \cdot \hat{I}_a \frac{C_1}{C_1 + C_a C_b/(C_a+C_b)}$$

$$\approx -\frac{1}{pC_p} \cdot \hat{I}_a \frac{C_1}{C_1 + C_p/4}, \quad (2.56)$$

where it is assumed that $C_a \approx C_b \approx C_p/2$. The coupling condensor C_1 must therefore be $C_1 \gg C_p/4$ if no pulse height loss is to result. With $C_1 \gg C_p/4$ the capacity C_1 can be neglected also in calculating the time constant τ ($R \cdot C$ in Fig. 2.38) which becomes

$$\tau = (R_a//R_b)(C_a+C_b). \quad (2.57)$$

Thus preferably $R_a \approx R_b$ is made.

Because of the condition $\tau \gg \tau_{fl}$, the relation (2.48) is reduced to

$$V_{out}(t) = -\frac{\bar{A}e\bar{N}}{C_p} \epsilon^{-\frac{t}{\tau}} \cdot H(t) = -V_0 \cdot \epsilon^{-\frac{t}{\tau}} \cdot H(t). \quad (2.58)$$

If τ is high and the pulse shaping is performed in the main amplifier first, many pulses do overlap and the well-known picture of pile-up results (Fig. 2.46). As will be shown in Chapter 3.1.4, the rms deviation of the output voltage from the zero value using (2.58) amounts to

$$\sqrt{\langle V_{out}^2 \rangle} = V_0\sqrt{\tfrac{1}{2}r\tau}, \quad (2.59)$$

where r is the mean pulse rate. In (2.59) all pulses are supposed to be of the same amplitude V_0. With $r=20$ cps, $\tau=100$ μsec, $\Delta E=1$ MeV,

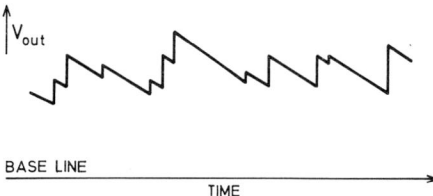

Fig. 2.46. The pile-up voltage at the photomultiplier output

$W=1000$ eV per electron, $\bar{A}=10^6$ and $C_p=16$ pF the voltage V_0 is $V_0=10$ V and $\sqrt{\langle V_{out}^2 \rangle}=10$ V. The linear range of the preamplifier must therefore be large enough (in the described case some 10 V). Vacuum tube preamplifiers have normally a linear range of more than

50 V. With transistorized preamplifiers the photomultiplier output pulse height must be matched to the available linear range of about 10 V by reducing the multiplier gain. Since because of (2.51) or (2.52) a high multiplier gain is desired, the gain reduction can be performed by artificial magnification of the capacity C_p or by using the current pulse of one of the last dynodes as the output signal.

The pulse height V_0 might be in itself high enough for further treatment in discriminators, analog digital converters etc. without any additional amplification. By appropriate choice of the multiplier gain A, only the voltage range of the analog digital converter need be matched to the desired range of the radiation energy. However, since a variation in the supply voltage V_s of the photomultiplier affects not only A, but also the energy and time resolution of the scintillation counter, the voltage V_s is kept constant and the desired variation of the total gain is accomplished by means of a variable voltage divider behind the preamplifier, the gain loss being compensated by a low-gain main amplifier. Usually the primary aim of the main amplifier is the pulse shaping, since its gain is only a secondary benefit.

It must be considered that by means of the voltage divider, the signal amplitude alone is reduced and not the noise of the first amplifier stage. Thus the signal-to-noise ratio can be affected adversely, especially if the pulse shaping occurs by means of the time constant τ (2.57) directly at the photomultiplier output and a broadband main amplifier is used at the same time. Therefore, in systems with single differentiators, the differentiating time constant must always be located behind the voltage divider and the first amplifier stage ($A_1 \approx 10$). In double differentiating systems the second time constant is commonly located in one of the last amplifier stages, thus enabling the photomultiplier circuit with τ to be used as the first differentiator.

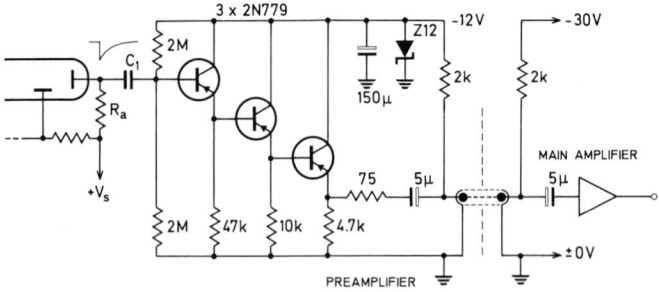

Fig. 2.47. Preamplifier for scintillation counters according to EMMER [2.119]

Reference is made only to a transistorized preamplifier by EMMER [2.119] for negative pulses, which does not need any explanation (Fig. 2.47). The R_b is about 1 MΩ. The use of npn transistors offers advantages, if a positive signal is taken from one of the dynodes. The supply voltage for the preamplifier is fed through the signal cable from the main amplifier. The two resistors of 2 kΩ each do not load the 75 Ω signal line noticeably.

In coincidence applications a signal is often needed which is proportional to the anode current $I_a(t)$. In Fig. 2.48 two variants of a fast signal output are shown. The last dynode—anode stage serves as a fast current generator with infinite internal resistance. In the variant (A) the loop D_n, A is closed by C and Z_0. A coaxial cable of the same characteristic impedance can be connected to Z_0. $C \cdot Z_0$ must be much longer than the maximum possible pulse length δ. The variant (A) offers the advantage of terminating the signal cable on the photomultiplier end by its characteristic impedance, thus avoiding multiple signal reflections even where the cable is not terminated correctly at the other end. In fast photomultipliers the system D_n, A forms a structure with a defined characteristic impedance Z_0 (mostly 50 Ω) which can be directly connected to a corresponding coaxial cable, as shown in the diagram (B). The anode voltage in this case must be applied by means of a resistor $R \gg Z_0$ (1 to 10 kΩ). Again it is $C \cdot Z_0 \gg \delta$. Here the cable must be terminated correctly at the output end, since it is completely open on the photomultiplier side which cannot thus absorb possible signal reflections.

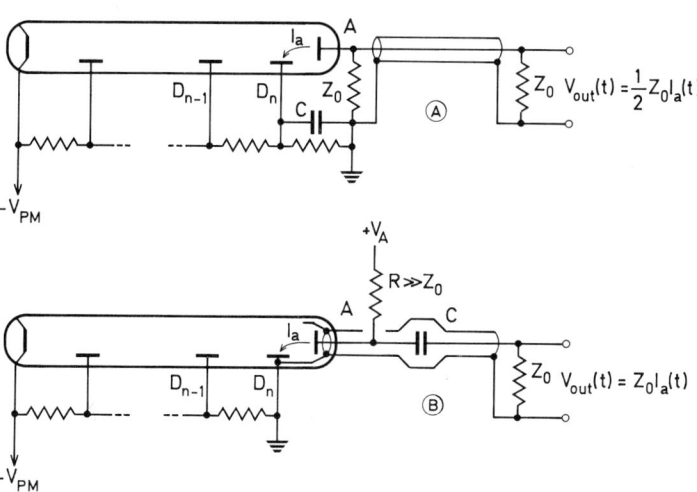

Fig. 2.48. Connection of a coaxial cable (impedance Z_0) to the photomultiplier output

If the fast photomultiplier signal is to be differentiated, e.g. by a short-circuited cable, the open photomultiplier end of the signal cable in (B) causes difficulties, since the reflection from the differentiator is again reflected, thus forming satellite pulses. HILL [2.179] reports a reflection-free differentiating system (Fig. 2.49). As can easily be seen, the incomming signal at the point (X) always meets with the same impedance Z_0 if both shaping cables have the length l and if all cables have the characteristic impedance Z_0. Only the output cable end must be terminated correctly. Of course, the absence of reflections is paid for by the loss of one half of the pulse height.

KREHBIEL [2.257] developed a simple fast preamplifier with an autotransformer and a common base transistor stage, with a gain of 8, which can be used for fast PM pulses, when the phototube gain is not sufficient.

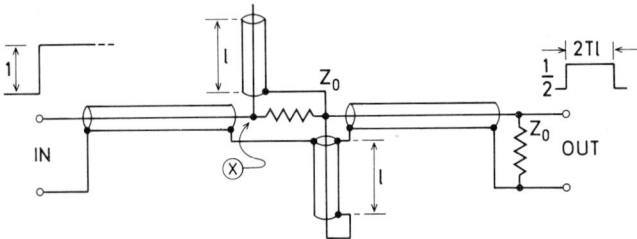

Fig. 2.49. Reflection free differentiator according to HILL [2.179]

Although the common operating mode of the photomultiplier is with constant high tension V_s, the dependence of the gain $\bar{A}$ on the voltage V_s can be used for control of $\bar{A}$ or even for gated operation of the scintillation counter. The gated operation offers advantages not only in coincidence applications, but also if the photomultiplier must be protected against extremely high pre-pulses (e.g. in pulsed accelerators), after-pulses (e.g. in spark chambers where triggering is accomplished by means of gas scintillations) or other time-correlated large signals. In gated operation either the whole voltage V_s [2.180], or merely the voltage of a single dynode section, or the potential of another electrode (e.g. the focussing or screening grid) can be controlled. The control pulse must be as high as 100 V. Because of this, difficulties arise from the capacitive pick-up of control pulses by the anode signal circuits. ROOSE [2.181] describes a dynode gate which is shown in Fig. 2.50. The gate pulse saturates the transistor BF 109 (normally cut off) via a transformer. The voltage between the dynodes D_2 and D_3 and thus the photomultiplier gain becomes zero. D_2 and D_3 exhibit about the same stray capacity relative to the anode A. Since none of the dynodes is blocked to ground, a positive pulse is formed on D_2 and a negative one on D_3. The parasitic anode signal caused by one of the pulses is compensated by the other. By means of R_T the two pulse heights can be equalized, and C_T allows for the parasitic capacity C_s. In a correctly adjusted circuit the interference with the anode signal lies below 0.1 mV.

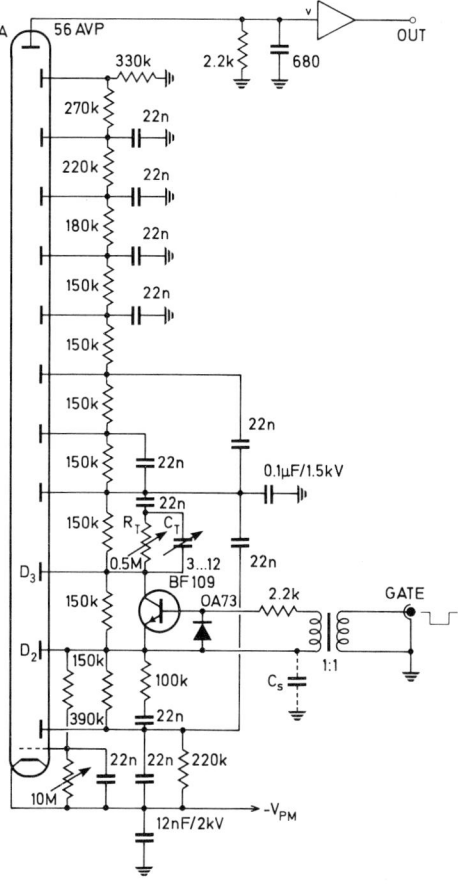

Fig. 2.50. Photomultiplier gating circuit according to Roose [2.181]

The transistor circuit of course can be modified for inverse operation (i.e. BF 109 normally saturated). A flip-flop control can also be foreseen.

GOERLICH (e.g. [2.182, 2.183]) developed a special phototube with a control grid shielded against the cathode for fast gating.

2.5.6. Auxiliary Circuits

The supply voltage to the particular dynodes usually is applied by means of a resistor voltage divider, as has been implied in the circuit examples of the Chapter 2.5.5. In this chapter an analysis of the voltage divider will be given.

The high tension applied to the voltage divider is provided by a high tension power supply which must be extremely stable against variations of mains voltage and of load, and which must have an excellent long-term stability, temperature independence and low hum. The description of suitable power supplies — which, of course, are also used for other applications besides nuclear ones — is outside the scope of this booklet. Modern transistorized high tension power supplies have been described e. g. by JOVANOVIĆ et al. [2.184], IZUMI and KOKUBU [2.185], SHEEN and RATCLIFFE [2.186] and FREVERT and KREISEL [2.187]. Concerning the general design criteria for power supplies, appropriate information can be found in the monograph by WAGNER [2.188] or any other similar text book.

The situation can be surveyed easily in the stationary case, viz. where the photocathode is sparsely illuminated, thus yielding a constant cathode current I_k=const. (Fig. 2.51). With zero light intensity $I_k=0$, and the whole current I_s flows through the resistor chain. Corresponding to the resistor values $R_0 \cdots R_n$ the particular dynodes have defined potentials dependent on the overall voltage V_s. For the sake of simplicity the typical case of such a voltage distribution will be assumed, viz.

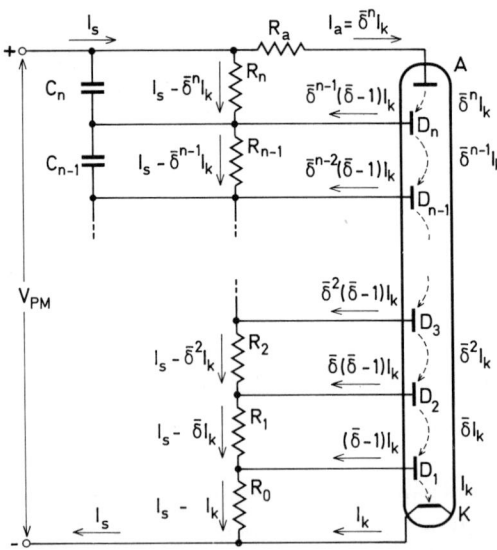

Fig. 2.51. Currents in the photomultiplier voltage divider ($V_{PM}=V_s$)

that all dynodes exhibit the same multiplication factor $\bar{\delta}$, resulting in $\bar{A} = \bar{\delta}^n$. If now a non-zero cathode current I_k is flowing, the resistor chain current I_s in the n-th dynode section is diminished by $\bar{\delta}^n \cdot I_k$. This lowers the potential differences $A - D_n$, $D_n - D_{n-1}$ etc. considerably, especially in the upper part of the resistor chain. Since the overall voltage V_s remains constant, the inter-dynode voltages in the lower dynode sections rise. The resulting increase in gain exceeds the gain loss in the upper dynode sections, and according to [2.163], the net increase of $\bar{A}$ can be approximated to

$$\frac{\Delta \bar{A}}{\bar{A}} = \frac{\bar{\delta}^n I_k}{I_s} \left(1 - \frac{1}{n+1} \cdot \frac{\bar{\delta}}{\bar{\delta}-1}\right). \tag{2.60}$$

The relative variation in gain $\Delta \bar{A}/\bar{A}$ is thus proportional to the ratio I_a/I_s. Hence, for stable operation, a chain current $I_s \gg I_a$ must be used.

The value of the anode resistor R_a is not critical, since the potential difference $A - D_n$ does not affect the gain severely. Nevertheless, $R_a \cdot I_a \ll R_n \cdot I_s$ must remain valid. A high resistance R_a may adversely affect the linearity of the photomultiplier, as will be shown later.

In pulsed operation the relations $I_s \gg I_a$ and $R_a \cdot I_a \ll R_n \cdot I_s$ are still valid, with I_a denoting the average anode current. In order not to disturb the voltage distribution and thus the gain, by the high peak pulse currents, the last dynode sections are blocked by the capacitors C_n, C_{n-1} etc. (cf. Fig. 2.52). In the case of gradually decreasing time constants $R_n C_n = \bar{\delta} \cdot R_{n-1} \cdot C_{n-1} = \bar{\delta}^2 \cdot R_{n-2} \cdot C_{n-2} = \cdots$ and with $\tau_{fl} \gg t_p$, the dependence of the gain variation on time after a pulse with $\bar{N}$ photo cathode electrons becomes [2.163]

$$\frac{\Delta \bar{A}}{\bar{A}} = \frac{\bar{A} e \bar{N}}{I_s R_n C_n} \frac{\epsilon^{-t/R_n C_n} - \epsilon^{-t/\tau_{fl}}}{1 - \tau_{fl}/R_n C_n} \cdot H(t). \tag{2.61}$$

Apparently $R_n C_n \gg \tau_{fl}$ must be chosen, so that (2.61) becomes

$$\frac{\Delta \bar{A}}{\bar{A}} = \frac{\bar{A} e \bar{N}}{I_s R_n C_n} \epsilon^{-t/R_n C_n} \cdot H(t). \tag{2.62}$$

As can be seen from (2.62), with a high time constant $R_n C_n$ the gain variation $\Delta \bar{A}$ is small, but it lasts correspondingly longer. If the mean pulse rate r is higher than $1/R_n C_n$, several pulses may overlap, yielding a higher gain deviation $\Delta \bar{A}$ as given by (2.62). For undisturbed statistics, the rms deviation of $\Delta \bar{A}$ can be derived in analogy with (2.59) to be

$$\sqrt{\left\langle \left(\frac{\Delta \bar{A}}{\bar{A}}\right)^2 \right\rangle} = \frac{\bar{A} e \bar{N}}{I_s R_n C_n} \sqrt{\frac{1}{2} r \cdot R_n C_n}. \tag{2.63}$$

With given maximum pulse rate r, maximum pulse height $\bar{N}$ and mean gain $\bar{A}$, $R_n C_n \approx 1/r$, and $I_s \gtrsim \bar{A} e \bar{N}/R_n C_n \cdot a_{max}$ is preferably chosen, where a_{max} denotes the maximum permissible rms photomultiplier gain deviation. For instance, with $r = 10^4$ cps, $\bar{A} = 10^6$, $\bar{N} = 10^3$ electrons and $a_{max} = 0.1\%$ the values $R_n C_n \approx 100$ µsec and $I_s \approx 1.6$ mA represent an optimum.

For the condition of equal dynode multiplication, a voltage divider with equal resistors $R_0 = R_1 = \cdots = R_n = R$ corresponds. This condition is usually fulfilled, with the sole exception of $R_0 > R$. For a given overall voltage V_s, this chain type yields the highest possible gain $\bar{A}$. The voltage between the cathode and the first dynode D_1 should not be lower than a value given by the manufacturer of the photomultiplier, if the collection losses of photo cathode electrons are to remain negligible, and hence $R_0 \approx 2 \cdot R$ is usual. Besides this, various focus electrodes must usually be supplied with specified potential; these are also taken from the resistor chain voltage divider.

However, PAGANO et al. [2.189] and WALTON [2.190] report that the thermal noise of the photocathode depends on R_0, increasing with increasing R_0. PAGANO loc. cit. has found a noise minimum for $R_0 \approx R$. This effect might be caused by the diminution of the Richardson potential of the photo cathode by the higher electrostatic field. In low noise systems this must be accounted for.

Due to the space charge effects in the last dynode sections the range of current pulse amplitudes, for which the photomultiplier operation is still linear, is limited. The linear range can be extended if the voltages of the upper dynode sections are made higher than in the lower sections. Such a voltage divider exhibits for instance the following resistor values: $R_0 = 2R$; $R_1 = R_2 = \cdots = R_{10} = R$; $R_{11} = 1.2R$; $R_{12} = 1.5R$; $R_{13} = 2.0R$ and $R_{14} = 5.0R$, in the case of a 14 dynode photomultiplier. Of course, for a given overall voltage V_s, the achieved gain A is much lower than the maximum with an equal-valued chain. Various authors have treated the optimum conditions for the resistor chain values ([2.191] to [2.193]).

BELLETTINI et al. [2.192] describe a high tension power supply where all particular dynode voltages can be adjusted separately, thus simplifying the search for optimum voltage distribution. DAVIS and SPORE [2.194] and NESS and SMITH-SAVILLE [2.195] separated the voltage divider (normally mounted close to the phototube socket) from the photomultiplier. The connection to the particular dynodes is made by means of coaxial cables, separately shielded in order to improve the insulation. The separate voltage divider with potentiometers allows the adjustment of all individual voltages.

If in fast-slow assemblies the slow signal is taken from one of the last dynodes, the situation shown in Fig. 2.52 arises. The voltage divider, including the capacitors $C_n \ldots$, exhibits values in agreement with the above-mentioned considerations, and the dynode D_{n-2} (or any other) is connected to the tapping point of the voltage divider via a resistor R_d which has the same purpose as R_a in Fig. 2.45 and which should therefore be relatively high. A mean current $\bar{\delta}^{(n-3)}(\bar{\delta}-1) I_k = I_{d,n-2}$ flows through R_d. The interdynode voltage is denoted by V_{dd}. The relationships $R_d \cdot I_{d,n-2} \ll V_{dd}$ or $R_d I_{d_{max}} \ll V_{dd}$ must hold, where $I_{d_{max}} \approx \bar{\delta}^{n-3}(\bar{\delta}-1) \cdot e \bar{N}/\tau_{fl}$ is the maximum amplitude of the signal current pulse. However, there is another criterion which often limits $I_{d_{max}}$, namely that the voltage pulse height $\bar{\delta}^{n-3}(\bar{\delta}-1) e \bar{N}/C_p$ remains small in comparison with V_{dd}, since the current pulse is integrated by the parasitic capacity C_p. Because the voltage of all other dynode sections remains constant, and the voltage D_{n-1}, D_{n-2} decreases by the same amount as the voltage D_{n-2}, D_{n-3} increases, the resulting gain varia-

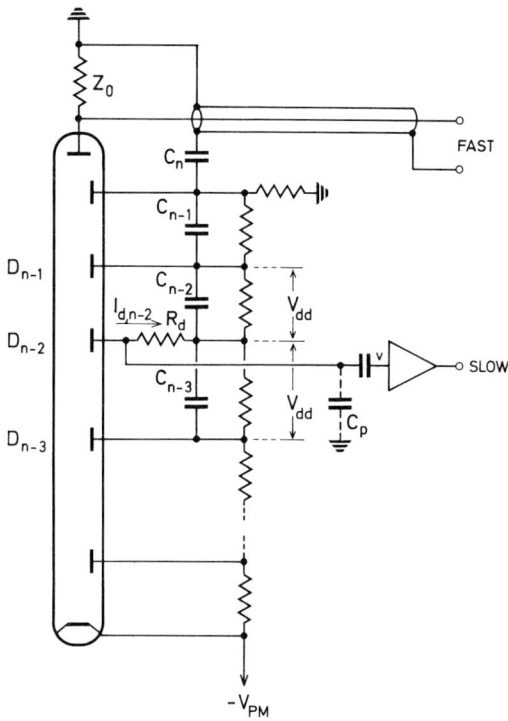

Fig. 2.52 Signal output from the dynode D_{n-2}

tions cancel each other out in the first approximation, giving the following relation

$$\frac{\Delta \bar{A}}{\bar{A}} = -\left(\frac{\Delta V_{dd}}{V_{dd}}\right)^2 \tag{2.64}$$

for the relative gain deviation. Thus R_d can be made much higher than the chain resistor R.

Where very high rates of extremely high current pulses must be dealt with, e.g. in particle accelerator experiments, it is recommended to supply the last 4–5 dynodes via cascaded emitter or cathode followers with the operating voltages. BARNA [2.196] reports an assembly with emitter followers for the dynodes D_{10} to D_{14} of the phototubes 6810A or 56AVP, having a total cross current of less than 10 mA, but being equivalent to a resistor chain with $I_s = 200$ mA up to dynode currents of 7 mA.

In fast applications (cf. Fig. 2.48) the positive pole of the high tension power supply is normally grounded, i.e. a negative voltage $-V_s$ is applied to the photocathode. The anode circuit of the phototube thus lies at a low potential, simplifying the signal output circuits. Because of the high negative potential of the cathode, and thus of the whole glass envelope, small discharges might occur between the envelope and the (grounded) phototube housing, resulting in increased phototube noise. For low noise applications the phototube must therefore be surrounded by a shield connected to the cathode potential $-V_s$, or the cathode must be grounded and a positive voltage $+V_s$ applied to the anode (Fig. 2.45). In the latter case C_1 must of course be a high tension capacitor.

2.5.7. Scintillation Counter Stabilizer Circuits

The photomultiplier gain A, its spectral response and the photon yield of the scintillator — all factors directly influencing the output pulse height — are relatively strongly temperature dependent ([2.197] to [2.199]). Moreover, the gain A also depends on the actual pulse rate r: variations of A by more than 10%, due to variations in r by a factor of 10^3, have been reported ([2.200] to [2.203], [2.258]).

> In the case of a sudden change of r, two components of the gain change can commonly be observed, which differ in time characteristics: a fast component with a time constant of about 1 minute, and a slow one with a time constant of some hours. Both components can exhibit an equal or a contrary sense. For instance, after an increase of r, the gain A at first rises fast by 1…2% and thereafter decreases by 5% with a time constant of 1 h, and vice versa. After a diminution of r to the starting value, the gain A also returns to the starting value. The mechanism of these reversal gain variations is not completely understood.

If the scintillation counter is to be used as a radiation spectrometer, or if a constant gain is desired for other purposes, the detector and the attached electronics must be stabilized, in so far as the calculation of corrections is not automatically done by a computer (COVELL [2.204]). The principle of a stabilizing loop is shown in Fig. 2.53.

The scintillation counter is supplied with a stable reference signal. Both the reference and the normal signal are amplified in the same manner by the photomultiplier and the amplifier system. Whilst the normal signal is further treated in a normal way, the reference signal is extracted and compared with a set nominal value. The deviation of the reference signal from its nominal value forms a correction voltage used for controlling the gain of the phototube or the amplifier. The whole device forms a simple control loop, the mathematical treatment of which is much the same as that of a negative feedback amplifier. The higher the stabilization factor (= gain variation without stabilizer/gain variation with stabilizer), the better is the control.

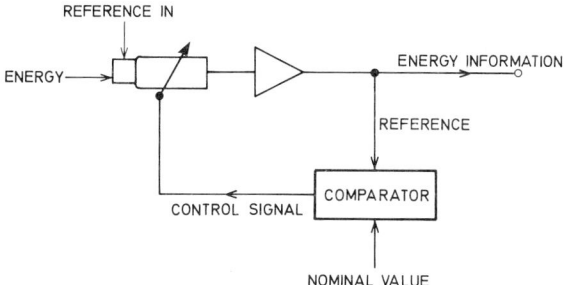

Fig. 2.53. Operating principle of a scintillation counter loop stabilizer

Different stabilizers are described in a vast number of papers [2.205] to [2.224], [2.259]. The particular devices differ in the origin of the reference signal, in distinguishing between the normal and the reference signal and in the control of the total gain by the correction signal.

Often a characteristic peak or Compton edge in the pulse height spectrum to be measured can itself be used as a reference signal ([2.205, 2.212, 2.215, 2.221, 2.223]). If the spectrometer must remain stable even during the measurement of different samples, a suitable reference signal must be formed by external means. Light pulses of constant amplitude fed into the photomultiplier via light pipes formed in cold cathode tubes (Z 70 U [2.206]) or in small indicator tubes (so-called "magic eyes" — i.e. triodes with scintillating coating of the anode, as e.g. Amperex 6977 [2.210], or the European equivalent Philips DM 160 [2.217, 2.224]) or even in gas discharge tubes [2.218] have been used as reference signals. Another possibility is to use auxiliary radioactive sources, possibly with a different scintillator ([2.207, 2.213, 2.219, 2.220, 2.222]), Finally, the pulse form of the reference signal can be relinquished and a constant [2.208] or a low-frequency modulated light source (50 Hz [2.209], 77 Hz [2.214]) may be used.

When no external reference source is used, it is not necessary to distinguish between the reference and normal signals. External light pulses are identified by their time coincidence with the triggering pulses, and a simple gate system can thus separate the reference light pulses from the scintillation pulses. DUDLEY and SCARPATETTI [2.213] use a $\beta-\gamma$ isotope as an auxiliary radioactive reference source. The β particles are detected in a 4π-plastic scintillator sandwich; the β pulses indicate the time instants of the emission of a γ quantum which is used as the reference signal, thus indicating the time intervals in which a reference pulse may occur. This method can also be modified for $\alpha-\gamma$ isotopes. COMUNETTI [2.220] described a device with a rotating lead shield having a narrow open slot, thus allowing the radiation of an auxiliary source to pass to the scintillator in a specified position of the rotor only. Therefore the reference signal is correlated in time with the movement of the shutter rotor. HINRICHSEN [2.219] distinguished his α reference pulses from the γ signal pulses with the aid of a pulse shape discriminator. Constant or low frequency modulated photocathode illumination is kept away from the "fast" signal channel by means of a frequency filter; however, it causes additional noise.

The method described by DE WAARD [2.205] is commonly used (Fig. 2.54) for producing the correction voltage. On both sides of a

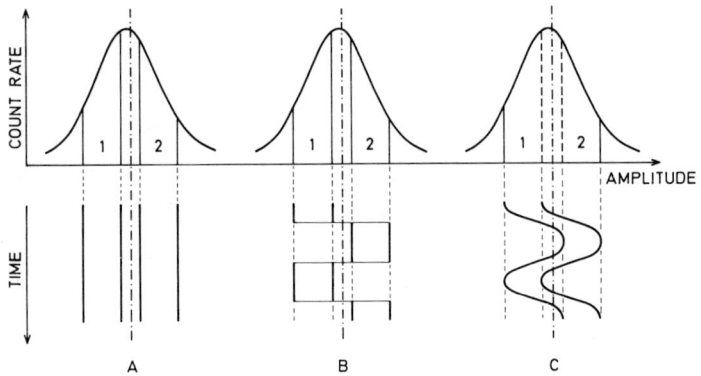

Fig. 2.54 A–C. Three methods for producing the correction voltage (according to DE WAARD [2.205])

reference peak, two discriminator channels of equal width are set (case A), and the count rate difference of pulses in channel "1" and "2" is formed by a difference ratemeter. The voltage which is proportional to the count rate difference "1"−"2" is used as the correction signal to directly control the gain. The peak centre is thus stabilized in the middle between the two discriminator channels (=nominal value). Instead of two discriminators, a single one can be used, with the channel wobbled periodically about the nominal value (case B and C). According to the actual channel position left or right of the nominal value the pulses are fed via a gate system to the " + " or " − " input of the difference ratemeter. The integrating time constant of the ratemeter must be high enough to avoid a gain variation due to the statistical deviations of the reference pulse rate (for a detailed treatment cf. DE WAARD loc. cit.).

In the case of constant peak height (i.e. constant rate of the reference pulses), a single discriminator channel can be set in the middle of the peak side [2.215]. The variation of the pulse rate at the discriminator output is a very sensitive measure of the displacement of the peak. A similar situation arises with an integral discriminator with the level set on the peak maximum [2.222].

As the windows "1" and "2", two channel ranges of a multichannel analyzer might also be used. Whether a pulse belongs to the range "1" or "2" is decided with the aid of a simple logical circuit, the output of which triggers a forward-backward digital counter. The state of this counter is converted in the control voltage in a digital-analog converter [2.221]. DIXON [2.212] describes another technique of forming the difference "1"−"2" which is especially useful with low count rates.

In earlier papers dealing especially with vacuum tube assemblies, the supply voltage V_s of the photomultiplier was used preferably as the controlling parameter of the total gain. Because of the high sensitivity of this control a small variation of V_s often suffices, so that V_s can be controlled directly by transistor circuits, too [2.213, 2.221]. If transistorized DC-converters with oscillators are used as high tension power supplies, the control voltage may also be applied in the low-voltage part of the converter [2.209, 2.217]. Since the paper by MARLOW [2.210], variable-gain attenuators between the preamplifier and the main amplifier are often used as the control element. One of the two voltage divider resistors is made from a Raysistor, which is a photosensitive resistor combined with a miniature bulb in a transistor case. Since the bulb is supplied with the control voltage, the attenuation factor becomes a function of the latter [2.215, 2.219, 2.223]. PATWARDHAN [2.216] used a Varicap diode in a capacitive voltage divider as the control element. DUDLEY and SCARPATETTI [2.213] allowed for the gain variations by controlling the high tension V_s of the phototube; the zero line drift of the main amplifier was compensated by changing the ADC threshold in the multichannel analyzer. Control of the high voltage V_s is not applicable where the phototube is used in fast coincidence assemblies. (For variable gain attenuators cf. Chapter 3.1.7).

The stabilization factors reported in the quoted literature vary between the tens and the hundreds.

BRIMHALL and PAGE [2.225] report the time shape of the gain change of RCA 6342-A to be almost the same for a step-like variation of the count rate and of the voltage V_s. If the voltage V_s is made dependent on the count rate, i.e. on the total anode current, in a suitable manner, the two variations cancel each other. However, this technique seems to be of very limited interest.

2.5.8. Čerenkov Counter

If the velocity v of a charged particle travelling through a transparent medium having a refractive index n is higher than the light velocity c/n in this medium, then

$$v > \frac{c}{n}, \qquad (2.65)$$

and the particle emits a continuous spectrum of photons all of wavelength λ, for which $n(\lambda) > c/v$. This well-known Čerenkov effect can be used for detection of relativistic particles if the radiator is coupled to a photomultiplier.

There is comprehensive literature on the Čerenkov counters, discussed in various reviews (e.g. [2.003, 2.226]). Since the photomultiplier is the actual signal-forming element of the Čerenkov counter, all the considerations in the preceding chapters remain valid, and a special discussion is hardly necessary, since this would only repeat the aforementioned conclusions. An electronic assembly designed especially for Čerenkov counters is reported by THORN [2.229].

Because there is no delay in the photon emission in the Čerenkov effect, the duration of the light pulse is given only by the light path in the radiator and is therefore extremely short. In a small (~ 25 mm) radiator, e.g. from lucite, the μ mesons of the cosmic radiation cause sub-nanosecond pulses with an amplitude of some tens of photo cathode electrons, which can be used for testing the time characteristics of fast phototubes (e.g. BIRK et al. [2.227]).

3. Analog Circuits

3.1. Linear Pulse Amplifiers

For further evaluation in analog-digital converters the radiation detector pulses must be suitably shaped and amplified. The most important function of the pulse amplifiers consists in the conservation of the energy information during the amplification process.

Probably the first application of electronics to the amplification of pulses from ionization counters with internal gas multiplication was described in 1919 by KOVARIK [3.001]. In 1924 GREINACHER [3.002] simplified his circuit and in 1926 increased the gain of a four-stage tube amplifier to such an extent that the ionization chamber pulses could be detected by means of ear-phones [3.003]. The pulse shape in these amplifiers was given by non-controlled parasitic properties of the circuits. The first pulse shaping by a differentiator with a relatively short time constant is described probably by WYNN-WILLIAMS and WARD [3.004] in 1931. A short outline of the early history of nuclear electronics is presented by FAIRSTEIN and HAHN [3.009] in an appendix to a review article on this subject.

The requirements for a pulse amplifier can be summarized as follows: the amplifier performance must be compared with the properties of the best detectors — the amplifier should not affect the energy resolution which can be reached by them. Since the semiconductor detectors exhibit an energy resolution (FWHM) of only a few tenths of a per cent, all contributions of the amplifier system to the width must remain well below 0.1 %. The following properties are of importance:
- The output pulse height must be a monotonous function of the energy (or of the input pulse height). When special characteristics, e.g. logarithmic or parabolic, are not required, *linear* amplifiers are used for the sake of simplicity, the characteristics of which can be easily checked. The integral and differential linearity (Chapter 3.1.1) influences the accuracy of the energy scale.
- The *temperature independence* and the *long-term stability* of the gain must be high.
- The noise should be low and the pulse shape selected must result in an optimum *signal-to-noise ratio*. In every case the noise of the main amplifier must be well below the contribution of the detector with the preamplifier. The amplifier must not pick up unwanted signals from external or internal sources (e.g. hum).

- Although there is a statistical time distribution of the pulses, the *sum effects* such as pile-up must remain low even for high pulse rates (Chapter 3.1.4).
- The amplifier must also amplify the small pulses correctly, even in the presence of overloading high pulses (Chapter 3.1.5).
- The *conservation of the time information* in the amplifier might eventually be of interest.

In the following chapters the particular properties are discussed in more detail, and in chapter 3.1.6 the design practice is instanced.

3.1.1. General Considerations, Linearity

Although the early amplifiers had vacuum tube circuits and even today some tube amplifiers are still in operation, transistor circuits offer so many advantages that almost all new apparatus is transistorized, and the tube circuits have become an anachronism. Hence a discussion of special tube circuits can be omitted. Nevertheless, since most of the following text is devoted to the discussion of passive coupling networks between the amplifying stages, the text is also valid for the interpretation of older tube circuits.

Strictly speaking, the historical situation is just the opposite: the "philosophy" of the pulse amplifier was developed for tube circuits and later adapted to transistor circuits, the transistor being a replacement for the tube. After all, most of the new transistor circuits are not mere transistorized equivalents of the tube ones, but circuits specially developed considering the specific transistor properties. Some of the well-known advantages of transistors over vacuum tubes will be mentioned briefly. The low power dissipation in transistorized circuits promotes the stability of the gain and allows a high degree of miniaturization. Transistors in general are faster than tubes and allow for a higher feedback to be applied to amplifier circuits, with a corresponding improvement in gain stability. The lifetime and the long-term stability of the transistors are much higher than those of vacuum tubes, due to the absence of heated cathodes suddenly changing their emission properties. The complementary npn and pnp transistors allow simplified circuit techniques with *dc* coupling. The low input impedances favor the use of current amplifiers.

It is hardly necessary to outline here a systematic theory of fundamental pulse and amplifier circuits, since these circuits are used also for other purposes and are described in other publications. From the vast number of excellent textbooks on this subject, only the standard book by SHEA [3.010], the comprehensive "Amplifier Handbook" edited also by SHEA [3.011] and a rather unorthodoxly written textbook by LITTAUER [3.012] can be mentioned by name. In this text the particular circuits will be analysed in detail only where this is necessary for an understanding of the principle of the amplifier system.

Voltage and current amplifiers. A voltage pulse measured on an ohmic resistance is always coupled with a current pulse through this resistance, thus making it unnecessary to distinguish between voltage and current pulses. However, the distinction proves to be of value in considering the two following extreme cases: 1) If a generator with signal voltage V_0 and with internal resistance R_i supplies a load $R_x \gg R_i$, then the load voltage V_x is nearly the same as the *signal voltage* V_0, independent of the actual value of R_x. 2) If on the contrary the internal resistance R_i is much greater than $R_x(R_i \gg R_x)$, the *signal current* $I_x = V_0/R_i$ trough R_x remains constant and independent of R_x. Hence four amplifier types will be distinguished: the voltage amplifier, the current amplifier, the voltage-to-current converter and the current-to-voltage converter. Their circuit symbols together with the idealized input and output resistances R_{in}, R_{out} are summarized in Fig. 3.01.

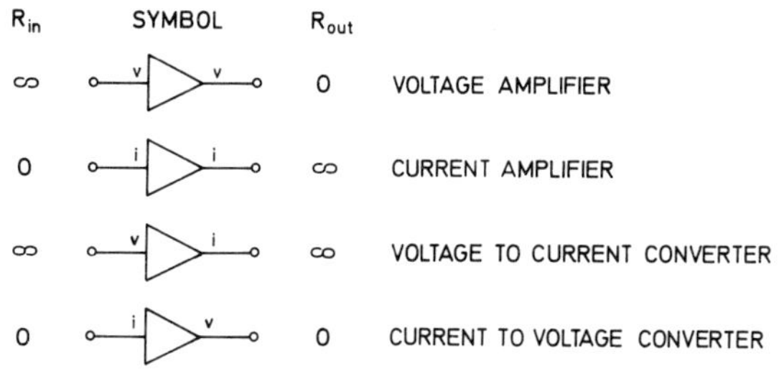

R_{in}	SYMBOL	R_{out}	
∞	v ▷ v	0	VOLTAGE AMPLIFIER
0	i ▷ i	∞	CURRENT AMPLIFIER
∞	v ▷ i	∞	VOLTAGE TO CURRENT CONVERTER
0	i ▷ v	0	CURRENT TO VOLTAGE CONVERTER

Fig. 3.01. Four basic amplifier types

The relation between the transient response and the frequency response of an amplifier. In Fig. 3.02 a simplified equivalent circuit of a real amplifier is shown: the upper cutoff frequency is obviously $f_2 = 1/2\pi R_2 C_2$, the lower cutoff frequency is $f_1 = 1/2\pi R_1 C_1$. (The three ideal amplifiers separate only the RC circuits; the input and output resistance, i.e. the type of the amplifier as a whole is irrelevant for the following considerations.)

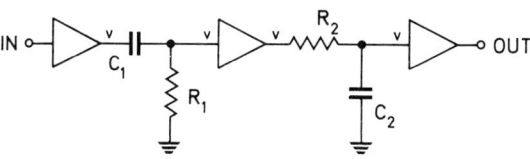

Fig. 3.02. The equivalent circuit of a real amplifier

On the other hand, the transient response function $F(p)$ of the amplifier can easily be calculated to give

$$F(p) = A \frac{p/R_2 C_2}{(p + 1/R_1 C_1)(p + 1/R_2 C_2)}, \qquad (3.01)$$

where A is the gain. An input voltage step $V_0 \cdot H(t)$ becomes shaped by (3.01) to

$$V(t) = A \cdot V_0 \frac{R_1 C_1}{R_1 C_1 - R_2 C_2} (\epsilon^{-t/R_1 C_1} - \epsilon^{-t/R_2 C_2}) H(t). \qquad (3.02)$$

The difference $\Delta f = f_2 - f_1$ denotes the bandwidth of the amplifier. In a wide-band amplifier there is $f_2 \gg f_1$, hence $\Delta f \approx f_2$. Because of $f_2 \gg f_1$ we have $R_1 C_1 \gg R_2 C_2$, the output pulse shape (3.02) is reduced to $A \cdot V_0 (1 - \epsilon^{-t/R_2 C_2}) H(t)$, and the rise time t_r (10% to 90% of pulse height) of the output pulse becomes

$$t_r = 2.2 \cdot R_2 C_2 \approx \frac{1}{3 \cdot \Delta f}. \qquad (3.03)$$

The time constant $R_1 C_1$ influences only the top of the pulse, introducing a droop distortion. Most of the amplifiers are dc coupled, thus having $f_1 = 0$, and the droop need not be considered.

Feedback. The properties of an amplifier are almost always stabilized by means of a negative feedback. The principle of the feedback is shown in Fig. 3.03. A part of the output voltage V_{out} of an inverting voltage amplifier with the gain $-A$ is fed back to the input by means of the attenuator $b < 1$ and added to the input voltage V_{in} in a mixer $(+)$.

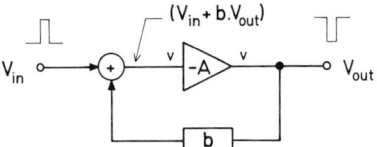

Fig. 3.03. Feedback principle

Since $V_{out} = -A(V_{in} + b \cdot V_{out})$, the output voltage is

$$V_{out} = V_{in} \frac{-A}{1 + bA}. \qquad (3.04)$$

The gain $-A_f$ of a feedback amplifier differs from the gain $-A$ by a factor of $(1 + bA)$

$$A_f = \frac{A}{1 + bA}. \qquad (3.05)$$

The product $b \cdot A$ describes the total gain in the feedback loop "amplifier A—attenuator b" and is often denoted by the term loop gain or feedback factor. If $bA \gg 1$, (3.05) reduces to

$$A_f \approx \frac{1}{b}, \qquad (3.06)$$

independent of the variations in amplifier gain A, and depending only on the values of the stable passive elements (e.g. metal film resistors) of the feedback network b. Relative variations in gain are reduced by the feedback factor

$$\frac{\Delta A_f}{A_f} = \frac{\Delta A}{A} \cdot \frac{1}{1+bA}. \qquad (3.07)$$

Of course, the same characteristic is obtained if a part $b \cdot V_{out}$ of the output signal of a *non-inverting* amplifier with $+A>0$ is *subtracted* from V_{in} in the mixer.

The influence of the feedback on the transient response $F(p)$ of the amplifier can easily be calculated from Fig. 3.04. The relations (3.04) and (3.05) obviously are valid for the Laplace transforms $\hat{V}_{in}(p)$ and $\hat{V}_{out}(p)$ too, if A is completed by $F(p)$. In an amplifier with only one integrator there is

$$A \cdot F(p) = A \frac{1}{pRC+1}. \qquad (3.08)$$

Introducing (3.08) in (3.05) we get the transient response function $F_f(p)$ of the feedback amplifier

$$-\frac{\hat{V}_{out}}{\hat{V}_{in}} = A_f \cdot F_f(p) = \frac{A}{1+bA} \cdot \frac{1}{pRC/(1+bA)+1}. \qquad (3.09)$$

Hence the amplifier becomes faster by a factor of $(1+bA)$.

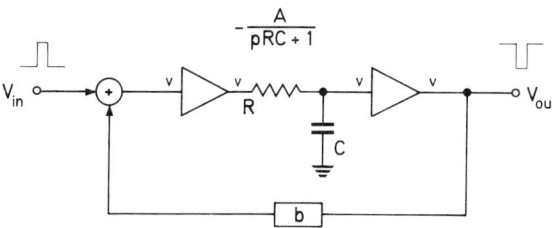

Fig. 3.04. Feedback amplifier with integrator

However, at the same time $\tau' = RC/(1+bA)$ becomes directly dependent on A and thus unstable to the same extent as A. Our discussion considers therefore only processes lasting much longer than $RC/(1+bA)$, which are not affected by variations in A and $F_f(p)$ (cf. [3.008] and FRANZ and PAUCKSCH [3.013], according to whom the feedback stabilization is always 2 to 3 times slower than the amplifier).

Analog Circuits

For the same reason the pulse shaping must always occur by means of passive networks outside the feedback loop or eventually in the feedback network b, but never in the amplifier A.

The output resistance $R_{out,f}$ of a feedback amplifier also becomes smaller by a factor of $(1+bA)$. The noise is attenuated to the same extent as the signal, yielding a constant signal-to-noise ratio independent of the feedback [3.014].

Parallel feedback. We have instanced the feedback principle on a configuration (Fig. 3.03) where a part of the output *voltage* is connected in *series* with the input voltage (so-called series voltage feedback). Corresponding to the four fundamental amplifier types (Fig. 3.01), there are still three other possible ways to realize a feedback loop. In Fig. 3.05

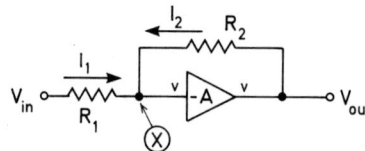

Fig. 3.05. Parallel voltage feedback (shunt feedback)

the so-called parallel-voltage feedback (sometimes called shunt feedback) is shown. A simple analysis of this circuit yields

$$V_{out} = -V_{in}\frac{R_2}{R_1}\cdot\frac{bA}{1+bA}, \quad \text{with} \quad bA = \frac{R_1}{R_1+R_2}A, \qquad (3.10)$$

where bA denotes the loop (voltage) gain once more. For $bA \gg 1$ the ratio $-V_{out}/V_{in} = R_2/R_1$, thus resulting in $I_1 = I_2$. Therefore the po-

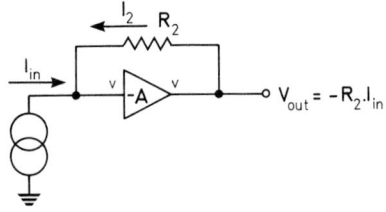

Fig. 3.06. A parallel feedback voltage amplifier serving as current-to-voltage converter

tential of the point (X) is independent of V_{in} or I_1 and equal to zero. The impedance of (X) being thus practically zero, (X) can serve as a current summing point. (X) is often denoted as a "virtual ground". A parallel feedback voltage amplifier can therefore be used as a current-to-voltage converter (Fig. 3.06). Because of $I_2 = I_{in}$, the output voltage $V_{out} = -R_2 I_2$ becomes

$$V_{out} = -R_2 \cdot I_{in}. \qquad (3.11)$$

R_2 is the conversion factor of the amplifier.

On the other hand a current-to-voltage converter with a given conversion factor of $-\rho$ can also be stabilized by a shunt feedback (Fig. 3.07) and converted into a voltage amplifier. The voltage gain becomes

$$V_{out} = -V_{in} \frac{R_2}{R_1} \cdot \frac{bA}{1+bA}, \quad \text{with} \quad bA = \rho/R_2 \qquad (3.12)$$

as the loop gain.

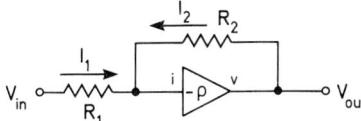

Fig. 3.07. Current-to-voltage converter stabilized by means of a shunt feedback

From the above it is evident that both circuits with a high input impedance (i.e. input stage in common emitter configuration) or with a low input impedance (i.e. input stage in common base configuration) can be used as amplifiers to which a parallel voltage feedback is applied.

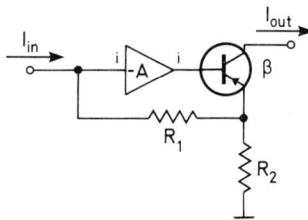

Fig. 3.08. Parallel current feedback

In Fig. 3.08 the stabilizing of a current amplifier by means of a parallel-current feedback is shown. In this circuit use is made of the fact that the collector and the emitter current of a transistor are approximately equal:

$$I_c = \frac{\beta}{1+\beta} I_e. \qquad (3.13)$$

Hence, while I_c serves as the output current I_{out}, I_e can be divided by R_1, R_2 and the fraction $I_e R_2/(R_1+R_2)$ can be fed back to the input. The output current becomes

$$I_{out} = \left(1 + \frac{R_1}{R_2}\right)\left(\frac{1}{1+1/\beta}\right)\frac{bA}{1+bA} \cdot I_{in}, \quad \text{with} \quad bA = (1+\beta)A\frac{R_2}{R_1+R_2} \tag{3.14}$$

as the loop gain. Since a high loop gain $bA \gg 1$ must be realized for the decided stable operation, it is $R_1 \gg R_2$, with $\beta \gg 1$, and equation (3.14) reduces to

$$I_{out} \approx \frac{R_1}{R_2}\left(1 - \frac{1}{\beta}\right) I_{in}. \tag{3.15}$$

Since I_e and not $I_c = I_{out}$ is used for the feedback, the output current (3.15) depends on the transistor current amplification factor β. Because of the temperature dependence of β ($\Delta\beta/\beta \approx 1\%/°C$), the current I_{out} is also a function of temperature; for $\beta \approx 100$, for instance, $\Delta I_{out}/I_{out} \approx 0.01\%/°C$. If this value is not acceptable, the circuit shown in Fig. 3.09

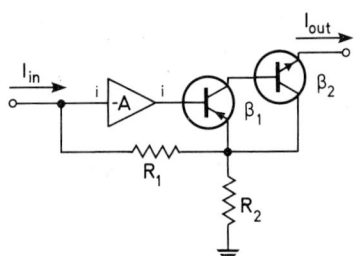

Fig. 3.09. Improved version of the parallel current feedback

must be used (e.g. ARBEL [3.015]). The amplification factors of the two transistors are denoted by β_1 and β_2, respectively. The relation between I_{out} and I_{in} in full detail is as follows:

$$I_{out} = \left(1 + \frac{R_1}{R_2}\right)\left(\frac{1}{1+1/\beta_1(1+\beta_2)}\right)\frac{bA}{1+bA} I_{in},$$

with

$$bA = [1+\beta_1(1+\beta_2)]A\frac{R_2}{R_1+R_2}, \tag{3.16}$$

yielding in the approximation of $R_1 \gg R_2$, $\beta_1 \gg 1$, $\beta_2 \gg 1$ and $bA \gg 1$

$$I_{out} \approx \frac{R_1}{R_2}\left(1 - \frac{1}{\beta_1 \cdot \beta_2}\right) I_{in}. \tag{3.17}$$

The configuration works as a transistor with overall factor $\beta = \beta_1 \beta_2$ and the temperature dependence is reduced by two orders of magnitude.

Because of the very simple relation between the output and the input signal ($\sim R_2/R_1$), which remains valid even if one or both feedback resistors are replaced by more general complex impedances, the parallel feedback amplifiers are often used as so-called operational amplifiers when carrying out arithmetic operations on analog signals. This will be discussed in more detail in Chapter 3.2.

Circuit diagrams of feedback amplifiers used in practice can be found in the rest of this chapter.

In all these considerations the gain A was assumed to be independent of the frequency. However, because of the dependence of transistor parameters on the signal frequency, A becomes an operator $A(p)$. Hence the gain of the feedback amplifier $A_f(p)$ is an operator too, and the whole system can become unstable and begin to oscillate. For oscillations of a given frequency f_0, the negative feedback must just be turned over to a positive feedback, i.e. there must be just a phase difference of π between the input and the output. Since the phase difference of a single stage is $<\pi/2$, single and double-stage amplifiers are always stable, amplifiers with three or more stages only under certain conditions. The different methods of circuit synthesis yielding an aperiodic behaviour of the amplifier are dealt with in the textbooks on this subject. ARBEL and BAR-DAVID [3.016] describe the application of the "root-locus" method used commonly in the control techniques (cf. [3.017]) in the design of nuclear pulse amplifiers. FRÄNZ [3.144] presented the feedback analysis of multistage pulse amplifiers.

Linearity. The desired linear relationship between the input and the output signals can be reached only approximately. An absolutely linear amplifier would exhibit a strictly constant gain A_0 independent of the output pulse height. The gain of an actual amplifier

$$V_{\text{out}} = V_{\text{in}} \cdot A_0 [1 + \varepsilon_i(V_{\text{out}})] \qquad (3.18)$$

however, depends on V_{out}, the function $\varepsilon_i(V_{\text{out}})$ being denoted as the *integral nonlinearity*. In analogy to (3.18) the *differential nonlinearity* ε_d is defined

$$dV_{\text{out}} = dV_{\text{in}} \cdot A_0 [1 + \varepsilon_d(V_{\text{out}})], \qquad (3.19)$$

which describes the distortion of small pulse height differences. As can easily be shown by differentiating equation (3.18), ε_d is always greater than ε_i. A large integral nonlinearity affects the energy axis calibration of a pulse height spectrum (i.e. the position of individual peaks), the differential nonlinearity manifests itself by modulation of the channel width (i.e. the distortion of the peak shapes).

However, the differential nonlinearity of the first amplifier stage can also influence the total gain, namely if the operating point of this stage is moved over a wide voltage range by the pile-up effect (cf. Chapter 3.1.4).

The amplitude dependence of the gain A is due to the dependence of the current amplification factor and the mutual transconductance of the transistor on the actual collector current. In order to obtain good linearity therefore, the collector current must be kept as constant as possible, e.g. by using a large collector load resistance R_c, by increasing the dynamic load with the aid of a bootstrap feedback or even by using a complementary transistor in constant current configuration as a load [3.006]. The bootstrap technique is discussed in detail by FAIRSTEIN [3.018].

The linearity can be improved substantially by means of a feedback. This can easily be demonstrated by introduction of the amplitude-dependent gain $A = A_0[1+\varepsilon_i(V_{out})]$ into (3.05)

$$A_f = \frac{A_0(1+\varepsilon_i)}{1+bA_0(1+\varepsilon_i)} = \frac{A_0}{1+bA_0}\left(1 + \frac{\varepsilon_i}{1+bA_0(1+\varepsilon_i)}\right). \quad (3.20)$$

Since $\varepsilon_i \ll 1$ the nonlinearity is reduced by a factor of $(1+bA_0)$.

3.1.2. The Transient Response of an Amplifier

Cascaded RC differentiators. Fig. 3.10 shows the equivalent circuit of a RC-coupled amplifier used in the following calculations of the shaping of a voltage step $V_0 \cdot H(t)$ (cf. for instance SCHLEGEL and NOWAK [3.019]).

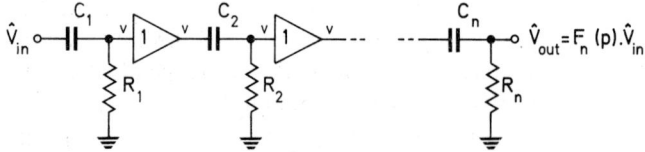

Fig. 3.10. Equivalent circuit of a RC-coupled n-stage amplifier

For the sake of simplicity the gain of the impedance decoupling stages is assumed to be unity. Further, all time constants $R_1 C_1 = R_2 C_2 = \cdots = R_n C_n = \tau_c$ are supposed to be equal. Since the transfer function of a differentiator is $p/(p+1/\tau_c)$, the final response function $F_n(p)$ of an n-stage amplifier comes to

$$F_n(p) = \left(\frac{p}{p+1/\tau_c}\right)^n. \quad (3.21)$$

With $V_{in}(t) = V_0 \cdot H(t)$, corresponding to $\hat{V}_{out}(p) = V_0 F_n(p)/p$, we get

$$V_{out}(t) = V_0 \xi_n\left(\frac{t}{\tau_c}\right) \cdot H(t) = V_0 \left[\sum_{\nu=0}^{n-1} \frac{(-1)^\nu}{\nu!}\binom{n-1}{\nu}\left(\frac{t}{\tau_c}\right)^\nu\right] e^{-t/\tau_c} \cdot H(t) \quad (3.22)$$

as the transient response of the amplifier. The term in brackets in (3.22) is known as the $(n-1)$th Laguerre polynomial [3.020]. The functions $\xi_n\left(\dfrac{t}{\tau_c}\right)$ are plotted in Figs. 3.11 and 3.12 for $t<\tau_c$ and $t>\tau_c$, respectively.

For time intervals $t \ll \tau_c$ the function ξ_n can be approximated by

$$\xi_n\left(\frac{t}{\tau_c}\right) \approx 1 - n \cdot \frac{t}{\tau_c}. \tag{3.23}$$

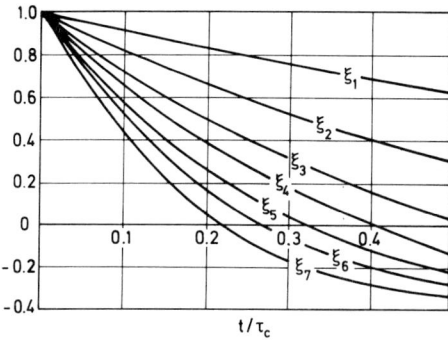

Fig. 3.11. The functions $\xi_n(t)$ for $t<\tau_c$ (according to [3.019])

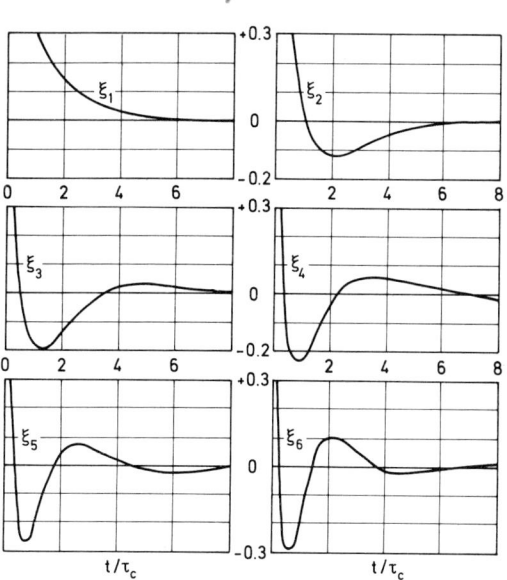

Fig. 3.12. The functions $\xi_n(t)$ for $t>\tau_c$ (according to [3.019])

Hence, the droop distortion of a voltage step applied to the amplifier input is proportional to the stage number n and inversely proportional to the coupling time constant.

Because of the polynomials in (3.22), the output voltage in general does exhibit a ringing, a sort of damped oscillation around the zero line. As will be discussed in more detail in Chapter 3.1.4, this ringing causes difficulties in measuring the pulse amplitudes at higher count rates, since the pulse height spectrum is disturbed by the pile-up.

Response to a very short pulse. A very short voltage pulse of the duration $t_0 \ll \tau_c$ can be represented by a delta function (Fig. 3.13): $V_{in}(t) \approx V_0 \cdot t_0 \cdot \delta(t)$. The Laplace transform of the response an amplifier characterized by $F_n(p)$ (3.21) to such a pulse is

$$\hat{V}_{out}(p) = V_0 t_0 \left(\frac{p}{p+1/\tau_c}\right)^n. \tag{3.24}$$

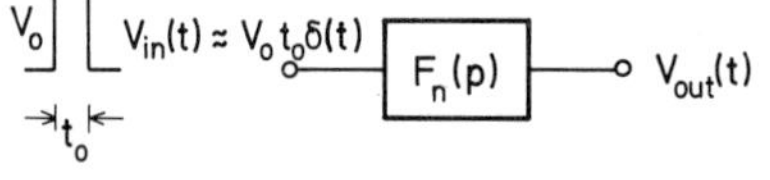

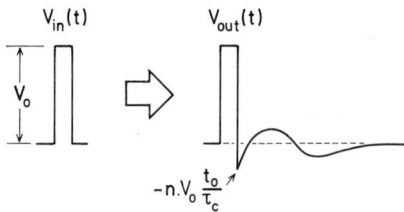

Fig. 3.13. Response of an amplifier to a very short pulse

Fig. 3.14. Response of a multi-stage amplifier to a short pulse

Transforming $\hat{V}_{out}$ into the time space yields

$$V_{out}(t) = V_0 t_0 \frac{d}{dt}\left[\xi_n\left(\frac{t}{\tau_c}\right) \cdot H(t)\right] = V_0\left[t_0 \delta(t) + \frac{t_0}{\tau_c} \xi'\left(\frac{t}{\tau_c}\right) H(t)\right], \tag{3.25}$$

where

$$\xi'\left(\frac{t}{\tau_c}\right) = \frac{d}{d(t/\tau_c)} \xi_n\left(\frac{t}{\tau_c}\right). \tag{3.26}$$

The function (3.25) is shown schematically in Fig. 3.14. Hence, the short input pulse ($t_0 \ll \tau_c$) is amplified with practically no distortion, except

The Transient Response of an Amplifier

for a subsequent ringing proportional to $\dfrac{t_0}{\tau_c}\cdot\xi'_n\left(\dfrac{t}{\tau_c}\right)$. Since $\xi'_n(0)=-n$ (as can easily be shown by introducing (3.22) in (3.26)), the maximum amplitude of the ringing which occurs for $t=0$, or exactly speaking for $t=t_0$, amounts to $-n\cdot V_0\dfrac{t_0}{\tau_c}$. The exact shape of $\xi'_n\left(\dfrac{t}{\tau_c}\right)$ for $n=1$ to 6 is shown in Fig. 3.15. As can be seen from (3.25) the ringing amplitude can be reduced to any size by raising the time constant τ_c. However, any reduction of the ringing amplitude increases its duration to the same proportion.

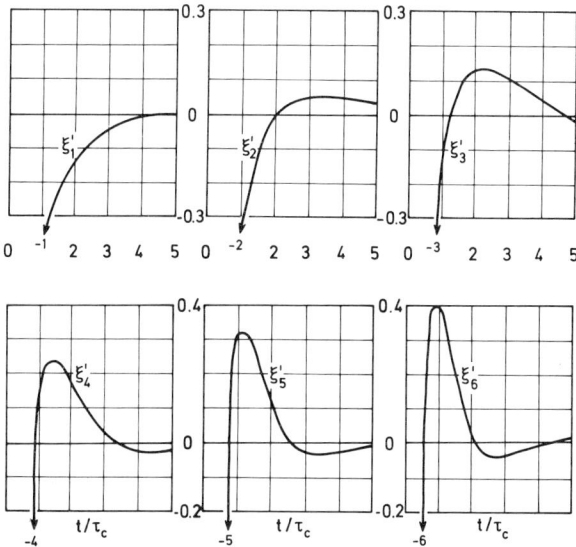

Fig. 3.15. The functions $\xi'_n(t)$ (according to [3.019])

Response to a short bipolar pulse. The behaviour discussed in the preceding chapter results from the fact that an RC-coupled amplifier cannot transmit any *dc* component. Hence the so-called unipolar pulse V_{in} in Fig. 3.14 must be shaped to an output pulse exhibiting the same area above as below the zero line. For minimum zero line shift even at high pulse rates, so-called bipolar pulses are often shaped and amplified (Chapter 3.1.3). The *dc* component of a bipolar pulse amounts to zero.

Without claiming a close mathematical correctness, the bipolar input pulse of Fig. 3.16 can be represented formally by

$$V_{in}(t) \approx V_0 t_0^2 \delta'(t), \qquad (3.27)$$

where $\delta'(t)$ is the time derivative of $\delta(t)$. Hence $\hat{V}_{in} \approx V_0 t_0^2 p$ and a calculation analogous to (3.24) yields the following response:

1. For $t_0 \ll \tau_c$ the pulse is transmitted without distortion.

2. The subsequent ringing comes to $V_0 \left(\dfrac{t_0}{\tau_c}\right)^2 \xi_n''\left(\dfrac{t}{\tau_c}\right)$, with ξ_n'' being the second derivative of ξ_n. The maximum of ξ_n'' occurs for $t=0$, or exactly speaking for $t=2t_0$, it amounts to $\tfrac{1}{2}n(n+1)$. The maximum ringing amplitude $V_0 \dfrac{n(n+1)}{2}\left(\dfrac{t_0}{\tau_c}\right)^2$ decreases with the square of t_0/τ_c and normally can be neglected if $\tau_c > 100 \cdot t_0$.

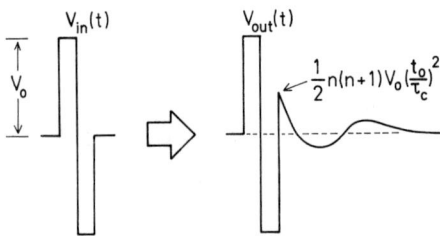

Fig. 3.16. Response of a multistage amplifier to a short bipolar pulse

Cascaded integrators. Circuits with integrating characteristics affect the shape of the pulse front (Fig. 3.17). Again, for the sake of simplicity, the equation $R_1 C_1 = R_2 C_2 = \cdots = R_n C_n = \tau_a$ is assumed. Since the response function of a single integrator is $1/(p\tau_a + 1)$, the overall response function $G_n(p)$ of a n-stage amplifier is [3.019]

$$G_n(p) = \left(\frac{1/\tau_a}{p + 1/\tau_a}\right)^n. \qquad (3.28)$$

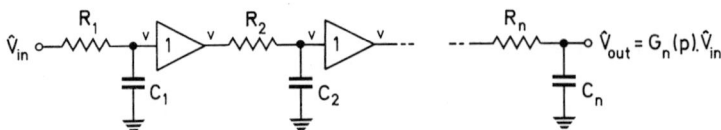

Fig. 3.17. Equivalent circuit of a n-stage amplifier (cascaded integrators)

With a voltage step $V_0 \cdot H(t)$ at the input we get $\hat{V}_{out} = V_0 G_n(p)/p$, yielding

$$V_{out}(t) = V_0 \eta_n\left(\frac{t}{\tau_a}\right) H(t) = V_0 \left[1 - \epsilon^{-t/\tau_a} \sum_{v=0}^{n-1} \frac{1}{v!}\left(\frac{t}{\tau_a}\right)^v\right] H(t). \quad (3.29)$$

The shape functions η_n are plotted in Fig. 3.18 for $n=1$ to 6 as a function of t/τ_a. Increasing n magnifies the rise time t_r which is roughly proportional to $\sqrt{n}$. Besides this the pulse front is delayed by a time proportional to $n \cdot \tau_a$.

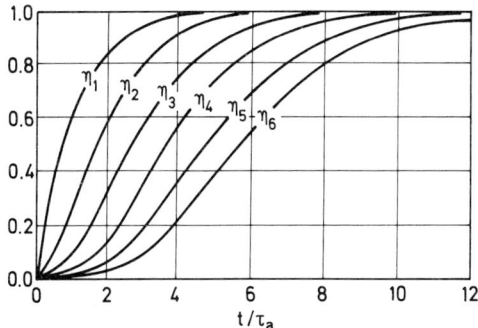

Fig. 3.18. The functions $\eta_n(t)$ (according to [3.019])

In the general case of transmitting a pulse with the rise time t_{r0} by a chain of integrators, each producing the rise time $t_{r1}, t_{r2}, \ldots$, respectively, the overall rise time t_r is given by a rule of thumb

$$t_r = \sqrt{t_{r0}^2 + t_{r1}^2 + t_{r2}^2 + \cdots} \quad (3.30)$$

The particular delays add linearly.

The actual time constants $R_1 C_1, R_2 C_2, \ldots$ can be given either by the load resistance and the parasitic capacities of the stage, or by the upper cutoff frequencies of the transistors being used. The transistor cutoff frequency, which is an inherent transistor property, depends on the operating point and on environmental conditions. Since the transistor cannot be represented by a simple integrator equivalent circuit, the amplifier design is substantially simplified if the time constants τ_a of the particular stages are given by passive circuit components (i.e. by the load resistance and parasitic or especially connected collector capacities).

To exclude changes in the pulse shape due to variations of the amplifier parameters, therefore, the amplifier must be much faster than it would otherwise have to be in order to correspond to the desired integrating time constant τ_a (cf. Chapter 3.1.3). Anyway, it must be taken into account that the particular time constants τ_a also delay the signal. If τ_a changes, for instance, due to variation of impedances when using an attenuator, or if the effective reduced τ'_a of a feedback amplifier varies due to changes in the open loop gain A, the corresponding signal delay may vary also, even if the pulse shape is approximately conserved. Where the signal is to be used for time analysis, e.g. making use of a zero crossing discriminator, the whole amplifier must be designed to be fast enough, and the expected delay variations must be lower than the desired time resolution.

3.1.3. Pulse Shaping

Normally the detector signal is a current pulse, the time integral of which — its total charge — is proportional to the energy ΔE. The circuit shaping this current pulse must exhibit the following characteristics:
   1. The shaping must allow a precise energy analysis,
   2. The signal-to-noise ratio must be high,
   3. The pulse must be of short duration and must allow high pulse rates,
   4. Occasionally, a precise time analysis is also desirable.

Optimum pulse shape. According to FAIRSTEIN [3.021] the optimum pulse shape for reliable triggering of a Schmitt trigger discriminator is the one shown in Fig. 3.19a. The pulse must exhibit a flat maximum,

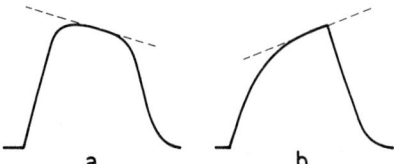

Fig. 3.19 a and b. Optimum pulse shape according to FAIRSTEIN [3.021]

at least long enough, to ensure charging the discriminator input capacity to the full amplitude value. The pulse top droop must remain lower than the hysteresis voltage before the trigger flips over in the metastable state. This also defines a minimum width of the pulse. The speed of the pulse voltage change during this time interval must be lower than the speed of the voltage changes at the parasitic capacities of the circuit. The discriminator will be triggered, even by pulses a little higher than the discriminating level, so long as these conditions are fulfilled. This behaviour is common to all level discriminators [3.007].

Analog-to-digital converters working on the principle of charging and subsequent discharging of a capacitor (cf. Chapter 4.2) require a pulse shape as shown in Fig. 3.19b. The pulse maximum should be reached slowly, in order to ensure that the converter capacitor is charged to the full pulse height. On the other hand the droop of the pulse should occur suddenly and should be fast, in order to release the discharging process with precision.

Current mode and voltage mode amplifier. If an amplitude proportional to the energy ΔE is desired, the detector current pulse must always be *integrated* irrespective of the actual pulse shape.

The integration can be accomplished in two different ways:

1. The current pulse is integrated by means of an RC circuit directly at the detector output and the resulting voltage signal is further amplified. This is known as *voltage mode* operation.

2. The current pulse is amplified prior to the integration which is accomplished in one of the last amplifier stages. This technique is called *current mode* operation.

Both amplifier techniques are shown schematically in Fig. 3.20.

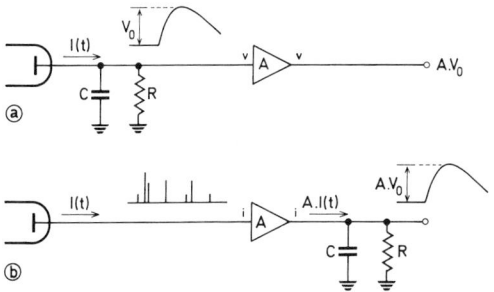

Fig. 3.20a and b. Current mode (bottom) and voltage mode (top) amplifier techniques.

In the current mode operation the input impedance of the preamplifier is made low, the input time constant resulting from parasitic capacitances thus being lower or at least of the order of magnitude of the current pulse duration (e.g. $\approx \tau_{fl}$ in a scintillation counter). Hence the shape of the detector current pulse is conserved in a first approximation during the process of amplification. Therefore the pulses remain very short, some 10 nsec, depending upon the actual detector used. Very high pulse rates can be tolerated. Various fast selection operations (coincidences, rejection of non-desired events) can be accomplished on these fast pulses. Only selected events are allowed to be integrated to slow voltage pulses. Obviously this technique strongly reduces the sum effects (Chapter 3.1.4).

It must be pointed out that alone the conservation of the detector current pulse shape and not the current amplification is characteristic of the current mode operation. The amplification itself can also be performed by voltage amplifiers. Therefore it is possible, for example, to convert the photomultiplier current pulse into a voltage pulse by means of a small resistance R_0 (termination of the coaxial cable etc.) and to handle the voltage pulse $V(t) = R_0 \cdot I(t)$ in a fast voltage amplifier. On the contrary, the voltage mode amplification in current amplifiers is possible if the integrated voltage pulse is converted into a current pulse by means of an appropriate resistor. The terms current mode amplification and voltage mode amplification have a different meaning from current amplification or voltage amplification.

The current mode amplification techniques are described in more detail by ARBEL [3.015, 3.043].

There is only one reason for using voltage mode amplification techniques instead of the current mode ones: the integrated signal, even in the preamplifier, possesses a "macroscopic" length of about 1 μsec and the amplifier system can be realized with fewer difficulties than the one for nanosecond pulses. Correspondingly the first current mode amplifiers appeared only recently and even today (1969) the majority of amplifying systems operates in voltage mode. Due to this situation the concept of the voltage mode operation is worked out much better than the one of the current mode amplification. In what follows we shall use the voltage mode operation for the discussion of the general principles of signal shaping and amplification.

Single differentiator. In principle a short integrating time constant RC at the detector output (Fig. 3.20) could be used to determine the pulse length. If an unipolar pulse is desired, the time constants τ_c of all coupling networks in the amplifier must be $\tau_c \gg RC$, or the amplifier must be dc-coupled, according to Chapter 3.1.2. In this case all the noise which is generated mainly behind RC is amplified with much smaller frequency limitations than the RC shaped signal and the signal-to-noise ratio is adversely affected.

Hence the short pulse shaping time constant τ_{diff} must be introduced behind the amplifier stage A_1 and a $RC \gg \tau_{\text{diff}}$ must be chosen (Fig. 3.21). The noise level which can be tolerated at the point (b) determines the necessary gain A_1. Since the maximum pulse rate is given by τ_{diff}, the much longer pulses at (a) overlap strongly and the well-known pile-up voltage picture is generated. Even if A_1 and A_2 are dc-coupled the circuit in Fig. 3.21 does not yield purely unipolar pulses, since the voltage "step" in (a) itself exhibits an exponential droop with the time constant RC. The relative undershoot is proportional to τ_{diff}/RC.

Pulse Shaping

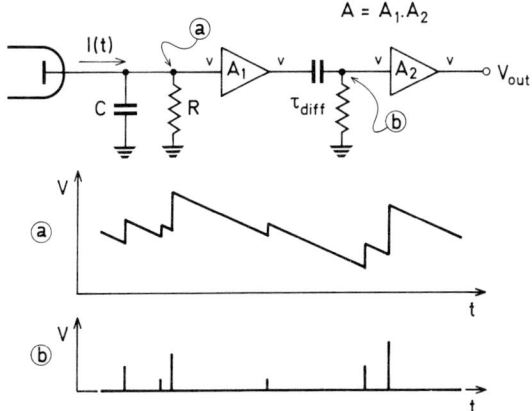

Fig. 3.21a and b. Amplifier with a single differentiator

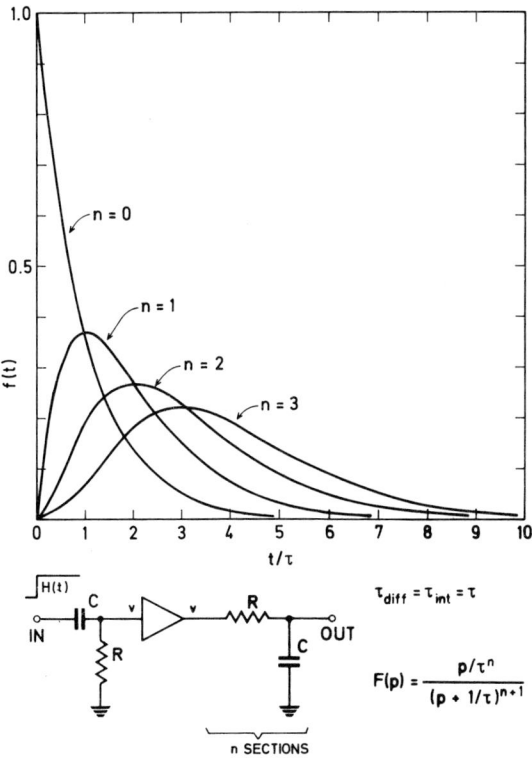

Fig. 3.22. Pulse shapes with one differentiator and n integrators (according to [3.022])

The pulse resulting from a single differentiation has a shape $V_0 \epsilon^{-t/\tau_{\text{diff}}} \cdot H(t)$ with a sharp maximum. It is unsuitable from the aspect of both pulse height analysis and signal-to-noise ratio, the latter being discussed in more detail later. Therefore the pulse front is faced additionally by one or more integrators with the time constant τ_{int}. The best signal-to-noise ratio is obtained when $\tau_{\text{int}} = \tau_{\text{diff}}$. The resulting pulse shapes are shown in Fig. 3.22. As can be seen, multiple integration in fact yields lower amplitudes but more symmetrical pulses.

For differentiating the voltage step, RC differentiators with inductive compensation according to Fig. 3.23 can also be used. The flattening effect of the inductance $L = R^2C/4$ on the pulse top can be seen clearly, especially in the case of $n=0$. For higher n the pulse becomes higher and at the same time narrower than without L. FAIRSTEIN and HAHN [3.008] report the RCL response $p(p+4/\tau)/(p+2/\tau)^2$ of differentiators

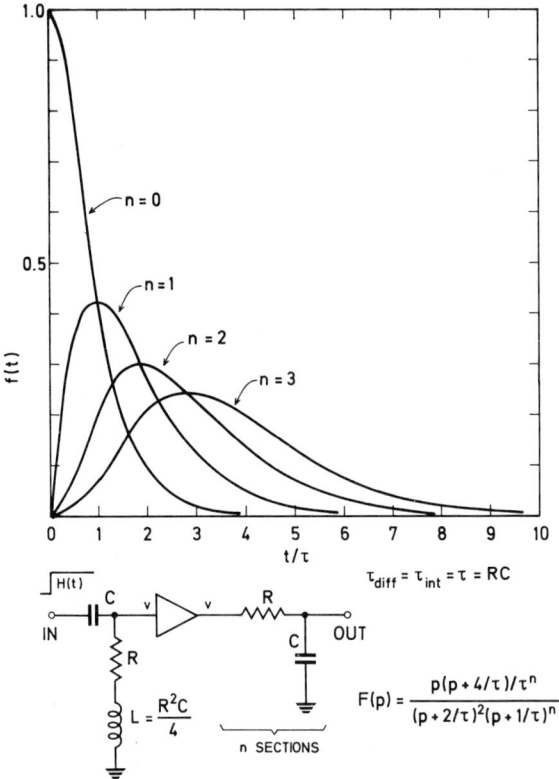

Fig. 3.23. Pulse shapes with one inductive compensated differentiator and n integrators (according to [3.022])

with inductive compensation (Fig. 3.23) to be duplicated by an active network consisting solely of resistors and capacitors in the feedback loop of an operational amplifier (Fig. 3.24).

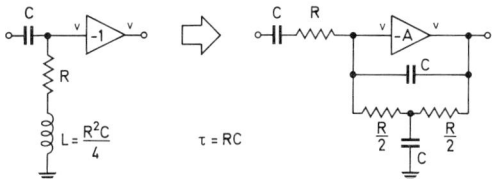

Fig. 3.24. Realization of the response function $F(p) = \dfrac{p(p+4/\tau)}{(p+2/\tau)^2}$ of the inductive compensated differentiator by means of an active network (according to FAIRSTEIN and HAHN [3.008]). An infinite gain of the amplifier is assumed

The *n*-fold integration can be performed by half of the passive networks only, if a critically damped RCL circuit with double integrating characteristics according to Fig. 3.25 is used [3.008, 3.015]. The integrator operates on current signals, so that a voltage pulse must be converted into a current pulse prior to integration. The arbitrary conversion factor of $1/R$ in Fig. 3.25 yields the same response function for the left-hand and right-hand circuits, respectively.

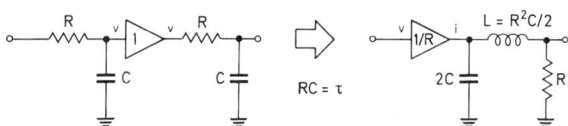

Fig. 3.25. Realization of the response function $F(p) = \left(\dfrac{1/\tau}{p+1/\tau}\right)^2$ of two integrators by means of a single passive network ([3.008, 3.015])

Delay line pulse shaping. Besides RC or RCL differentiators, the delay lines can also be used for determining the pulse duration. Fig. 3.26 shows the operating principle. The circuit (C) enables the mode of operation to be explained most simply. A voltage step is applied directly to the non-inverting input of a difference amplifier and via a delay line with the delay T_D to its inverting input. The amplifier output signal thus consists of the difference between the original input signal and the one delayed by T_D. The other circuits differ from (C) mainly in the way the difference between these two signals is formed. In circuits (A) and (D) the reflection at the open or short-circuited cable end is used, the delay therefore corresponds to twice the cable length: $2T_D$. If the cable is used for transmission only (case (B)) both ends can be terminated properly, thus reducing undesirable reflections. This is advantageous especially if low performace technical delay lines must be used [3.023, 3.024]. HAHN and GUIRAGOSSIAN [3.025] describe another circuit where the delay line is connected with proper terminations on both ends in one branch of a current bridge.

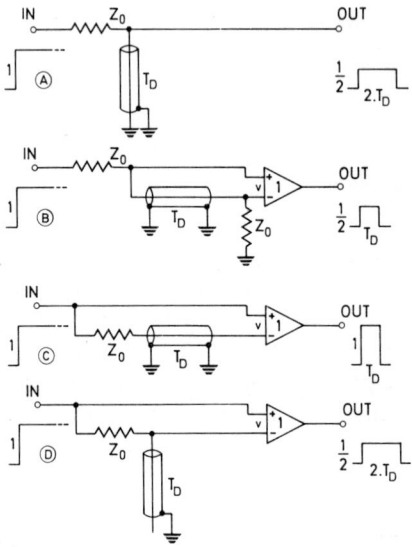

Fig. 3.26 A–D. Delay line pulse shaping

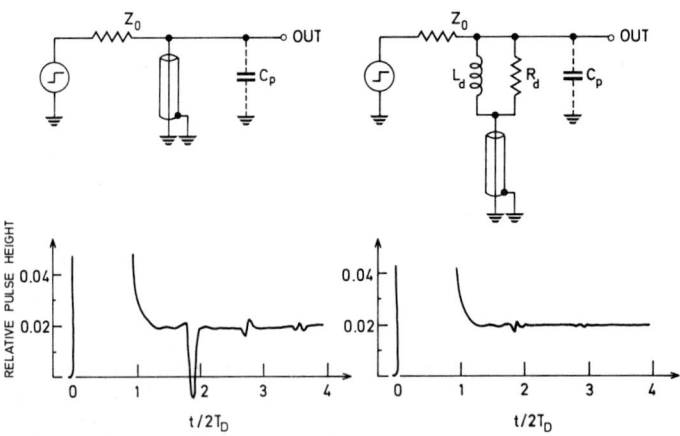

Fig. 3.27. Compensation network according to FAIRSTEIN [3.022] for reduction of ringing in delay line differentiator

The delay line pulse shaping and the experimental measures against parasitic effects in circuit (A) of Fig. 3.26 have been discussed in full detail by FAIRSTEIN [3.022, 3.026, 3.027].

The non-ideal properties of available delay cables cause a ringing due to various end effects and a pulse pedestal due to the finite ohmic resistance of the shorted line (Fig. 3.27). The ringing can be reduced by an appropriate compensation network. The pedestal is avoided if the droop of the input tail pulse during twice the transit time $2 \cdot T_D$ just compensates the voltage rise due to the cable resistance. The necessary droop can be achieved by proper adjustment of the time constant RC (NOWLIN and BLANKENSHIP [3.028]).

Depending on whether the integrator time constant is small ($\tau_{int} \leq T_D/5$) or comparable with the pulse length, the delay line differentiated pulse remains square or becomes triangular, respectively, when integrated (Fig. 3.28). Because of its flat top, the square pulse is well suited for level discriminators, though its signal-to-noise ratio is not optimum. The triangular pulse yields a high signal-to-noise ratio, but it can only be used in analog-to-digital converters based on capacitor charging.

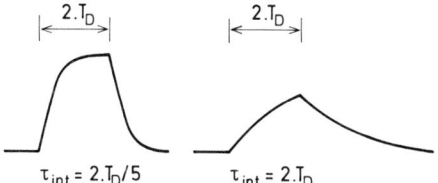

Fig. 3.28. Pulse shape with single delay line differentiator and single integrator

Pulse shaping according to the pole-zero compensation techniques. It was pointed out that single differentiating of a voltage step by a short time constant τ_{diff} always causes an undershoot, because of the intrinsic droop of the input voltage "step" resulting from detector current pulse integration $V_0 \cdot e^{-t/RC} \cdot H(t)$. Even if this undershoot does not disturb the normal operation, it causes difficulties in case the amplifier is overloaded by very high pulses (this will be discussed in Chapter 3.1.5). NOWLIN and BLANKENSHIP [3.028, 3.029] describe the application of the pole-zero compensation method, well known in control techniques, to the synthesis of pulse shapes. The principle is very simple: the input function $V_0 e^{-t/RC} \cdot H(t)$ corresponds to the Laplace transform

$$\hat{V}_{in} = V_0 \frac{1}{p + 1/RC}. \tag{3.31}$$

The response function of a differentiator is $F(p) = p/(p + 1/\tau_{diff})$. Hence its output voltage is

$$\hat{V}_{out} = \hat{V}_{in} \cdot F(p) = V_0 \frac{p}{(p + 1/RC)(p + 1/\tau_{diff})}. \tag{3.32}$$

Because of the two poles, (3.32) does not correspond to a simple exponential function. However, if the differentiator would have a response function

$$F^*(p) = \frac{p+1/RC}{p+1/\tau_{\text{diff}}}, \qquad (3.33)$$

the output voltage would become

$$\hat{V}^*_{\text{out}} = \hat{V}_{\text{in}} F^*(p) = V_0 \frac{(p+1/RC)}{(p+1/RC)(p+1/\tau_{\text{diff}})} = V_0 \frac{1}{p+1/\tau_{\text{diff}}}. \qquad (3.34)$$

Since the terms $(p+1/RC)$ in numerator and denominator cancel each other out, $V^*_{\text{out}}(t)$ becomes a simple exponential function with τ_{diff} having no undershoot. Thus the problem is reduced to the search for a network having a $F^*(p)$ according to (3.33). As can easily be shown the circuit Fig. 3.29 has

$$F^*(p) = \frac{p+1/R_1 C_1}{p+1/(R_1 /\!/ R_2) C_1}. \qquad (3.35)$$

The system of two equations $R_1 C_1 = RC$ and $(R_1 /\!/ R_2) \cdot C_1 = \tau_{\text{diff}}$ can be solved if $\tau_{\text{diff}} < RC$.

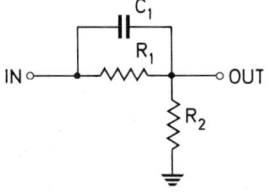

Fig. 3.29. Compensated differentiator according to the pole-zero compensation technique, having a response function

$$F^*(p) = \frac{p+1/R_1 C_1}{p+1/(R_1 /\!/ R_2) \cdot C_1},$$

where

$$R_1 /\!/ R_2 = \frac{R_1 R_2}{R_1 + R_2}$$

This design principle, namely to specify a desired pulse shape, to calculate its Laplace transform, to approximate it by a polynomial fraction as accurately as necessary and to synthesize a network corresponding to this expression, can also be applied to more complex pulse shapes than a simple exponential tail. BLANKENSHIP and NOWLIN [3.030] synthesized the network Fig. 3.30 in such a manner, yielding a very symmetrical pulse shape, truly unipolar without any undershoot, and with a high signal-to-noise ratio. The resulting pulse shapes are shown in Fig. 3.31. Even when overloaded 500 times no undershoot can be seen.

An amplifier system which is pole-zero compensated throughout has been described by STRAUSS et al. [3.145].

Double differentiation. Normally it is not possible to maintain a dc coupling from the detector up to the discriminator or *ADC* input. Any ac coupling by a capacitor, independent of its time constant, causes a base line shift as a function of the pulse rate, in order to equalize the pulse area above and below the base line. As has already been shown in Chapter 3.1.2, a sequence of bipolar pulses is transmitted without any noticeable base line shift if only the coupling time constants τ_c are high enough.

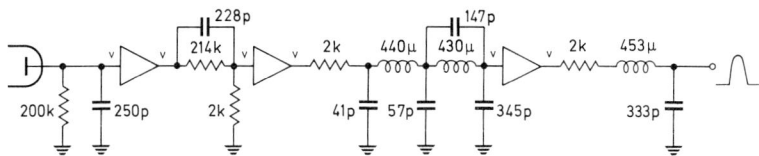

Fig. 3.30. Pulse shaping network synthesized by BLANKENSHIP and NOWLIN [3.030] for unipolar pulses of about 1.9 μsec width

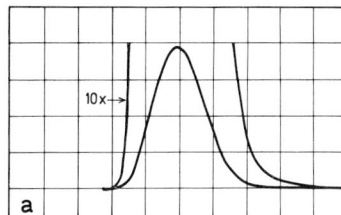

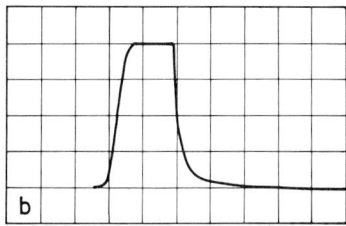

Fig. 3.31 a and b. Pulse shapes resulting from network Fig. 3.30. a) Non-overloaded pulses with and without 10× magnification, 2 μsec/division. b) 500× overload, 5 μsec/division

The length of an overshoot after a bipolar pulse is characterized by τ_c, its amplitude by $\left(\dfrac{t_0}{\tau_c}\right)^2$, where t_0 is the pulse duration. Thus the contribution of a particular pulse to the overshoot pile-up is proportional to $1/\tau_c$ and the base-line shift can be reduced by making τ_c very high, e.g. 10 msec or more.

Analog Circuits

However, the current integrating time constant RC (Fig. 3.21) plays the same role as τ_c so far as the overshoot is concerned. Since the value of RC is limited by the allowable pulse pile-up in the preamplifier, the first differentiation can for example be performed by a network which compensates for the pole $-1/RC$ (cf. Fig. 3.29) if necessary.

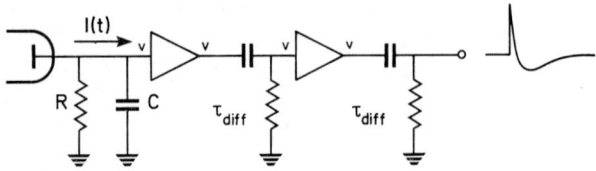

Fig. 3.32. Amplifier with two differentiators

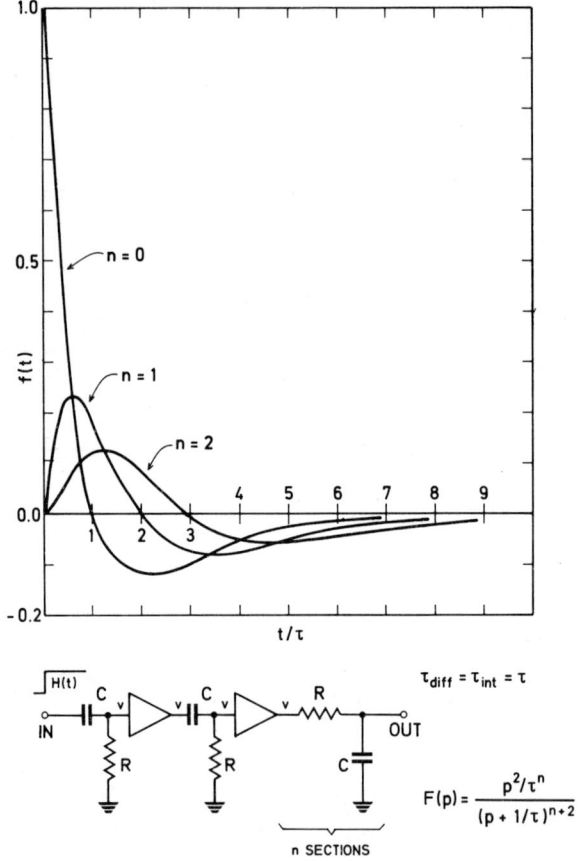

Fig. 3.33. Pulse shape with double RC differentiator and n integrators [3.022]

Pulse Shaping

Bipolar pulses are shaped by means of double differentiation of the integrated detector current pulse (Fig. 3.32). The double differentiation can be accomplished by two RC circuits, two critically damped RCL circuits, by some special networks like those in Fig. 3.30 or by two delay line differentiators. The pulse shapes of the first two cases are summarized in Fig. 3.33 and Fig. 3.34.

The zero crossing point of a bipolar pulse offers additional advantages. Because of the finite rise time of the pulses, a discriminator with

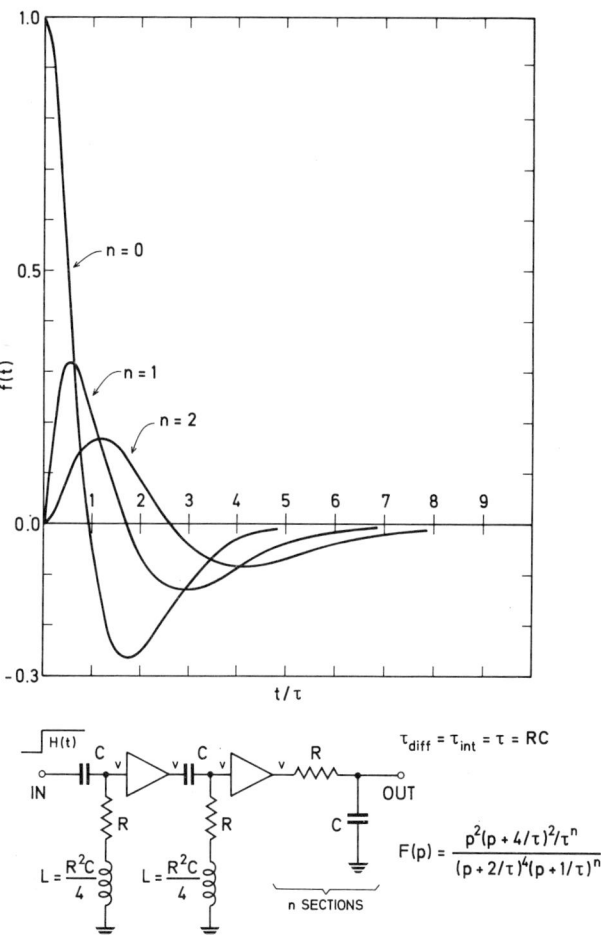

Fig. 3.34. Pulse shape with double inductive compensated differentiator and n integrators [3.022]

101

Analog Circuits

a non-zero level which is activated by the pulse leading edge, is triggered with a delay in relation to the releasing event. This delay is a function of the actual pulse height. A zero crossing discriminator watching the crossover point is triggered always with the same delay independent of the pulse amplitude, supposing the crossover point to be independent of the amplitude and given solely by the amplifier time constants. If the stability of all time constants in the amplifier is ensured, the time jitter of the zero crossing point is lower than 1% and in favourable cases even lower than 0.1% [3.030] of the pulse length.

The bipolar pulses in Fig. 3.33 and 3.34 are very unsymmetrical, the negative undershoot being much smaller and wider than the primary pulse. The symmetry can be improved by multiple integration, however, the pulse shape remains far from the optimum. A pulse shape which is very symmetrical is achieved by double delay line differentiation according to Fig. 3.35. For a small integration time constant τ_{int} the resulting pulse has an approximately square shape, both parts being equally high.

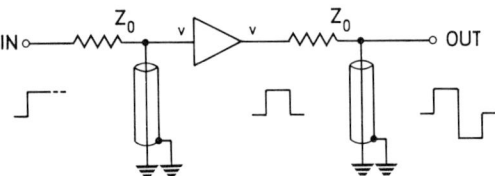

Fig. 3.35. Double delay line pulse shaping

Although double delay line pulse shaping is unsuitable from the viewpoint of high signal-to-noise ratio, it yields the shortest possible signal for a given length of the pulse top thus allowing high pulse rates. As a result it is very often used. However, recently in applications with high resolution semiconductor detectors, its adverse noise characteristics became prohibitive. In addition to FAIRSTEIN [3.026], who proposed the double line differentiation, a vast number of authors have dealt with these techniques (e.g. [3.016, 3.024, 3.025] and [3.031] to [3.034]).

GOLDSWORTHY [3.035] describes an interesting solution of double differentiation by one sole cable (Fig. 3.36) in which the signal is taken from the outer cable conductor terminated by the characteristic inpedance Z_0. At the amplifier output the cable is shorted by the zero output impedance, the other cable end is open. GORNI [3.147] described another method of bipolar pulse shaping by a single delay line and an active network.

If the unipolar signal is strongly symmetrical, i.e. if the maximum steepness of the leading and of the trailing edges are equal, a symmetrical bipolar signal can be obtained by mathematical differentiating of the unipolar one [3.030]. Therefore the pulse shaping network with the response function $F(p)$ has merely to be replaced by a network with the response function $p \cdot \tau_{ch} \cdot F(p)$. The constant τ_{ch} is a characteristic time of the unipolar pulse, and it determines the time scale only. If the unipolar pulse is described by $f\left(\dfrac{t}{\tau_{ch}}\right)$ the new network gives $\dfrac{d}{d(t/\tau_{ch})} f\left(\dfrac{t}{\tau_{ch}}\right)$. The circuit in Fig. 3.30 can easily be rearranged to give bipolar pulses as is shown in Fig. 3.37. The undershoot amplitude amounts to 99.4% of the primary pulse amplitude.

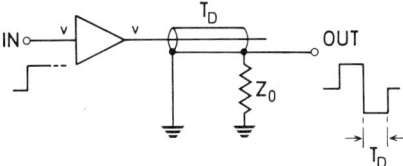

Fig. 3.36. Double delay line differentiator with one single delay line according to GOLDSWORTHY [3.035]

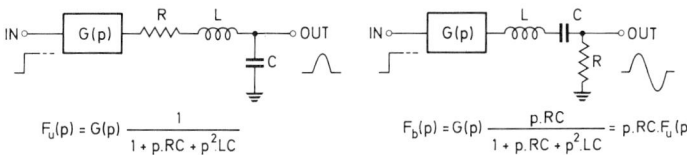

Fig. 3.37 Bipolar pulse shaping by mathematical differentiating an unipolar one (BLANKENSHIP and NOWLIN [3.030])

The noise. According to Appendix 8.2, the equivalent noise charge depends on the pulse shaping network. The ratio of parallel and series noise determines an optimum time constant characteristic for each particular network. At this time constant the contributions of parallel and series noises are equal and the total noise is a minimum.

As can be demonstrated [3.036], an optimum pulse shape exists for which the equivalent noise charge reaches a theoretical minimum. In Fig. 3.38 the relative noise charges $Q_{N,rel}$ for different pulse shaping networks are summarized and normalized to unity for the theoretical optimum, the so-called cusp pulse. The noise values are calculated assuming optimum time constants for each particular network (i.e. equal levels of parallel and series noise). Concerning the calculation of Q_N, refer to Appendix 8.2, to GILLESPIE [3.039], the papers [3.040, 3.041] and to the literature quoted in Fig. 3.38.

Two of the pulse shapes cannot be realized physically: the cusp and the Gaussian. (For approximations of the cusp pulse shape cf. [3.155, 3.161].) NOWLIN et al. [3.038] describe the realization of an almost

Analog Circuits

Pulse shape	Function	Network	$Q_{N,rel}$
 "Cusp"	$\epsilon^{-\frac{t}{\tau}}H(-t)+\epsilon^{-\frac{t}{\tau}}H(t)$	Not exactly realizable approximation cf. [3.037]	1.00
 "Triangle"	$\left(1+\frac{t}{\tau}\right)[H(t+\tau)-H(t)]$ $+\left(1-\frac{t}{\tau}\right)[H(t)-H(t-\tau)]$	Integrated bipolar DL-shaped rectangular pulse [3.038]	1.08
 "Gaussian"	$\epsilon^{-\left(\frac{t}{\tau}\right)^2}$	Not exactly realizable, asymptotic pulse shape for $n\to\infty$	1.12
	$\left(\frac{t}{\tau}\right)^n \epsilon^{-\frac{t}{\tau}} H(t)$	Single RC differentiator, n RC integrators $n=1$ $n=2$ $n=3$ $n=4$ $n=5$	1.36 1.22 1.18 1.17 1.16
	$\left[\frac{t}{\tau}-\frac{1}{2}\left(\frac{t}{\tau}\right)^2\right]\epsilon^{-\frac{t}{\tau}}H(t)$	Double RC differentiator, single RC integrator	1.88
		Unipolar pulse, network Fig. 3.30 [3.030]	1.14
		Bipolar pulse, network Fig. 3.37 [3.030]	1.78
 →|2.T$_D$|←	$\left(1-\epsilon^{-\frac{t}{\tau}}\right)H(t)$ $-\left(1-\epsilon^{-\frac{t-2T_D}{\tau}}\right)H(t-2T_D)$	Single DL differentiator, single $RC=\tau$ integrator $2T_D=\tau$ $2T_D=5\tau$	1.10 1.41
	$\left(1-\epsilon^{-\frac{t}{\tau}}\right)H(t)$ $-2\left(1-\epsilon^{-\frac{t-2T_D}{\tau}}\right)H(t-2T_D)$ $+\left(1-\epsilon^{-\frac{t-4T_D}{\tau}}\right)H(t-4T_D)$	Double DL differentiator, single $RC=\tau$ integrator $2T_D=\tau$ $2T_D=5\tau$	1.38 2.15

Fig. 3.38. Relative noise charges for various pulse shaping networks

exact triangular pulse by operational integration of a double delay line differentiated pulse. The triangular pulse can successfully be used in high-resolution spectroscopy with analog-to-digital converters based on capacitor charging. The common case of a single differentiator and a single integrator with equal time constants gives a Q_N only 36 % higher than the theoretical minimum. By means of an up to four-fold integration, the Q_N can be lowered markedly, though more than 4 integrators do not pay ($n \to \infty$ yields $Q_{N,\text{rel}} = 1.12$). The pulse shape corresponding to the network Fig. 3.30 yields a favourable Q_N, a single delay line differentiated and highly integrated ($2T_D \approx \tau_{\text{int}}$) pulse as well.

Bipolar pulses exhibit a high Q_N due to the signal amplitude loss in the second differentiator. The double delay line differentiated pulse with almost no integration ($2T_D/5 \geqslant \tau_{\text{int}}$) — which formerly was used very often — is especially disadvantageous concerning Q_N and cannot be used in high-resolution semiconductor detector spectroscopy.

The enhancement of the noise in delay line differentiators has the following qualitative reason [3.008, 3.042]: In delay line differentiators the output pulse is the *difference* of two mutually delayed input signals. Since the noise of the two input signals is not correlated, the high frequency ($f > \frac{1}{2} T_D$) noise components *add* quadratically and the original noise is enhanced by a factor of $\sqrt{2}$. BLALOCK [3.034] describes different circuits in which this effect is eliminated or reduced. The delay line, for instance, can be connected to the input by means of a linear gate which is normally closed. Thus the delay cable is free of noise. The gate is opened only if the cable is needed for delaying the signal. Hence no additional cable noise is added to the noise contained in the top of the input voltage step. The signal-to-noise ratio of bipolar pulses can be improved by means of amplitude measurement, both on the positive and negative part of the signal [3.149], or by peak to peak amplitude measurement [3.148], which moreover performs the effect of base line restoration.

Depending on the actual dominating noise component (parallel or series noise) each particular pulse shaping network will need another optimum time constant, i.e. another pulse length. In other words, for given particular noise conditions and desired pulse length there will be a particular optimum pulse shaping network. Therefore modern amplifiers must allow for the shaping network time constants, or even the shaping networks, to be changed easily.

Various authors ([3.150] to [3.161]) discussed recently the optimum pulse shape for accurate measurement of the pulse height in the presence of noise. BERTOLACCINI et al. [3.152] proved that no nonlinear shaping network can yield better signal-to-noise ratio than the optimum linear one. BERTOLACCINI et al. [3.154] presented an analysis of the optimum filter synthesis for statistically distributed pulse sequences rather than for single pulses. KONRAD [3.153], WEISE [3.157], SCHUSTER [3.158] and RADEKA [3.160] discussed signal processing with the aid of time variant filters. An improvement of pulse shaping techniques, especially for high-counting-rate measurements, can be achieved by means of pulse sampling rather than by simple differentiation and integration (GOLDSWORTHY [3.156], BERTOLACCINI et al. [3.155]). According to RADEKA and KARLOVAC [3.159] the optimum way to perform amplitude measurements is to evaluate the signal in a time-dependent filter performing the operation of "weighting" the observed signal according to the least-square-error criterion. DE WIT and WOLFF [3.161] report an approximate "cusp" realization by means of a matched filter consisting of a tapped delay line and an adder.

Analog Circuits

GOLDSWORTHY [3.163] proposed the linearity of an amplifier to be improved by means of integrating the voltage pulse area instead of measuring its amplitude. The area integration techniques are also discussed by CONNELLY and PIERCE [3.164].

3.1.4. Sum Effects

Due to the random spacing between successive pulses, two or more pulses often overlap if the product of the pulse rate r and the pulse length is not negligible with respect to 1. The overlap effects in a voltage mode amplifier affect the signal in two different ways.

Prior to the first differentiation the signal has the shape $V_0 \varepsilon^{-t/RC} H(t)$, where RC is high and $RC \cdot r$ cannot be neglected with respect to 1. The voltage signal thus exhibits the well known pile-up shape in Fig. 3.21a, the voltage $V(t)$ fluctuates around the mean value $V_0 r RC$, its root mean square deviation σ_V being given by

$$\sigma_V^2 = r \int_{-\infty}^{+\infty} |V_0 \varepsilon^{-t/RC} H(t)|^2 dt \tag{3.36}$$

according to CAMPBELL and FRANCIS [3.044]. (The so-called Campbell's theorem (3.36) of course also applies to pulse shapes other than $V_0 \varepsilon^{-t/RC} H(t)$.) The evaluation of (3.36) gives

$$\sigma_V = V_0 \sqrt{\tfrac{1}{2} r \cdot RC}. \tag{3.37}$$

With typical values of $RC \approx 200\,\mu\text{sec}$ and $r \approx 50\,\text{kcps}$, it is $r \cdot RC \approx 10$; the mean pile-up voltage is 10 times higher, as it would correspond to the single pulse amplitude V_0, and σ_V comes to about $2.2 V_0$.

The linear operating range of the amplifier stages before the first differentiation must be correspondingly high. In addition, their differential linearity must be good enough to avoid any gain variation dependent on the pulse rate. Otherwise the gain modulation corresponding to σ_V would add to the noise and adversely affect the energy resolution.

However, FERRARI and FAIRSTEIN [3.146] proved, that the system count rate limitation is never in the preamplifier, but always in the following amplifier with the shaping networks.

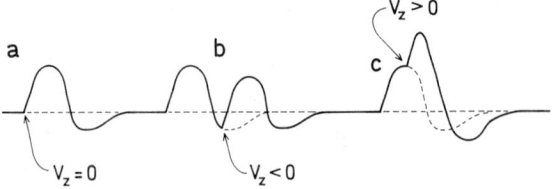

Fig. 3.39a–c. Baseline fluctuation

At the amplifier output, i.e. after the pulse shaping has been performed, the pulse length is surely small compared with $1/r$. Hence the overlap even of two pulses is rare. However, every pulse disturbs the baseline V_z for successive pulses due to its undershoot, ringing etc. (Fig. 3.39). Since most discriminator or analog-to-digital converters measure the pulse amplitude relative to the ground level, this causes a dispersion in the measured pulse heights. Hence the energy resolution is affected once more.

Calculation of the baseline fluctuation. The fluctuation of the baseline V_z (i.e. the starting voltage for an arbitrary pulse) can be analysed easily in the case of low probability of pulse overlap ($r \cdot RC \ll 1$). The distribution function of V_z is denoted by $g(V_z)$. Thus $g(V_z)dV_z$ denotes the probability that baseline voltage between V_z and $V_z + dV_z$ will occur within a given time instant. Apparently it is

$$g(V_z)dV_z = r \cdot dt \tag{3.38}$$

where r is the pulse rate, and dt is the time interval in which the pulse voltage $V(t)$ lies within V_z and $V_z + dV_z$ (cf. Fig. 3.40). Normally the interval dt is composed of the two contributions from the leading and from the trailing edges of the pulse respectively. Since

$$dV_z = \frac{dV}{dt} \cdot dt \tag{3.39}$$

we have

$$g(V_z) = r \frac{1}{\sum\limits_{V=V_z}(dV/dt)}, \tag{3.40}$$

where the sign $\sum$ symbolizes the addition over all points where $V(t) = V_z$. In practice the distribution $g(V_z)$ is determined approximately by dividing the pulse amplitude V_0 in convenient increments ΔV_0, determining the corresponding Δt and introducing these in (3.38) [3.045].

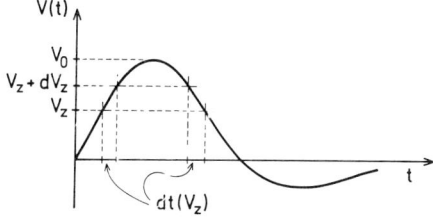

Fig. 3.40. Analysis of the baseline fluctuation

Analog Circuits

The calculation of $g(V_z)$ is discussed by SOUČEK [3.046, 3.047]. DE LOTTO and DOTTI [3.045] calculated and experimentally verified $g(V_z)$ for RC differentiated, single RC differentiated and integrated, double RC differentiated and single integrated, and for delay line differentiated pulses. Two of their results are shown in Fig. 3.41 and 3.42, namely the distributions $g(V_z)$ for single and double RC differentiated pulses with $\tau_{\text{diff}} = \tau_{\text{int}} = \tau$. As can be seen from these curves, for small values of $r \cdot \tau$ the distribution $g(V_z)$ retains the shape of a delta function at $V_z = 0$, i.e. the baseline is not shifted. For $r \cdot \tau \geqslant 0.1$ many pulses

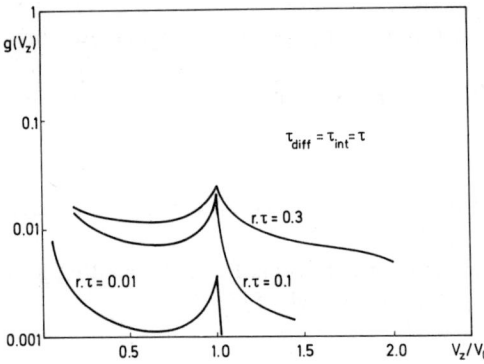

Fig. 3.41. The distribution function $g(V_z)$ for single RC differentiated and RC integrated pulses, having the pulse rate r, and an amplitude V_0, as measured by DE LOTTO and DOTTI [3.045]

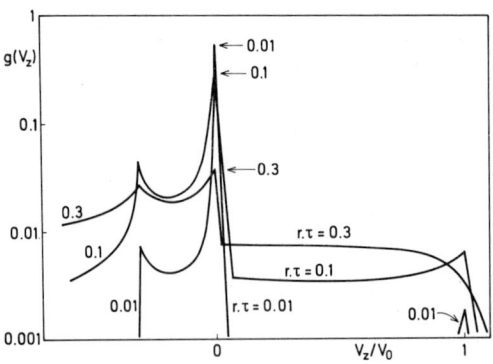

Fig. 3.42. The distribution function $g(V_z)$ for double RC differentiated and single RC integrated pulses, as measured by DE LOTTO and DOTTI [3.045]

overlap and the height of $g(0)$ is reduced. Due to the flat pulse top at $V = V_0$ the dt is here especially high and $g(V_z)$ exhibits a small maximum at $V_z = V_0$ (Fig. 3.41). Another maximum occurs at the amplitude of the negative undershoot, i.e. at $V_z \approx -0.3 V_0$ (Fig. 3.42).

Since V_z enters the pulse height determination as an additive error, the distribution function $g(V_z)$ describes directly the shape of the peak of a monoenergetic signal, as can be easily shown. In Fig. 3.43 a coincidence spectrum of a test pulse generator is shown, influenced by random spaced scintillation pulses of a Cs 137 source with a pulse rate of 114 kcps [3.030]. The true test pulse height is denoted by V_I, and the amplitude V_p corresponding to the photo peak of Cs 137 is $V_p = V_I/5$.

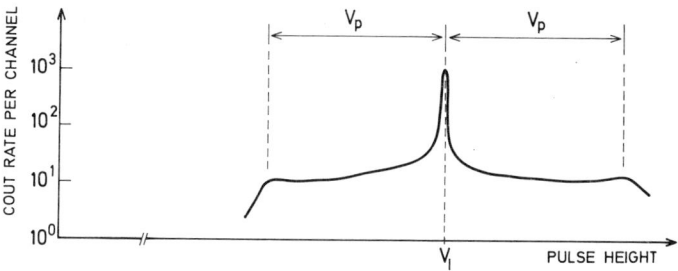

Fig. 3.43. Coincidence spectrum of a test pulse generator, influenced by random spaced scintillation pulses of a Cs 137 source, according to BLANKENSHIP and NOWLIN [3.030]

The pulses had the symmetrical bipolar shape according to Fig. 3.37. The spectrum shape obviously repeats the shape of the distribution function $g(V_z)$ from Fig. 3.42. However — due to the pulse symmetry — both the overshoot and the undershoot maxima are equally ($\pm V_p$) displaced from the main peak V_I.

From Fig. 3.41 and 3.42 the influence of the sum effects on the peak shape for different values of $r \cdot \tau$ can be seen. If the maximum pulse rate and the maximum permissible spectrum distortion are given, the maximum pulse length or the maximum time constant τ can be estimated. Contrary to the case of a monoenergetic signal of height V_0 or of a simple single peak pulse height spectrum, the calculation of $g(V_z)$ becomes very difficult in the general case. However, $g(V_z)$ can be estimated experimentally by periodic sampling of the amplifier output voltage $V_z(t)$ by short pulses of the duration $\delta \ll \tau$: the amplitude distribution of the sampled pulses corresponds precisely to $g(V_z)$ [3.048].

Analog Circuits

DE LOTTO et al. [3.048] describe a method of calculation for the computer correction of a measured real pulse height spectrum for the effects of sum events with known $g(V_z)$. A comprehensive analysis of the pile-up induced phenomena is given by AMSEL et al. [3.165].

Circuits eliminating the sum pulses. Circuits eliminating the sum pulses are described by different authors ([3.049] to [3.053] and [3.166] to [3.169]). Probably the most simple system is the one given by GUPTA et al. [3.052]. GUPTA measures the rate of events to be registered by an auxiliary detector. If this rate exceeds a preselected value the main measuring assembly is cut off. ROZEN [3.049] differentiates the input signal and measures the multiplicity of the differentiated pulses during the duration of the integrated voltage pulse in a simple coincidence circuit. Signals consisting of two or more fast components are eliminated by means of a linear gate. WEISBERG [3.051] (Fig. 3.44) describes a device, where a monostable multivibrator is triggered by the fast current signal or by the differentiated slow voltage signal. The constant amplitude monostable output pulses are integrated with the time constant τ_p corresponding to the duration of the slow voltage pulses. If two or more monostable pulses appear within τ_p, the integrated pulses overlap and exceed the triggering level of a discriminator. The discriminator output pulse controls a linear gate which eliminates the signals resulting from multiple pulses. Of course, the slow signal must be delayed by a time corresponding to the signal propagation in the multiplicity detector.

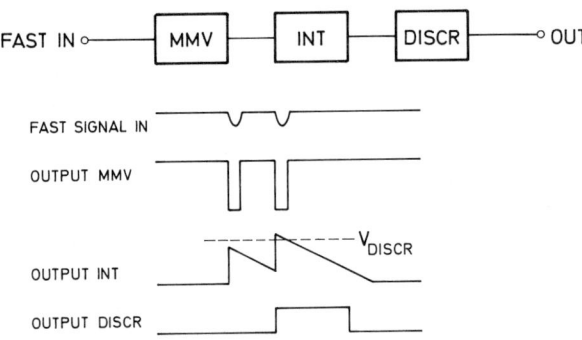

Fig. 3.44. Circuit for elimination of the sum pulses according to WEISBERG [3.051]

MONIER and TRIPARD [3.053] check the overlap of delay line shaped square pulses by measuring their length in the proximity of the base line. If this pulse length is greater than a known normal value the pulse obviously results from an overlap of at least two events and is eliminated by means of a gate system.

3.1.5. Overload Recovery

Very often small signal pulses must be detected in the presence of substantially greater pulses which cannot be avoided. Such a situation can arise for instance when measuring the weak X-rays from isotopes emitting high energy gamma quants at the same time. If the amplifier gain is set correctly for the small signal, the amplifier will be overloaded by the large pulses because of the limited linear range. The characteristic shape of an overload unipolar pulse is shown in Fig. 3.45. The amplitude of the overload pulse is limited by the amplifier nonlinearities, and the slow trailing edge of the original pulse is disproportionally amplified (in relation to the pulse height), making the pulse wider. Any small undershoot of the original pulse is also disproportionally overamplified (X). The baseline $V_z = 0$ is reached first after several coupling time constants and the correct function of the amplifier is disturbed for several tens to thousands of normal pulse lengths.

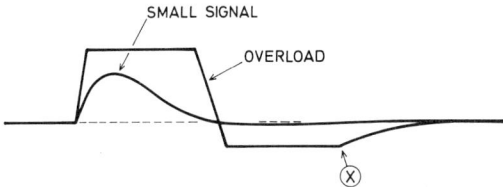

Fig. 3.45. Overload influence on the unipolar pulse shape

This effect may become magnified if the input impedance of the amplifier stages changes its value during the overload (e.g. by driving the tube grid into the positive voltage region). In this case the coupling capacitors are charged faster than they can be later discharged.

The amplifier paralysis by the overloading pulse can be avoided, or at least restricted and limited in its duration, if the following precautions are taken. The selected pulse shape must not exhibit a wide base. Therefore, delay line shaped pulses or pulses according to Fig. 3.31 are to be preferred, and pulses having a long exponential tail are to be avoided. The pulse must not exhibit an undershoot. Any undershoot must be small enough to remain in the linear range of the amplifier even for the maximum overload, in order to be able to restore the baseline by the following second differentiation. The input impedance of the amplifier stages must not change during the overload.

Analog Circuits

Circuits resistant to overload. CHASE and HIGINBOTHAM [3.054] introduced cathode-coupled difference amplifier stages ("long-tailed pairs"), in which none of the tubes could be driven into the grid current region, thus maintaining the input impedance constant under overload conditions. The transistorized pendant to this circuit, the emitter coupled difference amplifier shown in Fig. 3.46, has been investigated by COLLINGE et al. [3.055]. Positive and negative overload signals cut off the one or the other transistor SP 8303 and hence do not affect the impedance in point (X).

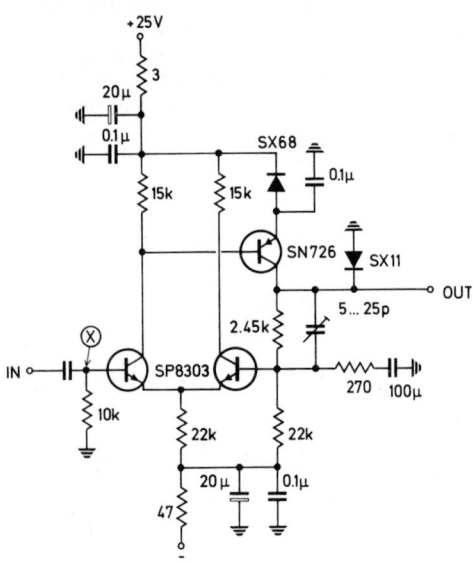

Fig. 3.46. Amplifier by COLLINGE et al. [3.055] resistant to overload pulses

As GOULDING et al. [3.032] pointed out, the bipolar signal should always be linearly amplified, since any overload may disturb the signal symmetry, thus shifting the baseline. In amplifiers with double differentiation the actual overload with the resulting amplitude limitation must occur immediately before the second differentiator and the following stages must not be overloaded by the limited bipolar pulses. The linear range of the first amplifier part is chosen advantageously much higher than necessary and the amplitude limitation, to a well defined maximum, is performed in a special limiter. The diode SX 11 in circuit Fig. 3.46 serves as an example of such a limiter by limiting the negative pulse height.

Limiters. An ideal limiter should be linear in the pass region and the limiting action should start suddenly after the input voltage has reached the chosen limit. Fig. 3.47 shows schematically the circuit diagram and the characteristics of a diode *voltage* limiter. The resistance R lies between $1\,\text{k}\Omega$ and about $10\,\text{k}\Omega$. The linear range extends up to the

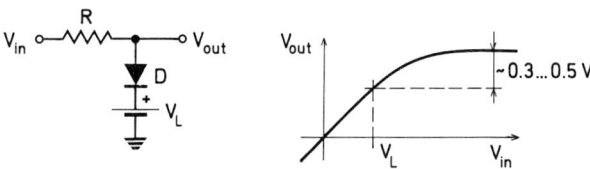

Fig. 3.47. Principle of a diode voltage limiter

limiting voltage V_L. However, the maximum output amplitude V_{out} is higher by about $0.3 \ldots 0.5\,\text{V}$ due to the diode characteristics, and there is only a gradual transition between the linear region and the saturation. As has been pointed out by LARSEN [3.056], the conventional semiconductor diode is more suitable for *current* limiters, since the high impedances of current sources virtually linearize the diode characteristics. The principle of a current limiter is shown in Fig. 3.48. Again,

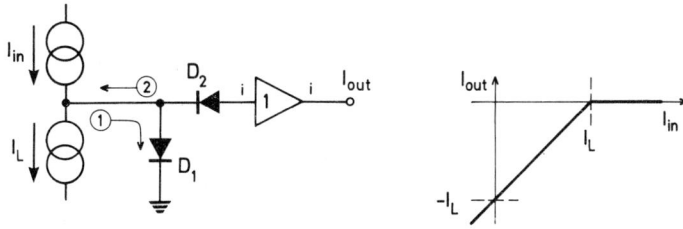

Fig. 3.48. Principle of a diode current limiter

I_L denotes the limiting current. As long as $I_{in} < I_L$, the difference $I_{in} - I_L$ is directed out of the limiter (2) and is drawn via D_2 from the current amplifier. As soon as I_{in} becomes $I_{in} > I_L$, the difference changes its direction (1) and flows via D_1 to the ground. A similar diode current limiter is described by GOULDING et al. [3.032], and a transistorized current limiter is discussed by HAHN and GUIRAGOSSIAN [3.025].

Analog Circuits

In Fig. 3.31 in addition to the normal non-overload pulse resulting from the shaping circuit by BLANKENSHIP and NOWLIN [3.030], the same pulse is shown under $500\,x$ overload conditions. Since there is no undershoot, the baseline is restored immediately after the voltage decline and the amplifier paralysis is longer than the normal pulse duration by a factor of only two.

LASCARIS and PAPADOPOULOS [3.057] avoid the overload of their amplifier system by an auxiliary discriminator measuring the input pulse amplitude — every pulse with an amplitude higher than an preselected value causes a linear gate to be cut off. Hence only small pulses may enter the amplifier.

3.1.6. Practical Design Criteria

Since different pulse shaping circuits, or at least different time constants, are needed for optimum results with different detectors, preamplifiers or, at different count rates, a modern main amplifier must exhibit a high degree of flexibility. This is especially true of semiconductor detectors: without the possibility of the selection of optimum pulse shaping parameters, the potentially high energy resolution of the detector cannot be utilized. In research establishments a system of amplifier modules with interchangeable passive or active [3.008] pulse shaping elements is preferred to a "wired" amplifier with fixed parameters. In this connection ARBEL [3.015] uses the term "nucleonic computer" which can be adapted (programmed) to the actual experiment.

In modern amplifiers with RC pulse shaping either the differentiating time constant or the integrating time constant can be selected by means of a panel switch (GOLDSWORTHY [3.058]), or the pulse shaping circuit is made as an interchangeable plug-in unit (BLANKENSHIP and NOWLIN [3.030]). Because of the limitations in the pulse length variation, the delay line pulse shaping is often avoided. GOULDING and LANDIS [3.059] describe an amplifier with optional RC or delay line pulse shaping.

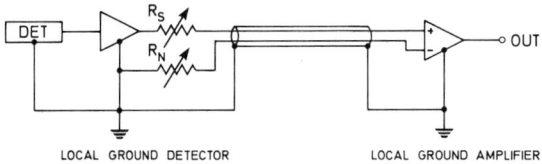

Fig. 3.49. Low-noise connection between the preamplifier and the amplifier by means of a twin shielded cable

It is recommended to lead all important signals (input of the main amplifier, output of the first differentiator, etc.) to small monitor connectors on the front panel, enabling the correct amplifier function to be tested by means of an oscilloscope.

Whenever sensitive amplifiers have to be used in an environment with high noise level, the input stage of the main amplifier is designed preferably as a difference amplifier and the signal from the preamplifier is fed by means of a shielded twin cable (Fig. 3.49, FAIRSTEIN and HAHN [3.009]). In the first approximation the noise induced in the ground connector cancels out the noise induced in the signal connector. The resistors R_S and R_N serve to match the impedances of the preamplifier and the cable. In practice the resistor R_N is adjusted for minimum noise pickup. The design criteria of the input difference stage are discussed by BLANKENSHIP and NOWLIN [3.030].

3.1.7. Amplifiers with Variable Gain

In amplifier systems with closed loop stabilization the output signal of the stabilizer must control the gain (or attenuation) of an amplifier stage. Since any control element adversely affects the linearity of the amplifier, the total gain is often controlled by means of the detector voltage variation. Of course, this can be done only when using scintillation or proportional counters, but not with semiconductor detectors. Moreover, changing the photomultiplier voltage also changes the signal delay in the multiplier. The working principle of a closed loop stabilization has been described in Chapter 2.57.

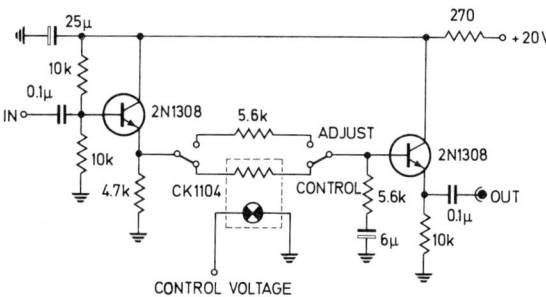

Fig. 3.50. Variable gain amplifier with Raysistor CK 1104 according to PAKKANEN and STENMAN [3.061]

In principle the amplifier gain can be controlled by the variation of a resistor in a voltage divider in the signal path or in the feedback loop. Potentiometers with a servomotor may represent a good solution, though this technique is rather slow and the necessary precision mechanics are expensive. Often a Raysistor[5] is used as a variable resistor according to a proposal by MARLOW [3.060]. The Raysistor consists of

---

[5] Raytheon Company, Newton/Mass., USA.

a photo-sensitive resistor and a light source which are encapsulated together, e.g. in normal transistor housing. The resistance of the photosensitive element depends on the light intensity and thus on the voltage applied to the light source (small incandescent bulb). Fig. 3.50 shows an attenuator circuit using the Raysistor type CK 1104 according to PAKKANEN and STENMAN [3.061]. Some of the details of the original circuit are omitted. The nominal Raysistor resistance is 5.6 kΩ; the fixed resistor 5.6 kΩ allows the control voltage to be adjusted correctly for nominal attenuation. Precautions must be taken to avoid any *dc* current through the photosensitive Raysistor resistor, which otherwise becomes heated and changes its value. The environmental temperature should also be kept constant.

Instead of Raysistors, the NTC resistors with external heating can be used, e.g. the type THERNEWID F 73[6]. The slow time response of the Raysistors or NTC resistor of some ten milliseconds is not normally prohibitive.

NTC resistors can also be heated directly by the control current if they are connected to the signal voltage divider by a decoupling capacitor. Any small incandescent lamp can also be used as a temperature-dependent element. However, the temperature coefficient is positive in this case. ARQUE ALMARAZ [3.062] describes circuits with different lamps, e.g. with the G.E. 344 having a resistance of 300 Ω at 3 mA control current. This value rises to about 1200 Ω at 25 mA. The lamp is connected in parallel to the emitter-resistor of a two-stage feedback amplifier (Fig. 3.51) via a 22 µF condenser, and the change in its re-

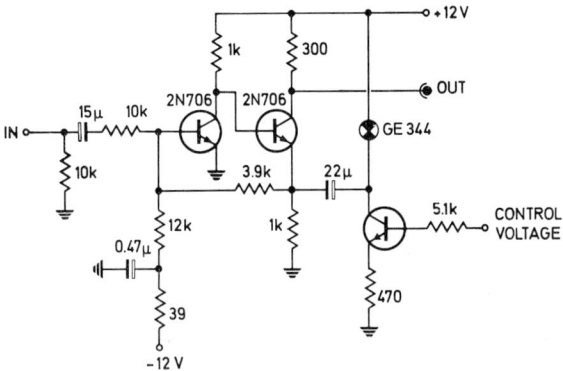

Fig. 3.51. Gain control using the incandescent lamp GE 344 [3.062]

---

[6] Siemens & Halske AG, Berlin, Germany.

sistance influences the feedback loop gain. The differential non-linearity of the circuit is better than 0.04%, the maximum gain variation is ±20% and the linear input voltage range is ±1 V. The thermal time constant is about 80 msec.

REMIGOLSKY and TEPPER [3.170] pointed out, that the variable element of a linear variable gain amplifier must be outside the feedback loop, if a constant rise time is required.

Field effect transistors offer an alternative as control elements and such attenuators are much faster, having a time constant of about 10 μsec [3.062]. PATWARDHAN [3.063] describes a charge-sensitive preamplifier, the sensitivity of which could be controlled using a voltage-sensitive Varicap diode capacitor as a part of the feedback capacity. The control voltage applied to this diode changes the integration capacity.

3.2. Arithmetic Operations on Analog Signals

In many experiments the physicist is interested in results which are given by arithmetic operations on different measured parameters—we need only to instance the identification of particles having different mass and charge by measuring the total energy loss E and dE/dx at the same time. Since these two magnitudes are represented as analog signals it is necessary to perform the arithmetic operations on these.

Of course, it is possible to digitize the pulse height information prior to performing the arithmetic operations and to allow the calculations be made by an on-line or off-line fast digital computer. However, both cases are much more expensive than a simple analog calculator unit. Even if the accuracy of the analog calculation is not high enough and a digital off-line data reduction is accomplished, the additional on-line analog treatment is still carried out because of the instant availability of results.

The techniques used correspond to those used in analog computer technology. In any case, the arithmetic operations on amplitudes of very short pulses cause some additional specific difficulties. Experiments which show the advantages of the analog reduction of measured data prior to the following digital treatment are reviewed by STRAUSS and BRENNER [3.064].

3.2.1. Operational Amplifiers

As already mentioned in Chapter 3.1.1., a parallel feedback amplifier is a suitable element for performing mathematical operations. The working principle can easily be seen if a current is chosen as the input signal and if the operational amplifier is regarded as a current-to-voltage converter (Fig. 3.52). We have

$$V_{out} = -I_{in} \cdot R \frac{A}{1+A}, \qquad (3.41)$$

and in the approximation of $A \to \infty$

$$V_{out} = -I_{in} \cdot R. \qquad (3.42)$$

The impedance at point (X) is

$$R_x = \frac{R}{1+A}. \qquad (3.43)$$

R_x becomes 0 (i.e. $R_x \ll R$) if $A \to \infty$. Because of its extremely low impedance, point (X) is often denoted as the "virtual ground". If more than one current generator is connected to (X), the partial currents add according to Kirchhoff's law. Hence the signal addition is performed by simply transforming them into currents flowing to point (X).

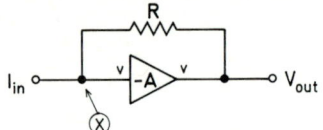

Fig. 3.52. Principle of the operational amplifier

The resistor R is a feedback element with an extremely simple relationship between the current I and the voltage V, namely $V = R \cdot I$. If a general component with the voltage-current characteristic described by $V = b(I)$ is used instead of R, the output voltage V_{out} is

$$V_{out} = -b(I_{in}) \frac{A}{1+A}, \qquad (3.44)$$

resulting in

$$V_{out} = -b(I_{in}) \qquad (3.45)$$

for $A \to \infty$. The function b can represent any time-dependent operator (e.g. $\sim \frac{d}{dt}$, $\sim \int_{-\infty}^{t} dt'$ etc.) or any time-independent function (exponential, logarithm, square, cube etc.).

Since the input signal exists mostly as a voltage, it must be converted into a current. The simplest way to do this consists of using an ohmic resistance (e.g. R_1 in Fig. 3.05). However, the conversion can also be performed by any other circuit element with more general voltage-current characteristic $V=a(I)$. Hence the general circuit consists of the two elements, a and b, and of the amplifier $-A$. Such a circuit shown in Fig. 3.53 will be discussed in more detail in the following paragraphs.

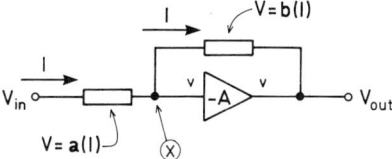

Fig. 3.53. Operational amplifier with two general components a and b

Denoting the potential of the point X by V_x we have $V_{out} = -A \cdot V_x$. The currents through a and b are the same (I). Since $V_{in} - V_x = a(I)$ and $V_x - V_{out} = b(I)$, the following equation is valid

$$b^{-1}\left(-V_{out} - \frac{V_{out}}{A}\right) = a^{-1}\left(V_{in} + \frac{V_{out}}{A}\right), \quad (3.46)$$

where $I = a^{-1}(V)$ and $I = b^{-1}(V)$ are the inverse functions of a and b respectively. An explicit solution of (3.46) can be given only for specified functions a and b. Nevertheless, in general it can easily be seen that in the approximation $A \to \infty$

$$V_{out} = -b(a^{-1}(V_{in})). \quad (3.47)$$

By an appropriate choice of the components a and b many functional relationships between V_{out} and V_{in} can be realized. Some of them are summarized in Fig. 3.54.

It is hardly necessary to comment on the particular circuits in Fig. 3.54. We need only point out that the given relationships between V_{out} and V_{in} are valid only in the approximation $A \ggg 1$, which is generally characteristic for the use of an operational amplifier. The exact relationships between V_{out} and V_{in} can be calculated from (3.46) by introducing the actual functions a^{-1} and b^{-1}. The equation (3.46), of course, also remains valid for the Laplace transforms $\hat{V}_{in}(p)$, $\hat{V}_{out}(p)$: the time-dependent operators a and b become algebraic expressions in p and the equation (3.46) can be solved easily.

Analog Circuits

Circuit	Equation	Function
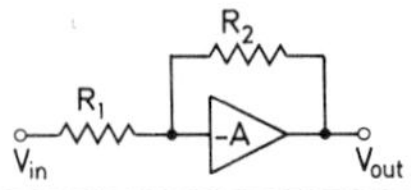 (op-amp with R_1 input, R_2 feedback)	$V_{out} = -\dfrac{R_2}{R_1} V_{in}$	Amplifier, multiplication by a constant factor
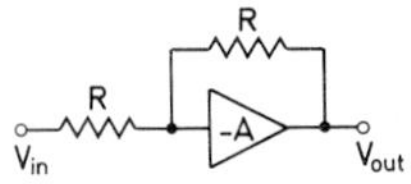 (op-amp with R input, R feedback)	$V_{out} = -V_{in}$	Inverter
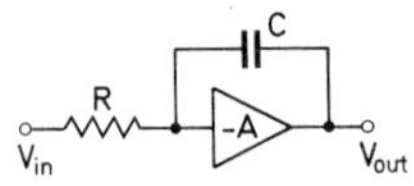 (op-amp with R input, C feedback)	$V_{out} = -\dfrac{1}{RC} \displaystyle\int_{-\infty}^{t} V_{in}(t')\,dt'$	Integrator (signal duration $\ll RC(1+A)$)
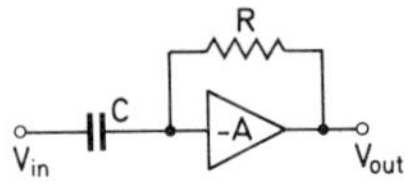 (op-amp with C input, R feedback)	$V_{out} = -RC\dfrac{d}{dt}V_{in}(t)$	Differentiator (signal duration $\gg RC/(1+A)$)
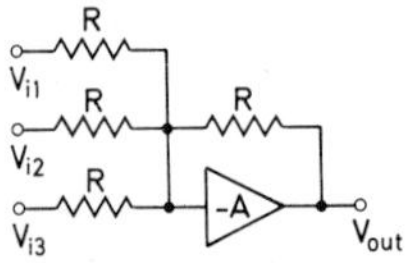 (op-amp with three R inputs V_{i1}, V_{i2}, V_{i3}, R feedback)	$V_{out} = -(V_{i1} + V_{i2} + V_{i3})$	Adder
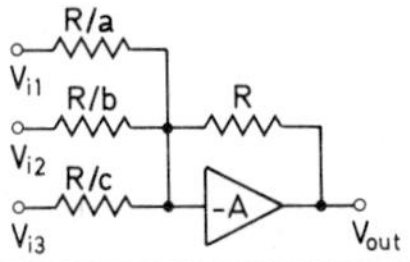 (op-amp with inputs $R/a, R/b, R/c$, R feedback)	$V_{out} = -(a\cdot V_{i1} + b\cdot V_{i2} + c\cdot V_{i3})$	Linear combination
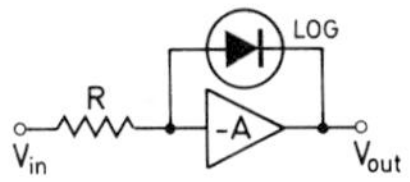 (op-amp with R input, LOG diode feedback)	$V_{out} = -V_0 \log \dfrac{V_{in}}{RI_0}$	LOG-converter
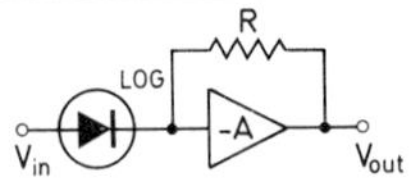 (op-amp with LOG diode input, R feedback)	$V_{out} = -RI_0\, e^{V_{in}/V_0}$	ANTILOG-converter

Fig. 3.54. Various operational amplifier circuits

3.2.2. Arithmetic Operations on Pulse Amplitudes

The true analog information is carried by the pulse amplitude, i.e. by the peak value of the pulse voltage or pulse current. Therefore, two pulses which are to be combined in an analog computer circuit must be exactly simultaneous. However, since the synchronism between two common RC-shaped pulses which frequently pass signal paths with

Arithmetic Operations on Pulse Amplitudes

different propagation times cannot be guaranteed, the pulses are shaped in a pulse stretcher prior to performing the arithmetic operation. The flat top of the stretched pulses must be longer than the maximum possible difference in signal delays. Normally pulse lengths of several microseconds are used.

In Fig. 3.55 a simplified functional block diagram of a stage performing the division operation is shown. Both input pulses, A and B, which may be delayed relatively, are first stretched in the pulse stretchers PS. The output signal of the division stage ($\div$) is proportional to the ratio of the instant values of the input signals. Hence the output signal OUT would have the shape shown by the dotted line if the stretched input pulses A' and B' were fed directly into the division stage: since B' is delayed with regard to A', the division $A':B'$ at first gives too high a value which decreases to the correct value A/B only after both pulses exhibit the flat top. In order to avoid this overshoot, a sampling pulse T is formed with the aid of a coincidence circuit CC and a monostable multivibrator MMV. A' and B' are fed into the division stage via two linear gates LG which are activated by the sampling pulse T.

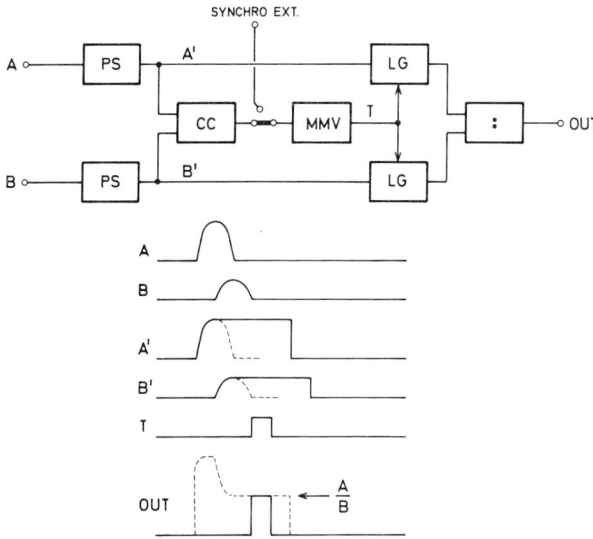

Fig. 3.55. Synchronizing of the arithmetic operations by means of a sampling pulse T

Sampling pulses limiting the arithmetic operation to the time during which all participating pulses exhibit a flat top are also used in operations other than division. Some of the circuits, such as for example the

121

LOG converter, are overloaded by negative signals V_{in}. The sampling pulse prevents the overshoots or pulse parts of false polarity being fed into the following operational stages. Besides the internally generated sampling pulse, external master pulses can be used. Moreover, the selection criterion for the generation of sampling pulses can be made more general as in Fig. 3.55, so that the arithmetic operation is performed on preselected pulse pairs or pulse groups only.

3.2.3. Practical Circuits

Normally the multiplication and the division are accomplished by adding or subtracting the logarithms of the original signals, which are formed in LOG converters. If necessary the result, which itself is logarithmic, can be linearized in an ANTILOG converter. However, it is somewhat difficult to find a circuit with a strong logarithmic characteristic.

Circuits using a logarithmic component. Although some vacuum tubes, such as pentodes and diodes, exhibit an approximately logarithmic characteristic (cf. [3.065]), the use of small semiconductor components which need not be heated offers many experimental advantages. Taking into account only the diffusion current in a semiconductor diode, the theory of SHOCKLEY [3.066] yields the following current-versus-voltage relation

$$I = I_{so}\left(\epsilon^{\frac{eV}{kT}} - 1\right) \qquad (3.48)$$

with I_{so} denoting the constant reverse saturation current. With forward bias, i.e. $V \gg kT/e \approx 25$ mV at room temperature, the term -1 can be neglected and I depends on V in a purely exponential manner. Various authors ([3.067] to [3.069]) use semiconductor diodes as logarithmic elements. However, the range of validity of the relation (3.48) is limited to one or two decades, since the diffusion current is only one of the components of I [3.070]. GIANNELI and STANCHI [3.071] reported the base-emitter pn-junction of a transistor to have logarithmic characteristics over much more than two current decades. Although the emitter current of a transistor consists of different components, all except one flow to the base and only the pure diffusion current flows to the collector, assuming $V_{CB} = 0$. This extremely precise logarithmic relationship between the collector current I_C and the emitter-base voltage V_{BE} up to nine decades [3.064] has been used by PATERSON [3.072], COOKE-YARBOROUGH [3.073], LUNSFORD [3.074], STRAUSS and BRENNER [3.064] and BYRD [3.171] in the design of LOG and ANTILOG converters.

In the LOG converter the "logarithmic" transistor is used in the feedback loop of an operational amplifier, the base being grounded and the collector being connected to the input "virtual ground" in order to fulfil the condition $V_{CB} \approx 0$ [3.072]. A simplified circuit diagram of a LOG converter according to STRAUSS and BRENNER [3.064] is shown in Fig. 3.56. By means of a variable resistor, the quiescent current I_{ref} (approximately 25 μA) of the silicon planar transistor 2N2219 is adjusted for operation in the logarithmic region of the I/V characteristic.

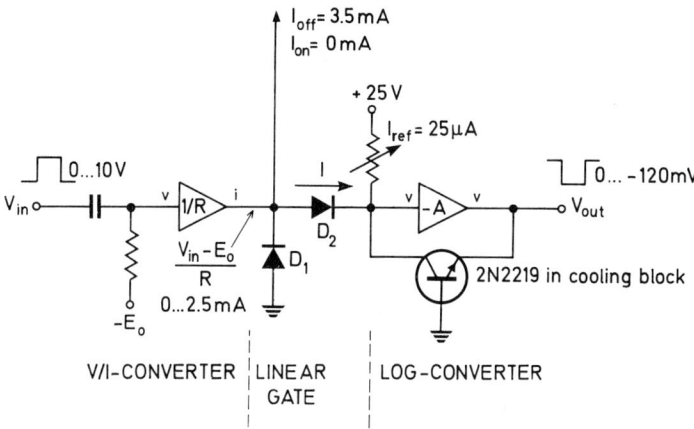

Fig. 3.56. Operating principle of the LOG-converter according to STRAUSS and BRENNER [3.064]

The output voltage amplitude V_{out} then becomes

$$-V_{out} = \frac{kT}{e} \log\left(\frac{I + I_{ref}}{I_{ref}}\right), \quad (3.49)$$

where I is the input current of the LOG converter. For V_{out} to be proportional to $\log(V_{in})$, the $(I + I_{ref})$ must be proportional to V_{in}. Therefore the voltage V_{in} is first diminished by a constant magnitude E_0, $V_{in} - E_0$ is converted into current I with the conversion factor $1/R$, and E_0 is adjusted to $E_0/R = I_{ref}$. Under these conditions

$$-V_{out} = \frac{kT}{e} \log\left(\frac{V_{in}}{R \cdot I_{ref}}\right). \quad (3.50)$$

The diodes D_1 and D_2 form a linear gate; this is explained in more detail in Chapter 3.4.

The LOG converter in Fig. 3.56 operates satisfactorily with positive input signal V_{in} only and must be protected against negative voltages V_{in}. PATERSON [3.072] reports a circuit accepting signals of both polarities (Fig. 3.57). The output signal is proportional to $\log|V_{in}/V_0|$ and exhibits a polarity opposite to that of the input voltage.

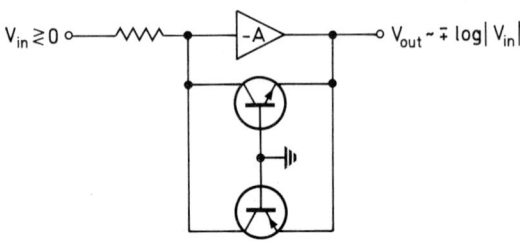

Fig. 3.57. LOG-converter for both positive and negative input pulses according to PATERSON [3.072]

The constant $V_0 = kT/e$ in (3.49) or (3.50) is a function of T, thus making the logarithmic conversion factor temperature-dependent. Hence the "logarithmic" transistor must be kept at constant temperature, e.g. with the aid of a small Peltier element [3.074].

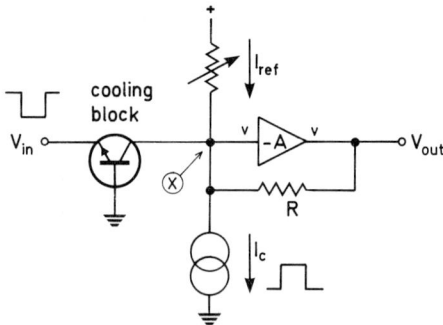

Fig. 3.58. The ANTILOG-converter

Fig. 3.58 shows the principle of an ANTILOG converter. The output pulse amplitude V_{out} is given by

$$-V_{out} = R \cdot I_{ref} \left(\epsilon^{\frac{eV_{in}}{kT}} - 1 \right). \tag{3.51}$$

The operating point of the transistor is maintained by I_{ref}. V_{out} differs from a purely exponential expression by the term $R \cdot I_{ref}$. Hence a correction current $I_c = -I_{ref}$ must be drawn from the point (X) during the signal pulse duration. Instead of a special additional current generator I_c, the reference current I_{ref} might also be interrupted during the pulse duration

Since I_{ref} must flow through the logarithmic transistor, the input voltage $V_{in} = -V_{BE}$ must be somewhat negative. The criterion for flowing in the transistor and not into the operational amplifier is of course $V_{out} = 0$. Hence the input bias voltage is advantageously controlled by a closed loop control circuit watching $V_{out} = 0$ (cf. STRAUSS and BRENNER [3.064]). These authors use the "logarithmic" transistor in common emitter configuration with a load resistance in the collector circuit. Obviously the condition $V_{CB} = 0$ given by PATERSON [3.072] for logarithmic operation of a transistor is not a strong one.

In both equations (3.49) and (3.51) the scale factor $V_0 = kT/e$ has the same temperature dependence. Hence, if by means of an appropriate mounting of all "logarithmic" transistors in the same cooling block local temperature differences are avoided, the scale factors of both LOG and ANTILOG converters become the same. Thus a series circuit consisting of a LOG and an ANTILOG converter is not temperature-dependent and very sophisticated thermostat equipment can be avoided.

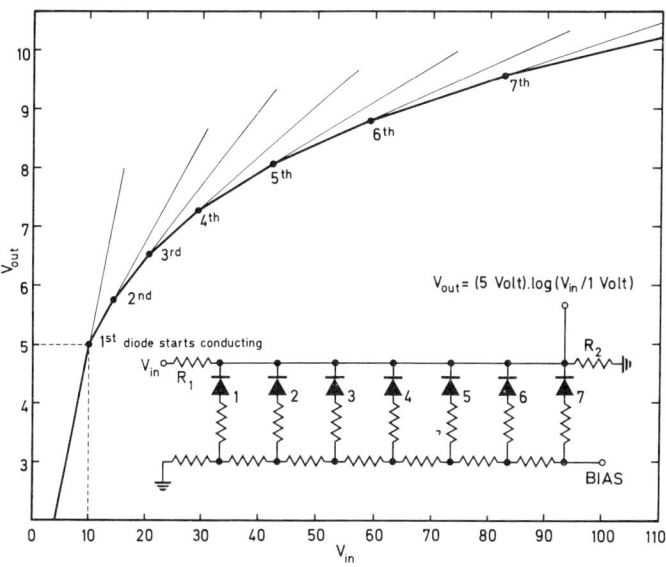

Fig. 3.59. Approximation of logarithmic characteristics by means of diode function generator according to WÅHLIN [3.075]

Analog Circuits

Approximation of logarithmic characteristics by means of nonlinear voltage dividers. Approximation of a logarithmic or any other function by means of short linear segments can also be used in operational circuits. For instance, Fig. 3.59 shows the approximate realization of the function $V_{out} = (5\text{ Volt}) \cdot \log(V_{in}/1\text{ Volt})$ according to WÅHLIN [3.075]. When the input voltage V_{in} is raised, more and more diodes start conducting, the input resistance R_1 is increasingly loaded and the V_{out} to V_{in} characteristic becomes respectively more flat. With appropriate resistor values it becomes approximately logarithmic. A similar circuit, however, using equally valued resistors in each attenuator stage, has been reported by VINCENT and KAINE [3.076]. GOLDSWORTHY [3.077] describes a pseudo-logarithmic amplifier in which the gradual nonlinear attenuation is performed by means of limiters placed between the particular amplifying stages: the last limiter is overloaded first, then the last but one, etc. A diode power function generator has been described by TUROS et al. [3.172].

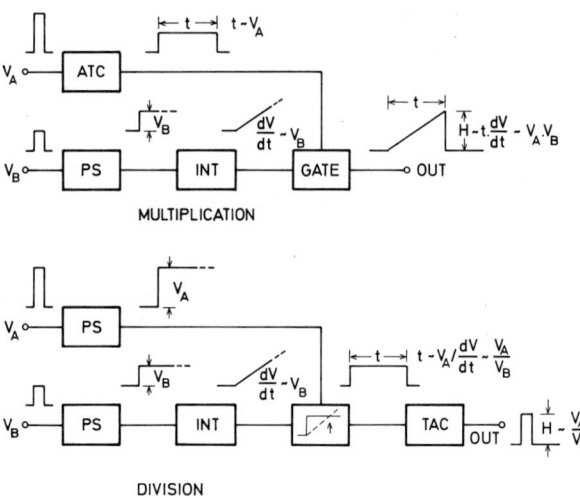

Fig. 3.60. Multiplication and division using the amplitude-to-time conversion

Multiplication and division using non-logarithmic techniques. Another possibility of performing the multiplication or division of two pulse amplitudes consists in using the subterfuge of pulse-height-to-time conversion (Fig. 3.60). For the purpose of multiplication, one of the two pulses (amplitude V_A) in an amplitude-to-time converter ATC is converted into a pulse of standard height and of duration $t \sim V_A$. From the other pulse (amplitude V_B) a linear ramp having the slope $\dfrac{dV}{dt} \sim V_B$ is produced e.g. by means of a pulse stretcher PS and an integrator INT. A linear GATE limits the length of the ramp pulse to t. Hence the amplitude H of the resulting saw tooth pulse becomes $H \sim t \cdot \dfrac{dV}{dt} \sim V_A \cdot V_B$, proportional to the product of $V_A \cdot V_B$.

In the division circuit a linear ramp with the slope $\dfrac{dV}{dt} \sim V_B$ is again produced. A difference discriminator compares the ramp voltage with the stretched pulse V_A and determines the instant t when both these voltages become equal. Since t is proportional to $V_A \Big/ \dfrac{dV}{dt} \sim V_A/V_B$, it is only necessary to convert the discriminator pulse in a time-to-amplitude converter TAC into a pulse of amplitude $H \sim t \sim V_A/V_B$.

Multiplication circuits based on this principle are described among others by AITKEN [3.078] and GRIFFITHS et al. [3.079]; division circuits by BAYER [3.080], TSUKUDA [3.081] and KUHLMANN et al. [3.082]. KONRAD [3.083] uses the pulse-height-to-time conversion with exponential wave forms (due to appropriate RC networks) for forming the logarithms of pulse signals.

Field effect transistors are majority carrier semiconductor devices in which the channel resistances are a reciprocal function of the gate voltage. Hence the channel current is proportional to the product of the channel and gate voltages. This effect can be used for signal multiplication. MILLER and RADEKA [3.084] describe a FET signal multiplier using a bridge configuration for improvement of the circuit linearity. GRUNBERG et al. [3.085] obtained a linear multiplication by FET transistors by means of a suitable feedback. Another FET multiplier has been described by FISHER and SCOTT [3.173], GERE and MILLER [3.086] used a double emitter transistor 3N64 as a multiplying element (such transistors normally are used in chopper circuits).

According to GRUETER [3.087] a parabolic characteristic can be achieved by feeding the input signal to the two grids g_1 and g_2 of a heptode if these are suitably biased. BRISCOE [3.088] who described one of the first multiplicating circuits used the identity

$$(A+B)^2 - (A-B)^2 = 4 \cdot AB. \tag{3.52}$$

The square of the signal sum and difference is formed in special tubes QK-329 with square characteristics.

CHAMINADE et al. [3.174] described a simple circuit for the approximation of the particle identification formula $(E + \Delta E)^\mu - E^\mu$.

An operational amplifier can linearize non-linear data if its characteristic is inverse to the non-linearity, which, of course, must be known [3.089]. This can be used in linearizing the energy-to-pulse-height relationship of detectors with non-linear response. However, non-linear amplifiers affect the pulse shape as well as the pulse height. These effects are discussed by HORN and KHASANOV [3.090].

3.3. Window Amplifiers

When using high response semiconductor detectors (or in other similar measurements), it may be necessary to spread a part of the pulse height spectrum over the entire range of a multi-channel analyzer. In this case window amplifiers are used. The gain of a window amplifier is zero for signal amplitudes below a well-defined level V_w, and increases (mostly to about 1 to 10) for $V_{in} > V_w$. The amplification characteristic for $V_{in} > V_w$ must be as linear as possible.

The linear range of a window amplifier is limited towards high input amplitudes, too, due to the inherent amplifier nonlinearities[7]. However this upper limit is not as pronounced as V_w. Fig. 3.61 shows a diode circuit exhibiting the desired characteristic. There is a negative bias $-V_w$ applied to the diode. Only that part of pulse which is higher than V_w is passed through the circuit. Due to the exponential form of the diode characteristic (3.48) there is a steady transition from cutoff to conduction. Hence the output voltage V_{out} depends upon V_{in} in a manner shown on the right in Fig. 3.61. The "break point" of the characteristic is displaced from zero by a voltage V_{in}^*. Input pulses with amplitudes of the order of magnitude of $2 \cdot V_{in}^*$ are passed through by the circuit, but the transmission is not linear.

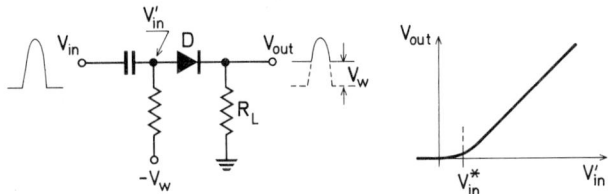

Fig. 3.61. Diode discriminator (diode window amplifier)

Of course, whether a diode is cut off or not must be decided by taking into account the load resistance R_L: the diode is cut off when its dynamic resistance $r_D \gg R_L$, and it is conducting when $r_D \ll R_L$. The break point V_{in}^* hence occurs at $r_D \approx R_L$. From (3.48) we have

$$r_D = \frac{dV}{dI} = \frac{kT}{e(I + I_{so})}, \qquad (3.53)$$

with $r_D \approx R_L$ giving

$$V_{in}^* = \frac{kT}{e}\left(\log \frac{kT}{eR_L I_{so}} + 1\right) - R_L I_{so}. \qquad (3.54)$$

Due to the low value of I_{so}, the term $R_L I_{so}$ can be neglected. At room temperature kT/eI_{so} with $I_{so} = 1\,\text{nA}$ amounts to approximately 25 MΩ, kT/e to 25 mV. $R_L = 2.5\,\text{k}\Omega$ yields for instance $2V_{in}^* = 250\,\text{mV}$.

---

[7] Therefore *window* amplifier!

According to (3.54) the non-linear range V_{in}^* becomes smaller if a high R_L is used. However, with a high R_L the parasitic diode capacity becomes prohibitive. A solution to this problem can be obtained if the diode (or any other non-linear component) is built into a feedback loop, e.g. as shown in Fig. 3.62. This circuit is cut off for $V_{in} < 0$ and transmits linearly when $V_{in} > 0$; the output voltage is inverted. The most convenient way to understand the circuit operation is to regard the amplifier and the diode D_1 as a new non-linear amplifier with an output resistance of r_D. Due to the feedback, r_D is reduced by $(1+bA)$, with the same result as if the load R_L were increased by $(1+bA)$. The diode D_2 further improves the operation by shorting the amplifier when $V_{in} < 0$. Hence the backward resistance of D_1 is not reduced by the feedback.

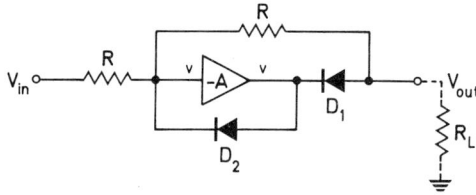

Fig. 3.62. Improving the linearity of a diode window amplifier by means of negative feedback

EMMER [3.091] described a window amplifier consisting of two biased transistorized difference amplifiers connected in series, the characteristic of which was linearized by feedback. By variation of the feedback intensity (i.e. the ratio of the two resistors R in Fig. 3.62), the gain of this amplifier can be adjusted between -1 and -16.

Instead of increasing R_L, a zero load $R_L = 0$ can also be used if a current generator with (almost) infinite internal resistance is connected to the signal input (Fig. 3.63). This has been proposed by LARSEN [3.056], ARBEL [3.043] and KANDIAH [3.092]. The bias current I_w delivered by a high ohmic resistance or by a pnp transistor collector flows via D_1 to the ground — D_2 is cut off. The signal current I_{in} at first decreases the D_1 current. Then at $I_{in} \geq I_w$, the total current $I_w - I_{in}$ changes its direction and flows via D_2 into the amplifier $-A$. The transition between the cutoff and linear range is very short and well defined. Since the currents are switched without changing the potential differences across D_1 and D_2, the parasitic capacities need not be recharged. Hence the circuit is a fast one.

In window amplifiers, only that part of a pulse is amplified which exceeds I_w or V_w. Thus the output pulse can be considerably shorter than the original pulse. Moreover, the output pulse length depends on its amplitude and on the window position. In order to conserve the

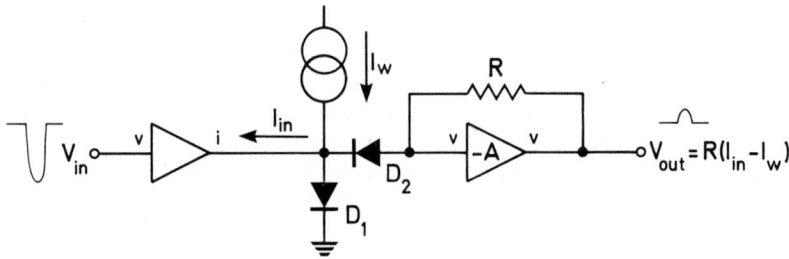

Fig. 3.63. Window amplifier consisting of voltage-to-current converter and a current discriminator according to LARSEN [3.056]

linearity, the window amplifier and the whole of the electronics following the non-linear element (i.e. the discriminator, analog-digital converter) must be very fast [3.042]. If the pulse shape is changed by circuits with time constants which are comparable to the pulse length, the amplitude of pulses of different lengths will be affected to different extents. Alternatively, the output pulses of a window amplifier could be shaped in a pulse stretcher [3.056].

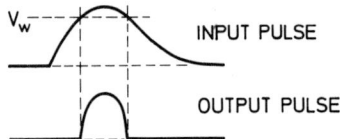

Fig. 3.64. The pulse shape at the output of the window amplifier

FABRI et al. [3.093] and BERTOLACCINI et al. [3.094] described another technique of spreading the energy scale which can be used in connection with charge-sensitive preamplifiers. According to their proposal the signal pulse from the preamplifier output triggers a standard pulse generator. The standard pulse is fed to the input of the preamplifier by means of a small coupling capacitor. Since the standard pulse exhibits opposite polarity, a charge of known and constant magnitude is subtracted from every signal pulse. The whole circuit is fast enough to allow the slow integrator in the main amplifier to form a pulse of amplitude proportional to the difference between the signal and compensation charges.

This method offers the following advantage: in a conventional window amplifier system the amplifier part preceding the non-linear element must exhibit an extremely high long-term stability, since any drift is magnified by the scale spread. Due to locating the "non-linearity" at the extreme input of the amplifier system by introducing the standard charge Q_w, there is, of course, no gain before the "non-linear element" and the accuracy of the assembly depends only upon the constancy of Q_w.

3.4. Linear Gates

The purpose of a linear gate is to disconnect the signal path when cut off and to transmit the signal with a minimum of distortion when open.

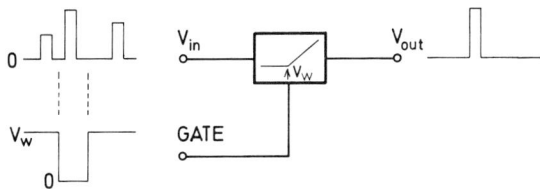

Fig. 3.65. Window amplifier used as linear gate

Window amplifier as a linear gate. A window amplifier can be used as a linear gate if the threshold is not kept constant but used as the gate control signal (Fig. 3.65). For a given threshold, e.g. V_w, all pulses with amplitudes lower than V_w are suppressed. The pulses are transmitted only during the time in which a zero threshold voltage is applied. Hence the circuit Fig. 3.65 operates as a linear gate, but only for pulses of a defined polarity and a defined maximum amplitude.

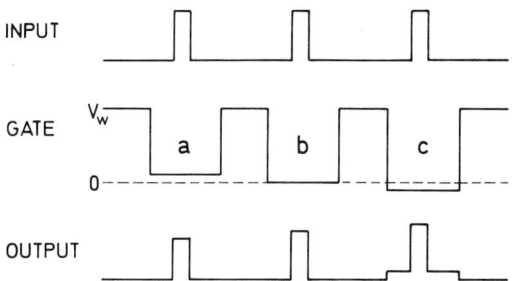

Fig. 3.66. Pulse shapes at a linear gate with too small (a), correct (b) and too large (c) gating signal, respectively

The gate control signal must amount to exactly zero for an "open" condition of the gate, otherwise either a small part of the pulse is lost (Fig. 3.66a) or the output pulse exhibits a pedestal (Fig. 3.66c). In the latter case the pedestal appears even if no input pulse is present and it can thus simulate a signal pulse.

MANFREDI et al. [3.175] report a method for compensating the pulse pedestal by means of a servo loop.

Linear gates for voltage pulses possess the same disadvantages as a corresponding window amplifier: If the characteristic is not linearized, e.g. by feedback [3.091], the break between cutoff and transmission is indefinite, small signals are nonlinearly distorted and a small pedestal is unavoidable. Moreover, the variation of threshold voltage V_w causes recharging of various parasitic capacities, thus resulting in a slow gate (cf. the analysis by MILLMAN and TAUB [3.095], Chapter 14).

On the other hand a current threshold amplifier according to LARSEN [3.096] (cf. Fig. 3.63) is particularly well suited as a linear gate — only the bias current I_w must be interrupted when opening the gate. Fig. 3.67 shows a simplified diagram of a circuit given by KANDIAH [3.092] operating on this principle. Instead of the diode D_2 in Fig. 3.63, the base emitter junction of the transistor Q_1 is used. The gate bias current of 10 mA flows through Q_3 and Z_2 in the diode AAZ 13, thus impeding any signal current $I_{sig} < 10$ mA from flowing into the operational amplifier $-A$. With the aid of R_3 an additional bias current can be applied to AAZ 13. Hence the circuit can serve as a window amplifier and a

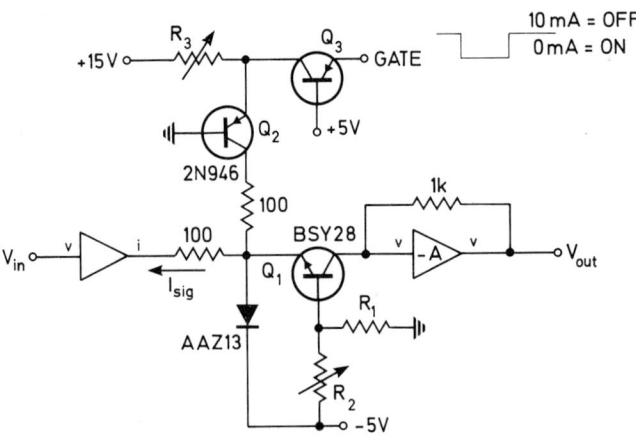

Fig. 3.67. Linear gate according to KANDIAH [3.092]

linear gate at the same time. When the gate bias current is disconnected (by cutting off the transistor Q_3) the gate is open for pulses axceeding the additional threshold bias current. For zero threshold ($R_3 \to \infty$) the height of the pulse pedestal can be adjusted to zero by means of R_2.

Switch-type linear gates. Linear gate circuits can be formed by networks with switches shown schematically in Fig. 3.68. Of course, the switches are replaced by electronic components in actual circuits[8]. In

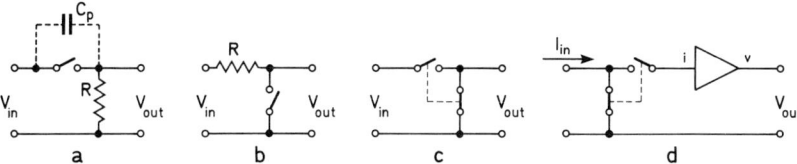

Fig. 3.68 a–d. Switch type linear gates

the case of the series gate (a) with closed switch (resistance R_{on}) the signal is transmitted with a low attenuation of $R/(R+R_{on})$ only, whereas with opened switch (resistance R_{off}) the signal is inhibited with the exception of a small part $R/(R+R_{off})$. Evidently $R_{on} \ll R \ll R_{off}$ must be valid. Signal transmission due to the parasitic capacity C_p causes difficulties.

In the parallel gate (b) the signal is inhibited with the exception of a small part $R_{on}/(R+R_{on})$ when the switch is closed, and transmitted with the attenuation $R_{off}/(R+R_{off})$, when the switch is off. Again $R_{on} \ll R \ll R_{off}$ must hold. A systematic analysis of both parallel and series gates is given by MILLMAN and TAUB [3.095].

The transmission and suppression factors of a series-parallel gate (c) or (d) are $R_{off}/(R_{on}+R_{off}) \approx 1$ and $R_{on}/(R_{on}+R_{off}) \ll 1$, respectively; because $R_{on} \lll R_{off}$, the on and off states of the gate are extremely well defined. A type (c) gate is described by GOULDING [3.096], and a schematic circuit diagram is shown in Fig. 3.69. In the gate state "OFF", the current I_{34} flows into point (1). Since $I_{34} > I_1$, the difference $I_{34} - I_1$ flows through the diode D_1 and the potential of point (1) is zero (or, more exactly, slightly negative). Hence the transistor Q_1 is cut off. On the other hand, I_2 flows into the base of Q_2 and saturates this transistor. Thus positive input signals are prevented from reaching the output. In the gate state "ON", Q_3 is cut off, the current I_1 flows into the

---

[8] Exactly speaking, the linear gates Fig. 3.65 are diode swich-type gates, too, the diode changing its switching characteristic (off — on) according to the signal polarity.

base of Q_1 and saturates Q_1. $I_{34} > I_2$ is now applied to point (2), the base potential of Q_2 is slightly negative and Q_2 is cut off. Hence the gate is open. The emitter potential of Q_2, which is approximately zero, can be finely adjusted for compensation of a possible pulse pedestal. Since the emitter current of Q_1 flows into the input signal generator, the output impedance of this must be very low.

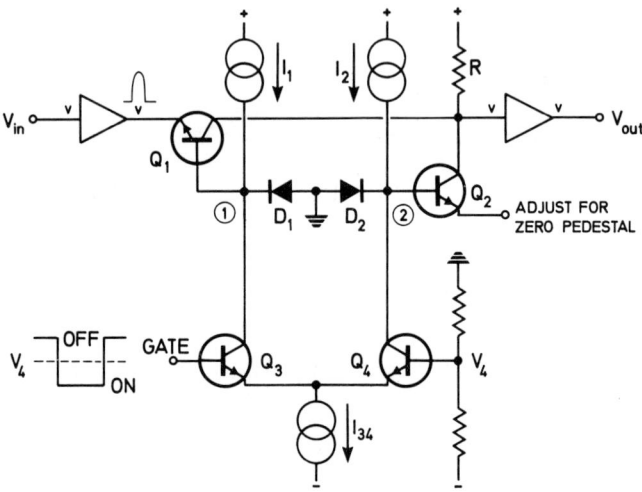

Fig. 3.69. Operating principle of a series-parallel gate with transistor switches

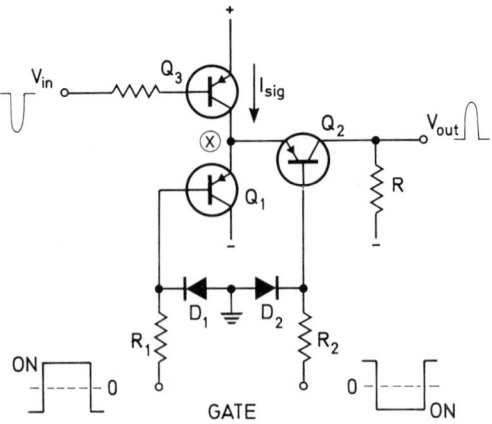

Fig. 3.70. Series-parallel gate for current input

BARNA and MARSHALL [3.097] used two transistors in the configuration (d) of Fig. 3.68. The principle of the circuit is shown in Fig. 3.70. Again two symmetrical gate control signals are needed. In the state "OFF", enough current is drawn by R_1 from the base of Q_1 to saturate Q_1. The transistor Q_2 is cut off. Signal current I_{sig} is fed through Q_1 to the ground (or exactly speaking to the nagative pole of the supply voltage). In the state "ON" Q_2 conducts and Q_1 is cut off. Hence an output voltage $V_{out} = R \cdot I_{sig}$ proportional to I_{sig} appears. The circuit is relatively insensitive, although it also works satisfactorily with gate control pulses of 50 nsec duration. FELDMAN [3.098] introduced an inductivity of about 1 mH between (X) and the ground, to avoid the pedestal due to the quiescent current of Q_3.

Some other papers dealing with linear gates are [3.176] to [3.178]. WHITE [3.162] described a fast linear gate for fast pulses in current mode operation, which integrates them when activated.

Linear gates for bipolar pulses. All the circuits discussed above operate with unipolar pulses only, or with one half of a bipolar pulse.

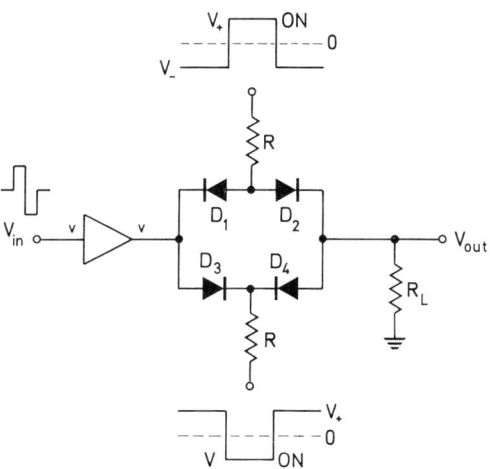

Fig. 3.71. Diode-bridge linear gate

If the gate is placed immediately before an ADC, the loss of the second half of bipolar pulse does not disturb since no count-rate-dependent baseline shift can occur. However, if the pulse shape must be conserved by the gate, special so-called bidirectional circuits must be used, which commonly operate on the principle of a diode bridge shown in Fig. 3.71.

In the state "OFF" all diodes are cut off, namely D_1 and D_2 by the voltage V_- and D_3 and D_4 by V_+. The state "ON" is obtained by applying symmetrical gate signals to the gating inputs. If $|V_+|=|V_-|$ holds, a current of the magnitude $V_+/R = -V_-/R$ flows through the bridge and all diodes conduct. When properly adjusted (i.e. equal currents through D_1 and D_3, and D_2 and D_4, respectively), no pulse pedestal appears at the output because of the symmetry. Bipolar pulses now are passed through. For the sake of small attenuation, the output impedance of the signal generator and the dynamic resistances of the diodes must be negligible against R, and $R \ll R_L$ should hold. (A more detailed analysis of these circuits is given in [3.095] and [3.099]). Because of the impedance conditions mentioned, an emitter-follower is normally connected to the input of the gate and another one to the output, to serve as impedance converters. The symmetrical gate pulses can be delivered by long-tailed pairs, as in Fig. 3.69 [3.100, 3.101], or by a transformer

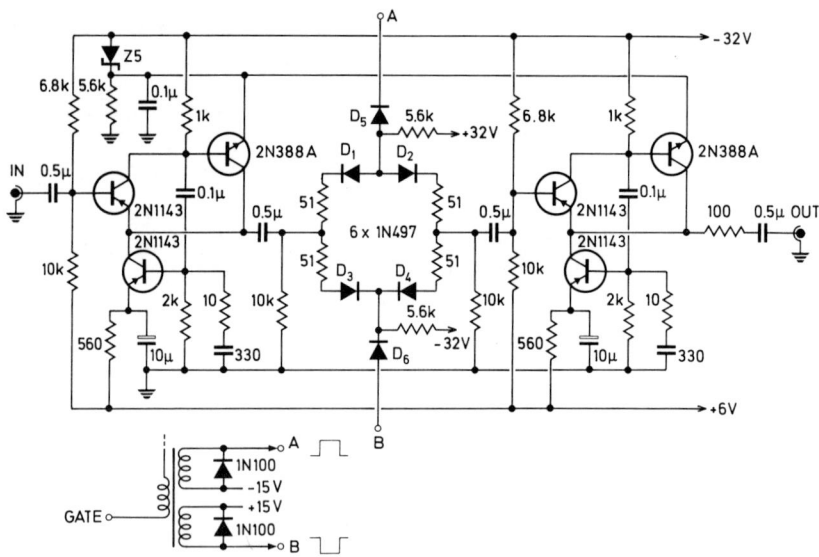

Fig. 3.72. Six-diode-gate according to GINGELL [3.103]

[3.102], as in Fig. 3.72. In the former case a *dc* operation is possible, but in the latter one the gate control pulse duration is obviously limited by the transformer. The bridge-type linear gates operate extremely fast and can be used down to gate pulse lengths of a few nsec.

To ensure that the gate control pulses need not exhibit accurate amplitudes without causing an output pulse pedestal, the resistor R is commonly connected to two fixed equal potentials V_+ and V_- of opposite polarity and the gating pulses are applied by two additional diodes. Such a circuit is known as a "six-diode gate" [3.095]. A circuit according to GINGELL [3.103], shown in Fig. 3.72, will serve as an illustration of the operating principle. In the OFF state potentials of -15 V and $+15$ V are applied to points (A) and (B), respectively, all the diodes D_1 to D_4 are cut off and the gate is consequently closed. The current through the two resistors 5.6 kΩ is fed by means of the diodes D_5 and D_6 to the -15 V and $+15$ V terminals of the power supply, respectively. In the ON state the two transformer windings deliver two pulses of about 30 V amplitude and of opposite polarity to the points (A) and (B), thus cutting off the diodes D_5 and D_6. The current applied by the two resistors 5.6 kΩ flows through the diode bridge and opens the gate. This current is given solely by the voltages of ± 32 V and by the exact magnitudes of both the resistors; it is independent of the actual height and symmetry of the control pulses. By adjusting the resistor values a pedestal can be avoided. Two white emitter-followers are used at the input and the output of the gate. Another diode-bridge linear gate with excellent linearity has been described by SCHUSTER [3.179].

Bidirectional gates can also be realized by means of pulse transformers. The signal is fed to the primary winding via diodes which are switched by a control signal between conduction and cutoff (CHAPLIN and COLE [3.104], VALCKX and DYMANUS [3.105]).

The output signal of a two-diode digital AND gate (cf. Chapter 6) is equal to that of the smaller input signal. This effect can be used for obtaining a relatively simple linear gate: The signal pulse is applied to one input of the AND gate, and the control voltage to the other. The control voltage must be zero for the OFF state of the gate, and higher than the maximum signal amplitude for the ON state. A diode gate operating on this principle is described by SASAKI and TSUKUDA [3.106], a corresponding transistor gate by LIU and LOEFFLER [3.107]. The AND gate configuration yields a relatively high pulse pedestal which must often be compensated by auxiliary circuits.

3.5. Pulse Stretchers

If the amplitude information of a pulse is to be analog stored for a given short time in order to enable the pulse height to be measured or compared with another, pulse stretchers are used. In a pulse stretcher a condenser is charged by a non-linear component with a short time constant to the full amplitude of the input pulse. Due to the high backward resistance of the non-linear component, the stored charge is conserved for a given time. There are three fundamental types of pulse stretcher, as shown in Fig. 3.73. The storage condenser C is discharged either exponentially by means of a resistor R (Fig. 3.73a) or linearly by a constant current generator I_C (Fig. 3.73b). The discharging can also be initiated by means of a switch (e.g. a transistor) activated by the input pulse which is delayed by a time interval of Δt (Fig. 3.73c). In the latter case an approximately rectangular pulse is produced, with an optimum aspect ratio (the ratio of pulse widths at the top and at the base).

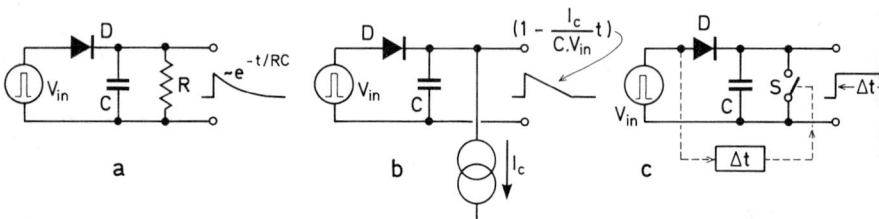

Fig. 3.73a—c. Pulse stretcher circuits

A diode (generally) or the base-emitter junction of a transistor is used as the non-linear component. The storage capacitor is charged with the time constant of $R_F \cdot C$, where R_F denotes the sum of the diode forward resistance and the output resistance of the signal pulse source. If C is to be always charged to the full input pulse amplitude, independent of the variations in the input pulse shape and width, the time constant $R_F \cdot C$ must be small compared to the minimum pulse width δ_{min}. Even if the output impedance of the signal pulse source is negligible, R_F remains equal to the non-zero diode forward resistance. Since C cannot be made infinitely small (cf. below), the minimum input pulse width is limited. The time characteristics of the charging process can be calculated more or less exactly according to KANDIAH [3.193].

Pulse Stretcher with Feedback. The resistance R_F can be diminished by introducing a negative feedback [3.042, 3.056]. Fig. 3.74 shows the principle of such a circuit. It is hardly necessary to explain the circuit operation in more detail. Since the resistor R normally terminates a coaxial cable, at the output of which the pulse to be stretched appears, the value of R is often $50 \ldots 100\,\Omega$. Hence the feedback loop must be separated from the storage capacitor by means of an impedance converter (emitter-follower) indicated by dotted lines. The discharging time constant is then $R' \cdot C$, where R' denotes the sum of the load and input resistance of the impedance converter.

It must be pointed out that the feedback loop is closed only during the charging process. After the end of the input pulse the diode D_1 is cut off. At the same time the whole output pulse V_{out} is fed back to the amplifier input (X) and normally overloads the amplifier. The overload can be avoided by another feedback loop with D_2. Nevertheless, the amplifier must not only be able to deal with the short input pulses, but must

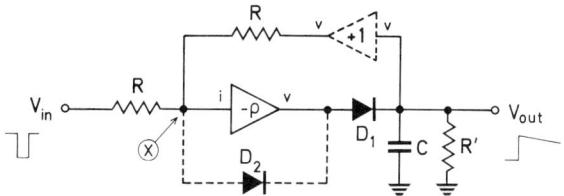

Fig. 3.74. Pulse stretcher with feedback

also withstand the long stretching period without any change in operating points. Hence a *dc* amplifier is preferably used. When R serves as a cable termination, the point (X) must remain a virtual ground independent of the state of the feedback loop. Therefore a current-voltage converter with a common base input stage offers advantages when used as the amplifier. Due to the extreme feedback situation with $b=1$, the system can exhibit some ringing and overshoots, which are avoided by proper layout of the circuit and by a current limiting resistor of a few ohms in series with D_1.

Parasitic properties of the charging diode. The non-ideal cutoff characteristic of the semiconductor charging diode D limits the maximum possible stretching of the pulse. Firstly, the backward resistance R_B of the diode always remains in parallel to C (Fig. 3.73), thus limiting the discharging time constant to $\leq R_B \cdot C$. Secondly, due to the parasitic capacity C_p of the diode, the charge stored in C is diminished by $C_p/(C+C_p)$ when the input voltage returns to zero. More exactly speaking, a stored charge q remains in the *pn* junction of the diode, the value of this charge depending among other things upon the actual charging current. This charge must first flow away before the junction becomes cut off (cf. e.g. SEILER [3.108] or other semiconductor text books, and especially KO [3.109]). The voltage across C is diminished by q/C by the charge q. Since q/C must remain negligible with regard to V_{in}, a lower limit for C results.

The non-ideal cutoff characteristic of a diode can be improved by a circuit, shown in Fig. 3.75, which has already been used by KELLY [3.110] in a vacuum tube pulse stretcher. The condenser C is charged via two diodes D_1 and D_2 connected in series. The voltage across C is fed back to the point (Y) by means of an impedance converter (emitter-follower) and another diode D_3. Since the potential of (Y) approximately corresponds to the output potential V_{out} during the charging process, no current flows through D_3. After the input voltage has returned to zero the diode D_1 is cut off, and the backward saturation current and the stored charge of D_1 are not drawn from C, but via D_3 from the low impedance output of the stretcher. Hence the potential at (Y) remains only slightly below the potential of the condenser C,

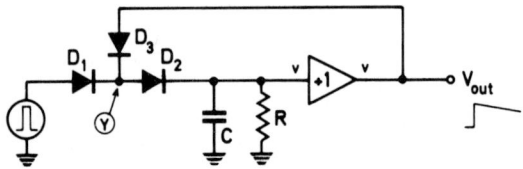

Fig. 3.75. Improving the diode cut-off in the pulse stretcher

and the diode D_2 is only slightly cut off and does not discharge C. Evidently the potential drop in the emitter-follower must be compensated by proper means.

Of course, any combination of the circuits shown in Fig. 3.74 and 3.75 can be used as a pulse stretcher.

Practical circuits. The older vacuum tube circuits are reviewed by MEYER [3.111]. SARAZIN [3.112] describes a pulse stretcher operating satisfactorily down to the minimum input pulse width of 5 nsec. The condenser $C = 27$ pF is charged through two cascaded cathode followers with tubes of very high transconductance ($g_m = 35$ mA/V). CUMMINS and BRANUM [3.113] used the parasitic input capacity of a normal CRT oscilloscope as the storage capacity C; the discharging of C is performed by the signal pulse itself, delayed by means of a delay line. The circuit served to facilitate the observation of very short pulses on the oscilloscope screen. KELLER [3.101] describes a circuit with a common base input stage designed for minimum input pulse duration of 20 nsec. The non-linear component used is either a fast diode (HD-5000) or a silicon transistor. With the latter the ratio of charging and discharging time constants is $1:10^6$. The circuit is linear within 5% in an amplitude range of 1:10. The stretched pulse is picked up by an emitter-follower with extremely high input impedance using a MOS transistor RCA TA 2330. WEDDIGEN and HAASE [3.114] report a pulse stretcher of the type Fig. 3.73c operating satisfactorily down to 6 nsec input pulse length. The principle of their circuit is outlined in Fig. 3.76. Normally both transistors Q_1 and Q_2 conduct. The potential at C is given by the base potential of Q_1, and the current through Q_2 by R_1, R_2 and R_3. The input pulse triggers a low delay tunnel diode monostable multivibrator TD, which delivers a negative pulse of about 1 μsec duration cutting off the transistor Q_2. At the same time the storage capacity $C=50\,\mathrm{pF}$ is charged to the full amplitude of the input pulse via transistor Q_1.

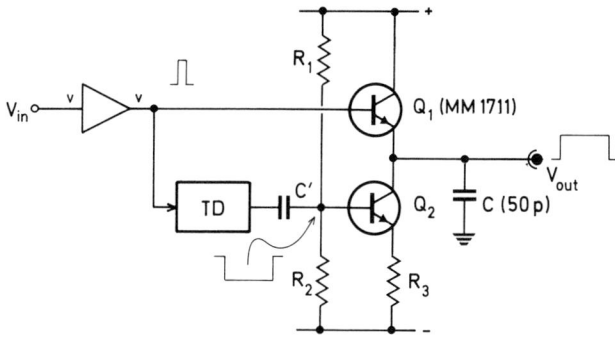

Fig. 3.76. Practical pulse stretcher circuit according to WEDDIGEN and HAASE [3.114]

After the input voltage has returned to zero, Q_1 is cut off too, and C stores the charge until it is discharged by Q_2 after 1 μsec. The stretched output pulse is picked up by cascaded emitter-followers with an input resistance of about 4 MΩ. The integral linearity of the stretcher is better than 2 % for input pulse heights varying between 0.1 and 2.5 V and pulse lengths varying between 6 and 300 nsec.

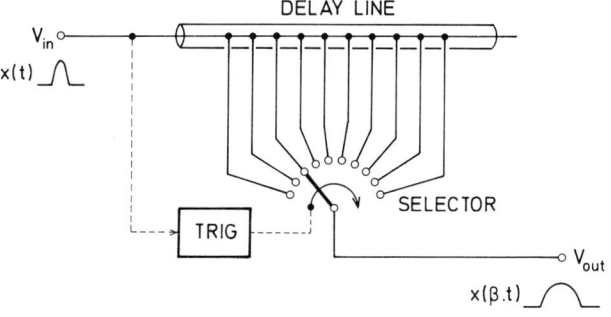

Fig. 3.77. Pulse stretching according to GERSHO [3.116]

Analog Circuits

A conventional nanosecond pulse stretcher is described by GOYOT et al. [3.180]. LOOTEN et al. [3.181] described a nanosecond pulse stretcher of the emitter-follower type, with output pulse amplitude proportional to the integrated input charge.

CRAIB [3.115] proposed the use of delay lines for the pulse stretching. He charges all capacitors of a lumped-constant artificial delay line at the same time. The stretched signal, the length of which corresponds to the line delay, appears at the properly terminated line ends. OWENS and WHITE [3.182] discussed the use of nonlinear delay lines consisting of chokes and biased diodes in nucleonic instrumentation. GERSHO [3.116] describes a general technique of stretching or shortening of signals without distortion of the signal shape. The principle can easily be explained using Fig. 3.77. The signal running down a delay line is picked up at different points and fed to the output via an (electronic) selector switch. The movement of the selector switch is initiated by the input pulse by means of a suitable trigger circuit TRIG. When the selector is fixed to one position, the output pulse corresponds to the input one $V_{out} = V_{in}$ (neglecting a constant delay). If the selector runs synchronously with the signal propagation in the delay line, the same voltage always appears at the output, and the input signal is stretched infinitely. A slower selector rotation in the direction of signal propagation (arrow) hence yields a finite pulse stretching: $X(t) \to X(\beta \cdot t)$ with $\beta < 1$. Selector rotation opposite to the direction of signal propagation results in a shortening of the input signal $(\beta > 1)$.

3.6. Fast Pulse Amplifiers

In high energy physics, especially in experiments on pulsed particle accelerators, pulses must often be processed at very high count rates. In these cases fast semiconductor detectors and scintillation or Čerenkov counters are used and the necessary selection operations (amplitude discrimination, coincidences) are performed on the fast signals, as has already been mentioned briefly in Chapter 3.1.3. Especially with semiconductor detectors, at high count rates because of the limited maximum anode current of a photomultiplier also with scintillation and Čerenkov counters, the detector current signal must be amplified. The amplifier must be fast (rise times of about 1 nsec) and must not distort the pulse shape. On the other hand the requirements concerning the linearity, gain, temperature stability etc. are somewhat reduced in relation to slow amplifiers (~ 1 μsec).

The rise time of an amplifier — or in other words its bandwidth — is limited on the one hand by the product of the actual load resistances and parasitic capacities which together form integrators, and on the other hand by the inherent response of the active component used. Since the voltage gain is proportional to the load resistor, a short rise time can be realized only at the expense of low gain. Also the forward current transfer ratio $\beta(=h_{fe})$ of a transistor becomes low at high frequencies, yielding an approximately constant gain bandwidth product. Hence the frequency response of a transistor can be sufficiently characterized by the upper cutoff frequency f_T at which $\beta(f_T) = 1$ becomes unity.

Distributed amplifiers. With vacuum tubes relatively high load resistors must be used for high voltage gain due to the low transconductance. Hence the rise time of the amplifier is determined by the parasitic capacities C_p between anode and ground and not by the inherent frequency response. In addition to the capacity of the external circuitry, especially the internal electrode capacities of the tube also contribute to C_p. Unfortunately, tubes with higher transconductance exhibit a higher C_p too. The situation is discussed in all textbooks on high frequency techniques (cf. too [3.019]). A shorter rise time can be obtained by connecting several tubes in parallel with the aid of lumped-constant delay lines (Fig. 3.78). The delay times of the particular line section in the grid and anode circuits are equal. Therefore the total delay between input and output is always the same, independent of the actual signal path through any particular tube. Hence the transconductances of the tubes add together. The parasitic capacities of the tubes, however, are separated by the inductances L and they merely increase the delay line section capacities C. By variations of C the proper value of the total capacity $C+C_p$ can be adjusted.

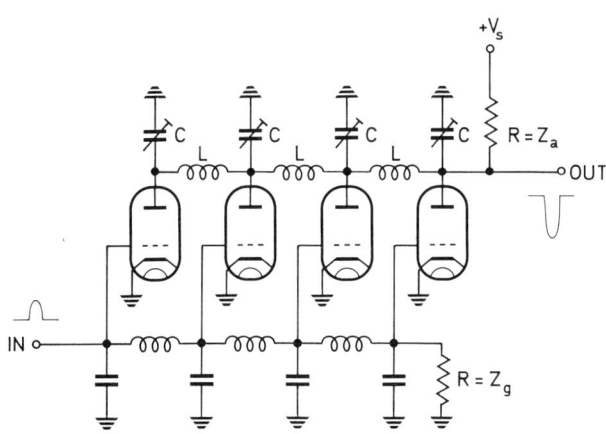

Fig. 3.78. Distributed amplifier with vacuum tubes

Such distributed amplifiers are described by different authors ([3.117] to [3.121]). Transistorized distributed amplifiers have been investigated, too (BENETEAU and BLASER [3.122]). Rise times of about 1 nsec can be achieved, but the amplifier must be carefully adjusted. A long-term stability of the gain is difficult to arrive at.

Transistorized fast amplifiers. With transistors the mutual transconductances are very high and the external circuits can easily be made of low-valued resistors so that the amplifier rise time is given by the inherent frequency response of the transistors used. There is a vast number of equivalent circuits [3.123], which describe approximately the physical behaviour of a transistor at high frequencies and which can be used as a starting point in a circuit synthesis.

An improvement of the frequency response by means of a feedback loop containing more than one amplifier stage is not practicable due to the long signal propagation times per stage. Hence feedback circuits with only one stage (i.e. one transistor) must be used such as shown in Fig. 3.79 (but cf. Fig. 3.82 with two transistor feedback stages).

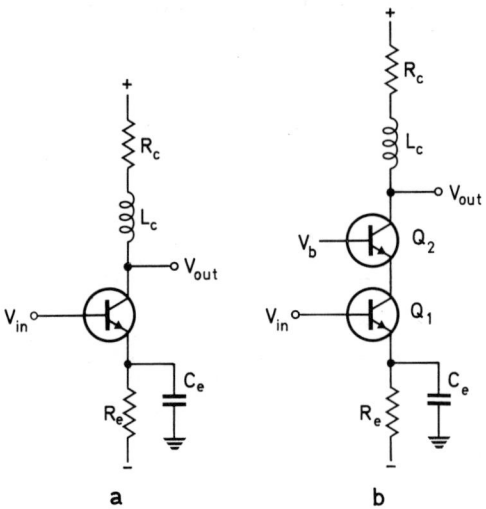

Fig. 3.79 a and b. Transistorized amplifier stages with negative feedback

The general case of circuit analysis proceeds as follows: With the aid of the transistor equivalent circuit used, the response function $G(p) = \hat{V}_{out}/\hat{V}_{in}$ of the stage is calculated. Normally $G(p)$ is a fraction of two polynomials such as

$$G(p) = \frac{1 + a_1 p + a_2 p^2 + \cdots + a_n p^n}{1 + b_1 p + b_2 p^2 + \cdots + b_m p^m} \tag{3.55}$$

with $m > n$. For stable operation of the amplifier the real part of all poles of $G(p)$ must be negative. Negative real poles yield an aperiodic response. According to Elmore [3.124], an amplifier with $G(p)$ from (3.55) exhibits a rise time t_r and a delay time (signal propagation time) t_d as follows:

$$t_r = \sqrt{2\pi[b_1^2 - a_1^2 + 2(a_2 - b_2)]}; \quad t_d = b_1 - a_1. \tag{3.56}$$

By introduction of suitably chosen compensation elements in the circuits of Fig. 3.79 the corresponding response function $G(p)$ can be influenced in order to yield the desired t_r. However, since the exact values of the equivalent circuit parameters are mostly unknown and moreover depend upon the amplitude, the optimum t_r can be achieved only by empirical adjustment of the actual circuit. For the comparatively simple but extremely tedious analysis of the particular fundamental circuits, see the original papers [3.125] to [3.130].

In the circuit of Fig. 3.79a the resistor R_e serves as a feedback element; the stage gain is approximately R_c/R_e, if $R_e \gg \dfrac{kT}{eI_E}$, where I_E denotes the emitter current. Normally R_c is chosen as $100\ldots 200\,\Omega$, $R_e \approx 50\,\Omega$, $I_E \gtrsim 10\,\text{mA}$. The inductance L_c (some tens of nH) and the "emitter peaking" capacitor C_e serve to compensate the gain loss at high frequencies. They must be adjusted experimentally for an optimum compromise between short rise time and low overshoot. The circuit of Fig. 3.79b makes use of the fact that the common base transistor stage is extremely fast. In the cascade circuit (b) the collector of the transistor Q_1 operates into the low emitter impedance of Q_2. Hence Q_1 has a "zero" voltage gain and its base-collector capacity is not magnified by the Miller effect. The rise time depends almost solely upon the properties of Q_1, the stage gain being given again by R_c/R_e. Since for Q_2 a transistor with a lower-upper cutoff frequency f_{T2} can be used, just as f_{T1} must be, a transistor type with a higher maximum collector dissipation and a higher collector base breakdown voltage can be chosen to extend the output range to well over 10 V.

Since the internal transistor capacities depend on the signal amplitudes, the compensation (L_c and/or C_e) in amplifiers having large dy-

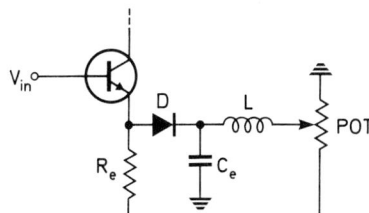

Fig. 3.80. Pulse-height-dependent compensation network according to ALBERIGI-QUARANTA and MARTIN [3.126]

namic amplitude ranges may be properly adjusted for a given signal amplitude. However, for higher or lower amplitudes the stage might be overcompensated or undercompensated. ALBERIGI-QUARANTA and MARTIN [3.126] proposed a circuit, shown in Fig. 3.80, in which the compensating capacity C_e is connected to the emitter via a diode biased to cutoff. Hence the emitter-peaking condenser starts compensating first for higher signal amplitudes. The bias is adjusted experimentally by the potentiometer POT.

Circuits with separate current and voltage amplification. Although the common emitter stages shown in Fig. 3.79 permit current and voltage amplification at the same time, are simple and inexpensive, and — if fast transistors are used — are fast too[9], preference is often given to the combinations of the a priori faster common base stages and emitter-followers. COLI et al. [3.131] divide the voltage and current amplification in the circuit shown in Fig. 3.81 between the common base stage Q_1 and the emitter follower Q_2, respectively. Another common base stage Q_3 serves as a voltage-to-current converter and yields a high impedance current output. At $A=10$ the rise time of such a configuration with twice the 2N918 (npn) and once the 2N976 (pnp) comes to 2.5 nsec.

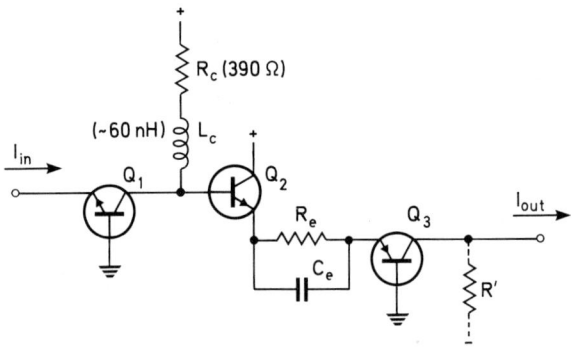

Fig. 3.81. Amplifier stage according to COLI et al. [3.131]

By slightly modifying the diagram shown in Fig. 3.81, a fast amplifier circuit is obtained (Fig. 3.82) which has been analyzed in detail by RUSH [3.128]. The transistor Q_1 of the common base stage operates directly into the input resistance of the emitter-follower Q_2, which itself is loaded by a very low input impedance of another common base stage Q_3. The resistors R_3, R_4, R_5 and R_6 serve only for adjustment of the particular operating points of the transistors, their values being much higher than the corresponding transistor impedances. Hence almost no

---

[9] LUNSFORD [3.132] describes an amplifier with $A=100$ and $t_r=1$ nsec!

signal current flows through R_3 or R_4. Therefore the circuit operates as a pure current amplifier, and the potentials at collector of Q_1 and emitter of Q_2 do not change, thus avoiding the slow charging and discharging of parasitic capacities. The gain is stabilized by the feedback loop R_1, R_2. Although a voltage signal is generated at the collector of Q_2 the parasitic collector capacity C_c can easily be compensated by means of C_2 (0.5 to 5 pF).

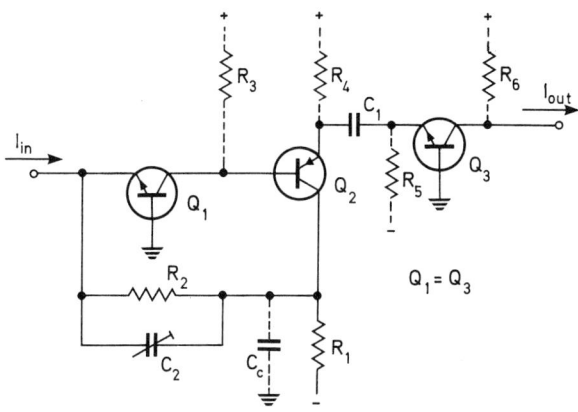

Fig. 3.82. Fast amplifier according to RUSH [3.128]

Using some plausible simplifications RUSH [3.128] calculated the rise time of this circuit to be

$$t_r = \frac{1}{f_{T1}}, \qquad (3.57)$$

where f_{T1} denotes the upper cutoff frequency of Q_1. For aperiodic response (no overshoot) the gain $A=(R_1+R_2)/R_2$ must be equal to

$$A=(R_1+R_2)/R_2 = 4(f_{T2}/f_{T1}), \qquad (3.58)$$

where f_{T2} denotes the cutoff frequency of Q_2. Together with the condition of R_1+R_2, which is high with respect to the input impedance of the common base stage Q_1, the relation (3.58) gives an equation for calculating the values of R_1, R_2. Fig. 3.83 shows the detailed circuit diagram of an amplifier having a gain of $A \approx 1500$ and a rise time of $t_r \approx 3$ nsec. These experimentally measured values agree very well with the estimates calculated by using the relations (3.57) and (3.58) and the data for f_T specified in manufacturers' catalogues(!) for the transistor types used. Besides the fast transistors 2N709 ($f_T = 800$ Mcps) and 2N976 ($f_T = 900$ Mcps), the type 2N2368 ($f_T = 650$ Mcps) was used in the end stage, in order to extend the linear amplitude range to about 3 V at the expense of the rise time.

Some more practical circuits are described in the papers [3.183] to [3.187].

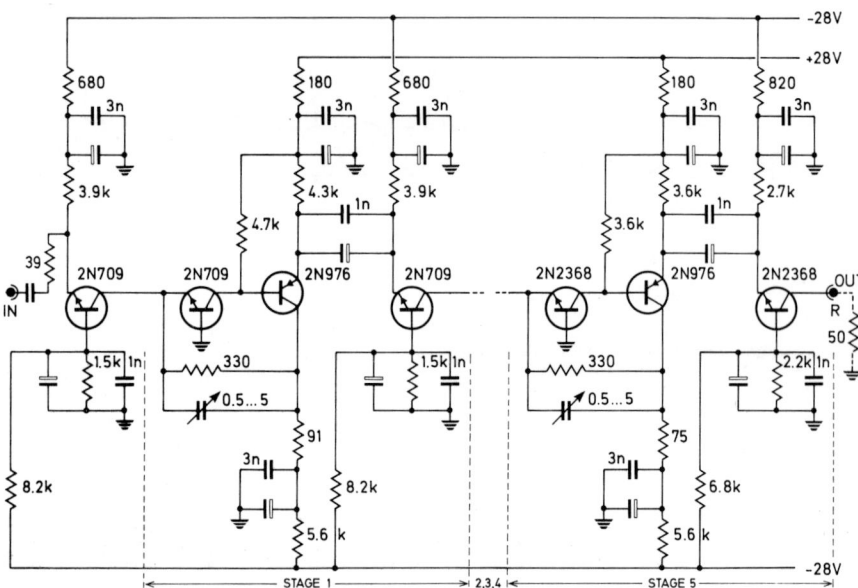

Fig. 3.83. Five-stage amplifier according to RUSH [3.128]. Gain $A = 1500$, $t_r = 3$ nsec (stages 2, 3, 4 = stage 1)

Amplifying of long pulse bursts. In experiments on pulsed particle accelerators the amplifiers are exposed during short periods of time (a few milliseconds) to extremely high pulse rates. In order to avoid a base line shift or saturation effects, direct coupling throughout the amplifier is necessary. The bias networks consisting of resistors and blocking capacitors, which are appropriate for adjusting the transistor operating points at low pulse rates, may also cause difficulties.

In Fig. 3.84a the situation in a common emitter stage with a biasing network R_1/C_1 is illustrated. When the pulse rate is raised suddenly, the potential of the point (1) rises too, causing a shift of the baseline at the collector. According to VERWEIJ [3.133, 3.134], this baseline shift can be compensated by introducing another biasing network R_4/C_2 in the collector circuit. If the time constants relating to the points (1) and (2) are equal, the potential decrease in (2) just compensates the potential increase in (1), yielding a negligible baseline shift at the collector. Of course, the voltage rise at point (1) must not be so high as to cut off the transistor (the so-called suicide effect). Hence the maximum duration of the pulse burst is limited.

VERWEIJ loc. cit. describes another technique for stabilizing the base line, using a non-linear feedback with a clamping diode D (Fig. 3.84c). The negative output pulses pass through without affecting D, though any positive undershoot is integrated by C_2 and fed via the amplifier A_s back to the input, causing the standing transistor current to remain constant. The time constant of the feedback loop of course must be much longer than the width of single pulses, but on the other hand the feedback loop can comprise more than one dc coupled amplifier stage. The high time constant of the point (1) and the value of C_2 influence the low-frequency response of the amplifier: Although a burst of short pulses of any duration can be transmitted, the width of a single pulse is limited.

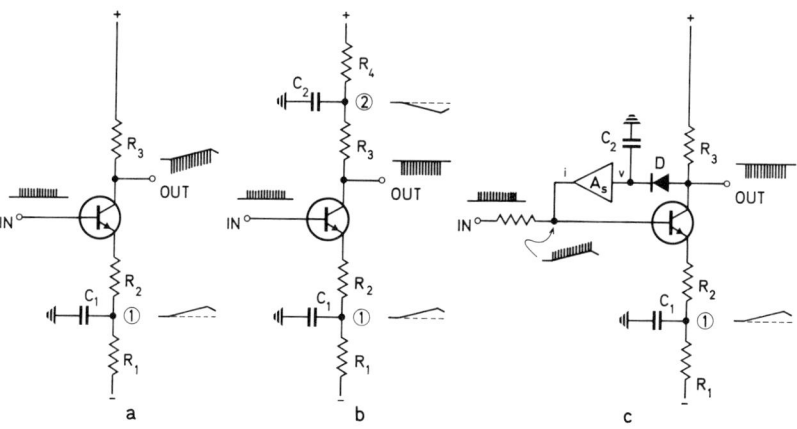

Fig. 3.84a–c. Amplification of a pulse burst in fast amplifiers [3.134]

ALBERIGI-QUARANTA and MARTINI [3.126, 3.135, 3.136] recommend the use of Zener diodes instead of blocking condensers for adjusting the operation points of the transistors. Other authors, too, use Zener diodes in fast amplifier circuits for burst applications. The Zener diodes eliminate virtually any pulse rate dependence of the operating points; however, supplying the transistors from voltage sources with very low internal resistance adversely affects the temperature stability of the circuits, since there is no *dc* feedback as in Fig. 3.84. In practice a decision must be made in each particular case as to which technique offers more advantages.

EPSTEIN [3.137], LAVAILLE [3.138], SCHAPPER [3.139] and JACKSON [3.140] have described various other fast transistorized amplifiers. AGENO and FELICI [3.141] separate the current and the voltage amplification, and use fast pulse transformers for voltage amplification. The current amplification between two transformers is performed by emitter-followers. A multistage amplifier using this curious principle exhibited a gain of $A=8$ and a rise time of $t_r=4.2$ nsec. DUBROVSKIJ [3.142] reviewed the fast pulse and wide band amplifiers.

4. Analog-to-Digital Converters

As we have seen in Chapter 1, the registration of pulses corresponding to selected events is performed in digital devices. Therefore it is necessary to digitize the analog information carried by the pulse, i.e. to decide whether or not and where a given pulse should be registered. A circuit performing the described operation is denoted as an analog-to-digital converter, shortened to ADC.

In an analog-to-digital converter a part of the information carried by the analog pulse is lost. This loss of information, of course, concerns the uninteresting part of the information only. It is necessary for the extraction of the relevant information. Two cases are possible:

1. Only those pulses which obey a certain criterion (e.g. the amplitudes of which are greater than a preselected value) yield normalized digital pulses, and the rest of analog pulses are suppressed. Circuits of this type are known as discriminators.

2. The pulse parameter to be measured (above all the pulse amplitude) is converted into a digital coded signal. A corresponding digital signal is associated with every analog pulse: essentially none of the analog pulses are suppressed. Circuits of this type are analog-to-digital converters in the stricter sense of the word.

The amount of information lost is different in each case. While a simple scaler or single memory cell suffices for the registration of the output pulses from a discriminator, in the second case as many memory cells are needed as the number of intervals into which the possible range value of the analog magnitude is divided. Strictly speaking the digital signal represents the address of the memory cell in which the corresponding event is to be registered.

However, there is no qualitative but only a quantitative difference between the two types of circuits. A given range of e.g. the pulse amplitude can be divided into n intervals, and one discriminator can be selected for each of these intervals. Thus none of the analog pulses is lost, since it always appears as an output pulse of one of the discriminators. On the other hand, a digital-to-analog converter accepts pulses only of higher amplitude than some minimum value[10], thus suppressing a part of the pulse height spectrum. Despite these relationships, a n-fold multiple discriminator becomes unwieldy even for $n \gtrsim 10$ and the possibility of a direct digitization of pulse height must be sought.

4.1. Pulse Height Discriminators

With the aid of the *negative* feedback an especially efficient conservation of the analog information could be achieved in the process of the pulse amplification, as has been demonstrated in Chapter 3.1. Hence we can expect intuitively that the desired reduction of the analog information can be performed by means of a *positive* feedback [4.001]. In fact, amplifiers with extreme positive feedback produce multivibrator circuits of threshold character which can be used as discriminators. The general properties of multivibrators as far as they concern their use as discriminators are discussed in the following chapter.

4.1.1. The Principle of a Multivibrator

Fig 4.01a shows an idealized characteristic of an amplifier. The circuit under consideration exhibits a linear range of the output voltage V_{out} where the gain is constant. The linear range limits V_1 and V_2 correspond approximately to the saturation and cutoff of the last stage transistor. For V_{in} outside the linear range, the characteristic $V_{out} = f(V_{in})$ takes a horizontal course. It must be pointed out that V_1 and V_2 are inherent constants of the amplifier $+A$, independent of any applied linear feedback.

With negative feedback (Fig. 4.01b) the slope of the linear part of the characteristic is lower, so that the linear range of the input voltage V_{in} is greater. The gain dV_{out}/dV_{in} from (3.05) amounts to

$$\frac{dV_{out}}{dV_{in}} = \frac{A}{1+bA}, \tag{4.01}$$

where the meaning of A and b is clear from the figure.

The feedback signal $b \cdot V_{out}$ can also be added to the original input signal thus magnifying the amplification. The gain of an amplifier with such a positive feedback (Fig. 4.01c and d) can be estimated immediately by taking into account that the only difference between a negative feedback and a positive one is an additional reversal of polarity. Hence, replacing b in (4.01) by $-b$ we get

$$\frac{dV_{out}}{dV_{in}} = \frac{A}{1-bA} \tag{4.02}$$

---

[10] This must be provided for because of noise, hum, etc.

The Principle of a Multivibrator

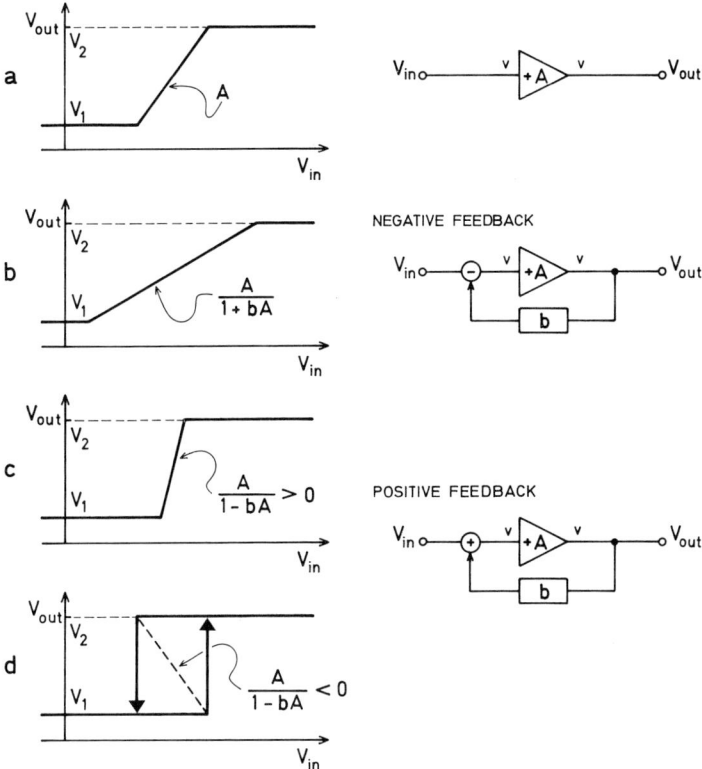

Fig. 4.01 a–d. Characteristics of an idealized amplifier with negative and positive feedback, respectively

for the gain of a positive feedback amplifier. So long as the loop gain bA remains less than 1, the gain (4.02) remains positive but higher than A. The linear range of the input voltage diminishes and vanishes when $bA=1$. With $bA>1$ the gain dV_{out}/dV_{in} becomes negative and the characteristic exhibits a shape according to Fig. 4.01 d. To simplify the discussion this characteristic is redrawn in Fig. 4.02.

For an input voltage within the interval V_{i1} and V_{i2}, there are three possible values of the corresponding V_{out}, one of which lies on the (dotted) negative part of the characteristic. Since this state is not stable the circuit assumes a bistable character: when V_{in} passes *up* the threshold V_{i1} the V_{out} changes suddenly from V_1 to V_2; when V_{in} passes *down* the threshold V_{i2}, the output voltage V_{out} changes back from V_2 to V_1. The difference between the thresholds is known as the hysteresis V_H:

$$V_H = V_{i1} - V_{i2}. \tag{4.03}$$

Analog-to-Digital Converters

From (4.02) the hysteresis, V_H can be calculated to give

$$V_H = \frac{V_2 - V_1}{A}(bA - 1). \tag{4.04}$$

The ratio of the maximum possible amplitude $V_2 - V_1$ and the gain A of a simple amplifier stage is in general constant and produces a proportionality between V_H and the loop gain $(bA - 1)$.

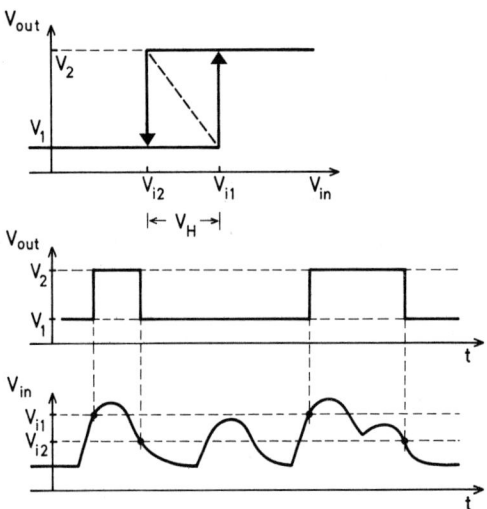

Fig. 4.02. Characteristic of a positive feedback amplifier

The lower part of Fig. 4.02 indicates a possible input signal voltage $V_{in}(t)$ and the corresponding output pulses $V_{out}(t)$ of the multivibrator.

Fig. 4.03 shows a transistorized dc amplifier with positive feedback in a configuration which was proposed for the first time[11] by SCHMITT [4.002] and which is consequently known as a Schmitt trigger. The principle of operation can easily be understood supposing $R = 0$ (i.e. the emitter of both transistors Q_1 and Q_2 interconnected) and if R_E is very high. In this case R_E can be regarded as a constant current source $I_T = $ const. If the voltage V_{B1} is low, transistor Q_1 is cut off. Hence I_T flows through Q_2, and when the base current is neglected the base voltage of Q_2 becomes $V'_{B2} = V_s R_b/(R_{C1} + R_a + R_b)$. The transistors Q_1 and

---

[11] Of course with vacuum tubes only.

The Principle of a Multivibrator

Q_2 form an emitter-coupled difference amplifier. Therefore if V_{B1} is raised, transistor Q_1 starts conducting as soon as V_{B1} reaches the proximity of V'_{B2} and, due to the feedback R_a, R_b, the circuit goes over into the second stable state: Q_1 conducting, Q_2 cut off. In this state the collector voltage of Q_1 is diminished by $R_{C1} \cdot I_T$ and the base voltage of Q_2 becomes $V''_{B2} = (V_s - R_{C1} I_T) R_b / (R_{C1} + R_a + R_b)$. Obviously $V''_{B2} < V'_{B1}$. If now V_{B1} is again decreased, the circuit reverts to the original state as soon as $V_{B1} \approx V''_{B2}$. Hence the voltage values V'_{B2} and V''_{B2} correspond approximately to the critical values V_{i1} and V_{i2} of the input voltage $V_{in} = V_{B1}$ as defined in Fig. 4.02. The difference $V'_{B2} - V''_{B2}$ is equal to the hysteresis V_H.

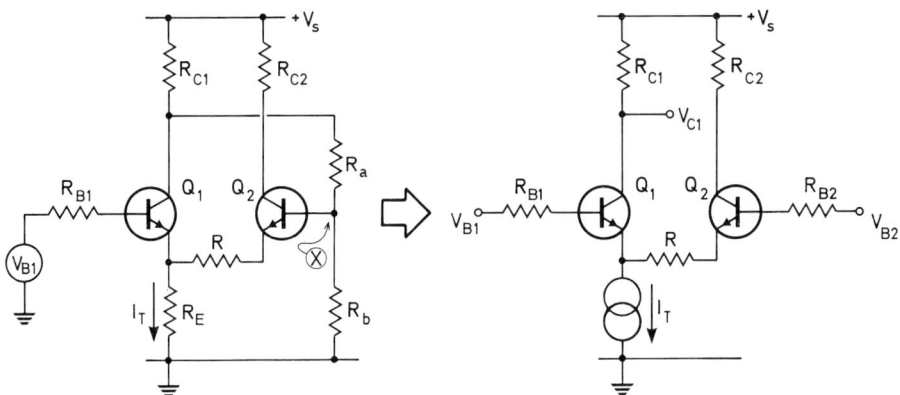

Fig. 4.03. The Schmitt trigger (left side) and an equivalent circuit for calculation of the loop gain (right side)

The loop gain bA can easily be calculated if (according to LITTAUER [4.001]) the loop is interrupted at the point (X) (cf. the right-hand diagram of Fig. 4.03). The voltage gain dV_{C1}/dV_{B2} is

$$\frac{dV_{C1}}{dV_{B2}} = \alpha_1 \frac{R_{C1}}{R + R_{B1}/(1+\beta_1) + R_{B2}/(1+\beta_2) + r_{E1} + r_{E2}}, \quad (4.05)$$

where α and β are the current amplification factors of the common base and common emitter configurations, respectively. The index 1 and 2 denotes the parameters of the transistors Q_1 and Q_2, respectively. The dynamic emitter resistance r_E is defined as

$$r_E = \frac{dV_{BE}}{dI_E} \approx \frac{kT}{e} \cdot \frac{1}{I_E}. \quad (4.06)$$

Analog-to-Digital Converters

Since $V_{B2} = V_{C1} \cdot R_b/(R_{C1}+R_a+R_b)$ we get

$$bA = \frac{R_b}{R_{C1}+R_a+R_b} \frac{dV_{C1}}{dV_{B2}}. \tag{4.07}$$

Obviously $R_{B2} = R_b/(R_a+R_{C1})$. The role of the resistor R can be understood from (4.05) and (4.07). Higher R decreases the loop gain bA. Hence in addition to the voltage divider R_a, R_b and the load resistor R_{C1}, the resistor R offers another possibility for influencing the loop gain and thus the hysteresis V_H (4.04). If R dominates the other terms in the denominator of (4.05) the gain dV_{C1}/dV_{B2} is given simply by R_{C1}/R.

The "non-involved" collector of Q_2 offers the possibility of picking up the digital output signal at R_{C2}. The feedback loop $V_{B2} \circlearrowleft V_{C1}$ is not affected by R_{C2}.

Another presentation of the situation is given in Fig. 4.04. The amplifying characteristic of the amplifier between V_{B2} and V_{C1} (i.e. with the loop interrupted at (X)) obeys the curve A, the position of which is still a function of V_{B1}. The attenuator b (R_a, R_b etc.) produces a second relationship between V_{C1} and V_{B2} which is strongly linear. The middle point of the three common points of these two characteristics b and A is not stable. If V_{B1} is decreased the characteristic A is shifted to the left until the position (K) (marked by a dotted line) is reached. Here the point of intersection V''_{B2} disappears and the only remaining possibility is $V_{B2} = V'_{B2}$. On the contrary, when shifted over the position (L) by increasing V_{B1}, the point of intersection V'_{B2} disappears suddenly and $V_{B2} = V''_{B2}$ remains. This form of presentation is especially useful if an attenuator b having a general non-linear characteristic is used (cf. Fig. 4.08).

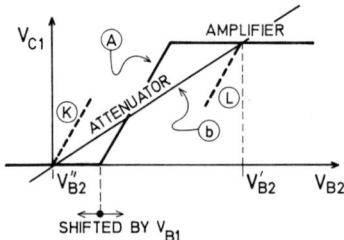

Fig. 4.04. Operating characteristic of a Schmitt trigger

In the preceding text we have assumed the amplifier A and the attenuator b to have an extremely wide band frequency response. In this case the flip-over from one stable state to the other would occur without any time delay at the instant of reaching the input voltage threshold V_{i1} or V_{i2}. However, the actual situation of an amplifying system with limited frequency response must be taken into account. Such a system can be approximated by means of an additional integrating circuit $1/(1+p\tau)$ within the loop (Fig. 4.05). The output voltage V_A of the ideal amplifier A remains well described by the static amplifier characteristic of Fig. 4.04. The attenuation relation $V_{B2} = b \cdot V_{C1}$, however, is valid only for V_{C1} given by

$$\hat{V}_{C1} = \hat{V}_A \frac{1}{p\tau+1}. \tag{4.08}$$

From (4.08) we get

$$p\tau \cdot \hat{V}_{C1} = \hat{V}_A - \hat{V}_{C1}, \tag{4.09}$$

which, transformed into the time space, gives

$$\tau \cdot dV_{C1} = (V_A - V_{C1})dt. \tag{4.10}$$

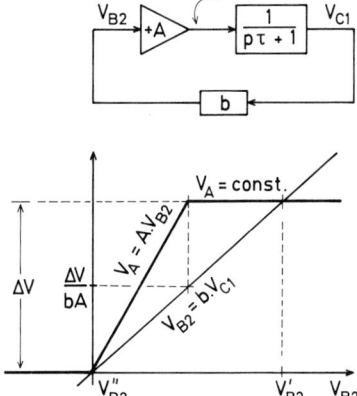

Fig. 4.05. Positive feedback amplifier with an integrator within the loop

The transition time t_t (i.e. the duration of the transition process from one stable state to the other) can be calculated by integration of dt from (4.10)

$$t_t = \int_{V_{C1}=0}^{V_{C1}=\Delta V} dt, \tag{4.11}$$

Analog-to-Digital Converters

where ΔV is the amplitude of the output signal V_{C1}. Introducing (4.10) into (4.11) and using the relation between V_A and V_{C1} from Fig. 4.05, we have

$$t_t = \tau \left(\int_0^{\Delta V/bA} \frac{dV_{C1}}{(bA-1)V_{C1}} + \int_{\Delta V/bA}^{\Delta V} \frac{dV_{C1}}{\Delta V - V_{C1}} \right). \tag{4.12}$$

The integration limits $V_{C1}=0\ldots\Delta V$ result if the input voltage just reaches the threshold. However in this case (4.12) yields $t_t=\infty$. Hence the input voltage must exceed the threshold by a small amount of say $\varepsilon \cdot V_H$ in order to obtain a finite transition time ($\varepsilon \ll 1$). The integration of (4.12) within the limits $V_{C1} = \varepsilon \cdot \Delta V \ldots (1-\varepsilon) \cdot \Delta V$ gives

$$t_t \approx \tau \frac{bA}{bA-1} \log \frac{1}{\varepsilon}. \tag{4.13}$$

The transition time is thus given mainly by the characteristic time constant of the loop. The slight dependence on the amplitude and shape of the input voltage becomes more pronounced only for signals just reaching the threshold. A small loop gain bA adversely affects the transition time. The relation (4.13) does not hold for an input amplitude comparable with or higher than the hysteresis (i.e. $\varepsilon \gtrsim 1$). In this case t_t is given merely by τ.

A detailed analysis of the transistorized Schmitt trigger circuit has been given by NIZAN and ELAD [4.120], who also proposed a method for the reduction of the hysteresis by means of an "intertriggering" action between two trigger circuits [4.121].

4.1.2. Integral Discriminators

The term "integral discriminators" denotes circuits producing a normalized digital output pulse for every input pulse, the amplitude of which exceeds a preselected threshold. Smaller pulses are suppressed.

Obviously a Schmitt trigger can be used as an integral discriminator. In any case, the following questions must be answered: How can the desired threshold be adjusted? What is the smallest acceptable input pulse height, i.e. the sensitivity of the circuit How is the preselected threshold influenced by the temperature variations etc.? Is there a threshold shift when changing the input pulse rate? What is the discriminator response to overloading input pulses? etc.

Fig. 4.06 shows the diagram of a simple integral discriminator circuit. The standing voltage V_{B1} can be adjusted by means of the potentiometer POT to V_{io}. As can easily be seen, all positive pulses with amplitudes higher than $(V_{i1} - V_{i0})$ trigger the circuit. $V_{DISCR} = V_{i1} - V_{i0}$ is thus the adjustable discriminator threshold.

The condensor C_a compensates for the parasitic capacity C_b of the voltage divider R_a, R_b only. It can be omitted in the following discussion.

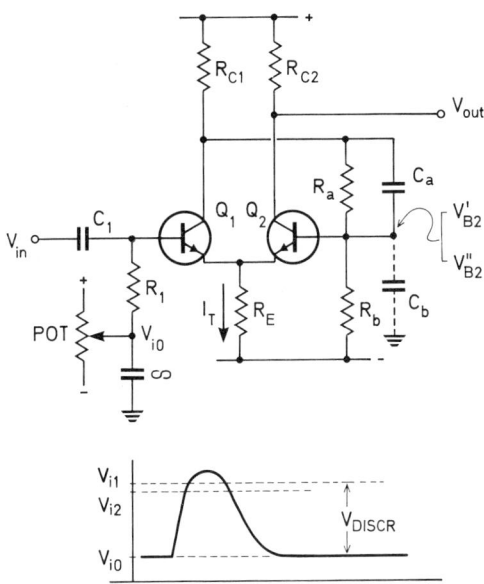

Fig. 4.06. Schmitt trigger type integral discriminator

Except for the difference in base-emitter voltages V_{BE} of Q_1 and Q_2 respectively, the critical voltage value V_{i1} is equal to V'_{B2}. Assuming V'_{B2} to be well-defined and constant, the temperature dependence of V_{i1} is given only by the temperature-dependent change in V_{BE} of the two transistors. In a difference amplifier with approximately equal emitter currents from both transistors, the relation

$$V_{BE} = \frac{kT}{e} \log\left(\frac{I_E + I_{E0}}{I_{E0}}\right) \tag{4.14}$$

holds for both V_{BE}[12]. Hence the temperature-dependent changes in V_{BE} just compensate one another. Although the circuit in Fig. 4.06 is in principle a difference amplifier, $I_{E1} \ll I_{E2} \approx I_T$ is valid at the begin-

---

[12] I_{E0} is the saturation cutoff current of the emitter-base junction.

ning of the transition. Therefore almost the whole temperature-dependent change in V_{BE} of Q_2 influences V_{i1} and V_{DISCR} directly. The resulting drift of V_{DISCR} amounts to $0.1 \ldots 1 \, \text{mV}/^\circ\text{C}$, and pulse amplitudes of $100 \, \text{mV}$ can be digitized with an accuracy of about 1% only when the temperature of the circuit is stabilized to within $1 \ldots 10 \, ^\circ\text{C}$.

V_{DISCR} results from the difference of two independent voltages V_{i1} and V_{i0}, which can be relatively high and which (at least in Fig. 4.06) are determined by completely different circuit components and voltage sources: $V_{i1} \approx V'_{B2}$ by R_{C1}, R_a and R_b; V_{i0} by POT. Small relative changes in values of these components or in the supply voltages may heavily influence the discriminator threshold V_{DISCR}. Therefore both voltages V'_{B2} and V_{i0} are best derived from the same voltage divider and the same voltage source. This can be realized easily in the case of RC coupling between Q_1 and Q_2 (Fig. 4.07). If $R_b \cdot C_a$ is chosen high enough in relation to the maximum pulse length, the differentiating circuit C_a, R_b forms an attenuator with $b=1$. The clamp diode D restores the baseline in a known manner, so that the standing potential of the Q_2 base corresponds to the potential of the point (X) independent of the pulse rate. Changes in the component values and in the supply voltage now influence V'_{B2} as V_{i0} as well. For the potentiometer POT a precision type is normally used with a resistor helix (e.g. "HELIPOT"). The selected threshold V_{DISCR} can be read directly from the potentiometer dial with an accuracy of better than 0.1%.

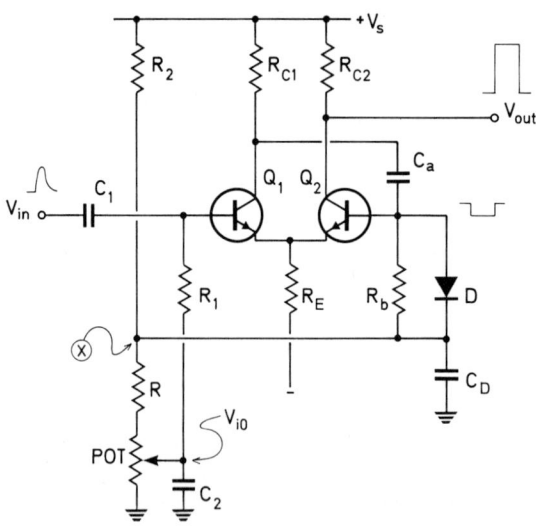

Fig. 4.07. RC coupled trigger circuit

From Fig. 4.06 and 4.02 the smallest detectable amplitude of unipolar input pulses can be seen to be higher than the hysteresis V_H, since for $V_{i0} > V_{i2}$ the circuit does not flip back when once triggered. Hence, in sensitive discriminators a small V_H is desired. On the other hand, a high hysteresis is a welcome safeguard against the circuit instability resulting in multiple satellite pulses or in free oscillations. It is not recommended that V_H be lower than a few 100 mV. A hysteresis of 100 mV already needs a very careful circuit layout. According to (4.04) a small V_H is obtained with a small loop gain $bA \gtrsim 1$. However, due to (4.13), a small bA yields a very slow circuit with a discriminator threshold depending largely upon the shape of the input pulses. A small V_H can be obtained without reducing bA if the voltage change at the base of Q_2 is limited by means of two diodes (Fig. 4.08). The circuit can be

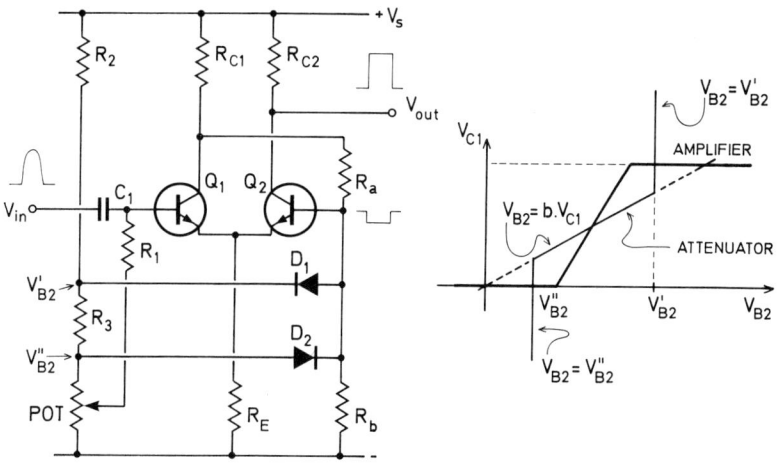

Fig. 4.08. Limiting the voltage excursion at the base of Q_2 by means of two diodes

investigated using the presentation of Fig. 4.04, the two diodes producing a non-linear attenuator characteristic. During the transitions from one stable state to the other bA is high, despite the small hysteresis, and the transition time is reduced. The parasitic capacities of the diodes D_1 and D_2 as well as the temperature dependence of the potential difference over D_1 (which directly influences the threshold) present some difficulties. The signal limitation can also be performed at the collector of Q_1 — where the voltage to be limited is somewhat higher.

Analog-to-Digital Converters

All the circuits discussed exhibit excellent overload characteristics. Assuming a constant current I_T, the current through Q_1 remains constant after the transition in the state "Q_1 conducting—Q_2 cut off", and the input is loaded only by the base current of Q_1, if $V_{CB} \neq 0$ (i.e. if Q_1 does not become saturated). Hence only the difference between the collector voltage $V_{C1} = V_s - R_{C1}(I_T + I_{a,b})$ of the conducting transistor Q_1 and the threshold voltage V_{i1} must be higher than the maximum possible input amplitude. $I_{a,b}$ denotes the current drawn by the attenuator R_a, R_b. Alternatively, the input pulse height must be limited to a value smaller than $V_{C1} - V_{i1}$.

Since the discriminator threshold is controlled by means of the bias of Q_1 base, the input pulses must be coupled to the base via a condenser C_1. As outlined in more detail in Chapter 3, any RC coupling introduces a pulse-rate-dependent baseline shift, which corresponds to a change in the effective threshold V_{DISCR}. With strongly unipolar pulses the bias V_{i0} at the base of Q_1 can be stabilized using a simple diode clamp, shown in Fig. 4.09. Due to the curvature of the diode characteristic this

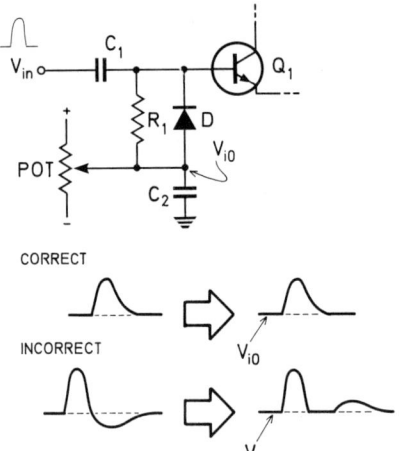

Fig. 4.09. Clamping diode for restoring baseline

circuit operates satisfactorily only with pulse amplitudes much higher than kT/e, i.e. $\gtrsim 200$ mV. The voltage divider POT must have sufficiently small impedance, and C_2 must be sufficiently high, in order to form a virtually constant voltage source V_{i0}. Diode clamp circuits are discussed in full detail, e.g. by MILLMAN and TAUB [4.003], p. 119. The simple circuit Fig. 4.09 fails to operate with bipolar pulses: although the negative undershoot is suppressed, the charged capacitor C_1 must subsequently be discharged through R_1 thus resulting in a positive satellite pulse.

ROBINSON [4.004] (cf. also GOULDING [4.005]) described a clamping circuit for bipolar pulses of not too small amplitudes (Fig. 4.10). The base potential of Q_1 is "anchored" with the aid of the diodes D_1 and D_2 at V_{i0}. Since the resistors R_1, R_2 are high, the currents $2 \cdot I_D$ and I_D can be assumed as constant. In the quiescent state the currents through both diodes are equal ($=I_D$). For small deviations of the base potential from V_{i0} the condenser C_1 operates in the very low resistance of $r_{D1}+r_{D2}$ $\approx 2\,\dfrac{kT}{eI_D}$ lying between the base of Q_1 and the voltage source V_{i0}. Hence small positive or negative undershoots of amplitude comparable to kT/e are paralysed with the small time constant $C_1(r_{D1}+r_{D2})$. High pulses of each polarity cut off one of the diodes D_1 or D_2 and are thus passed through with only a small distortion. The temperature dependences of the potential differences across the diodes, which normally would affect V_{DISCR}, fortunately compensate each other.

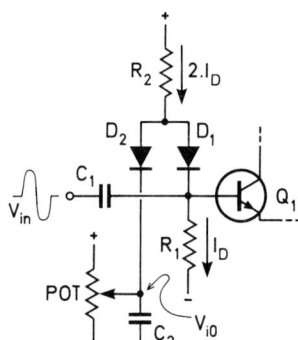

Fig. 4.10. Baseline restorer for bipolar pulses according to ROBINSON [4.004]

CHASE and POULO [4.122] described a variation of the Robinson dc restorer, in which the r_D is decreased by an operational amplifier. PATZELT [4.123] reports another method of baseline stabilization. GERE and MILLER [4.125] and WILLIAMS [4.126] introduced the so-called "active dc restoration" by a circuit producing a signal proportional to the peak-to-peak excursion of the input voltage.

GOULDING and McNAUGHT [4.006] compensate the pulse-rate-dependent baseline shift by means of a complicated control system.

When discussing Fig. 4.06, we stated that a Schmitt trigger with hysteresis V_H operates as an integral discriminator only when the input pulse amplitudes exceed the hysteresis: $V_{in} > V_H$. For a given hysteresis

Analog-to-Digital Converters

$V_H = V_{i1} - V_{i2}$ the circuit sensitivity can be improved only by adjusting the bias V_{i0} between the values V_{i1} and V_{i2}. Since the circuit in this case does not flip back automatically, a special reset device must be provided. For analysis of such circuit let us consider the basic Schmitt trigger configuration of Fig. 4.06, assuming for simplicity that the tresholds are defined exclusively by the base voltage V_{B2} of Q_2 (i.e. $V_{i1} = V'_{B2}$ and $V_{i2} = V''_{B2}$). In Fig. 4.11a the situation is shown where the bias voltage lies outside the hysteresis interval ($V_{i0} < V''_{B2}$). The voltages V_{out}, V_{B2} and V_{in} are plotted as a function of time t. If is hardly necessary to discuss these curves once more. The output pulse length is defined by the shape of the input pulse and by the actual threshold values V'_{B2} and V''_{B2}.

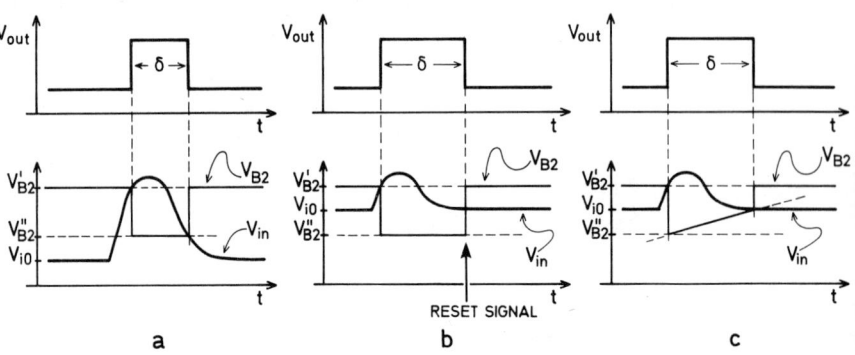

Fig. 4.11a–c. Trigger circuits biased within the hysteresis voltage range

In the case of $V''_{B2} < V_{i0} < V'_{B2}$ the transition $V''_{B2} \to V'_{B2}$ back into the original state must be initiated by an auxiliary reset signal (Fig. 4.11b) which is fed preferably into the base of Q_2 and the amplitude of which exceeds $(V_{i0} - V''_{B2})$. The pulse length δ is than given solely by the delay of the reset signal. Of course, δ must be longer than the maximum input pulse duration — or else multiple output pulses are produced. The discriminator circuit operating as shown in Fig. 4.11b represents a bistable multivibrator (flip-flop) which will be discussed in more detail from another point of view in Chapter 6.1.3. The reset signal is generated, for example, in an auxiliary monostable multivibrator triggered by the leading edge of V_{out}.

The introduction of a component with time-dependent characteristic (condenser, coil or delay cable) into the feedback loop offers a simpler way of resetting the discriminator (Fig. 4.11c). The flip-back occurs in the instant when $V_{B2}(t) = V_{i0}$. Of course, such a discriminator is none other than a monostable multivibrator. One way of realizing this is shown in Fig. 4.12.

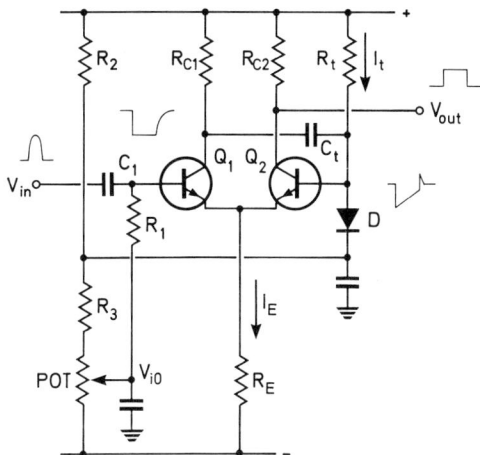

Fig. 4.12. Monostable multivibrator used as integral discriminator

The purpose of this somewhat unfamiliar presentation of the three types of discriminator circuits is to indicate the close interrelation between them. Hence, with the exception of the definition of the output pulse length, the discrimination properties of a Schmitt trigger, a bistable multivibrator or a monostable one are the same. All considerations concerning the temperature dependence of the threshold V_{DISCR} etc. already made for the Schmitt trigger therefore apply equally to the other two discriminator circuits.

Fig. 4.12 shows a monostable multivibrator with the time-defining components R_t and C_t. Assuming the current I_t through R_t to be constant, the output pulse length becomes $\delta \approx (V_{i0} - V''_{B2}) C_t / I_t$. Hence δ depends on the discriminator threshold V_{DISCR} (manifest in V_{i0}). The diode D stabilizes the standing potential V''_{B2} to the value defined by the voltage divider R_2, R_3, POT. The condenser C_t is discharged by R_t during the pulse duration. Due to the necessary recharging of C_t by R_{C1} and D after the flip-back, the collector voltage of Q_1 does not rise suddenly but exhibits a slow exponential rise with the time constant $R_{C1} \cdot C_t$. The characteristic pulse shapes are indicated in the circuit diagram. Since during this non-equilibrium of circuit voltages the threshold V_{DISCR} is not well defined, a short recovery period $\sim R_{C1} \cdot C_t$ is desired. Often an emitter-follower Q_3, acting as impedance converter, is introduced between R_{C1} and C_t. The corresponding circuit diagram in Fig. 4.13 needs no further discussion.

The temperature dependence of the potential difference across the diode D in Fig. 4.12 and 4.13 partially compensates the temperature-dependent changes in V_{BE} of Q_2, thus yielding a better constancy of V_{DISCR}. Monostable discriminators are especially suitable if a high sensitivity and a large hysteresis are desired at the same time.

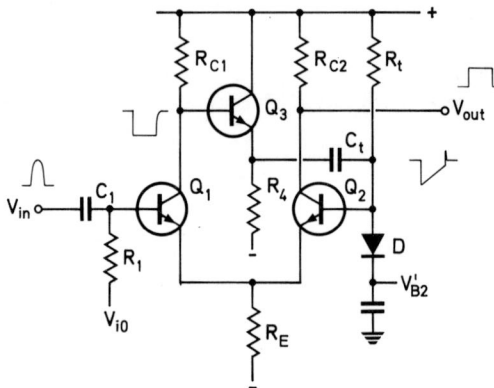

Fig. 4.13. Monostable discriminator with improved recovery time

As has already been pointed out, the output pulse length depends upon V_{i0} and thus upon the selected discriminator threshold V_{DISCR}. This difficulty can be avoided by using a variable-threshold window amplifier in front of a discriminator with constant bias V_{i0}. A window

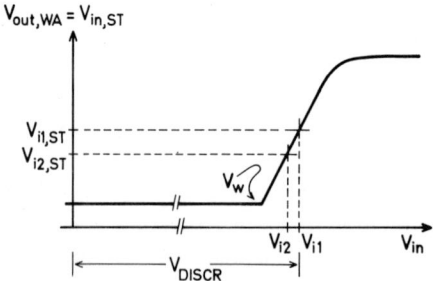

Fig. 4.14. Characteristic of a Schmitt trigger (index ST) preceded by a window amplifier (index WA)

Integral Discriminators

amplifier for the definition of V_{DISCR} offers advantages not only when used together with monostable discriminator, but also with dc-coupled Schmitt triggers: The stability of the Schmitt trigger threshold is improved by the gain of the amplifier, and the hysteresis is reduced by the same amount (Fig. 4.14). Exactly speaking, the demands on the discriminator threshold stability are transferred to the stability of the window amplifier threshold V_w.

The window amplifier used may consist of simple biased diode circuits such as those shown in Fig. 3.61 or of emitter-coupled biased difference amplifiers (long-tailed pairs [4.007] to [4.009]). KANDIAH [4.010] and CHASE [4.007] describe a current-sensitive discriminator based on the combination of a window amplifier analysed by LARSEN [4.011] (Fig. 3.63) and a simple Schmitt trigger. The principle of this circuit is shown in Fig. 4.15. In the absence of any input signal I_{in} the threshold current I_{i0} flows through the diode D. From the voltage divider R_2, R_3 another small current flows into D, biasing D slightly forward even when $I_{i0} = 0$. Hence because $V_{B1} > V_{B2}$, Q_1 conducts and Q_2 is cut off. However, this additional current from R_2, R_3 should be negligible compared with normal values of I_{i0}. An input current $I_{in} > I_{i0}$ then cuts off the diode D and the difference $I_{in} - I_{i0}$ is drawn from the base of Q_1, thus triggering the circuit.

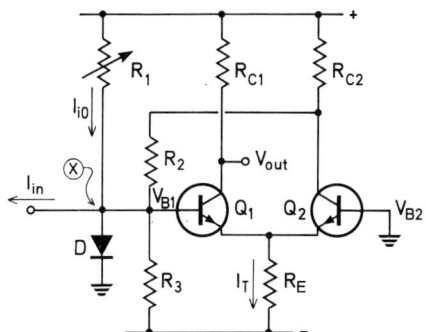

Fig. 4.15. Current sensitive transistorized discriminator

KANDIAH [4.012] proposed a very sensitive version of the trigger circuit making use of the strong relationship between the dynamic resistance and the current in a diode. Many of the modern precision discriminators for pulse lengths in the range 0.1 to 1 μsec use the operating principle of the basic circuit shown in Fig. 4.16 (cf. [4.010, 4.013,

4.014, 4.126]). In the quiescent state the standing current I_T is equally divided between the two transistors Q_1 and Q_2: $I_{C1} = I_{C2} = I_T/2$. The load of the collector circuit of Q_1 consists of R_{C1}, the input resistance of Q_2 (approximately $\beta_2 \cdot R$) and the dynamic diode resistance r_D. According to (4.05), the loop gain bA of the circuit can easily be calculated. Assuming R to be large in relation to the other terms in the denominator of (4.05) (i.e. $R \gtrsim 200\,\Omega$) and assuming $r_D \ll R_{C1}$ and $r_D \ll \beta_2 \cdot R$, the loop gain is

$$bA = \frac{r_D}{R}. \qquad (4.15)$$

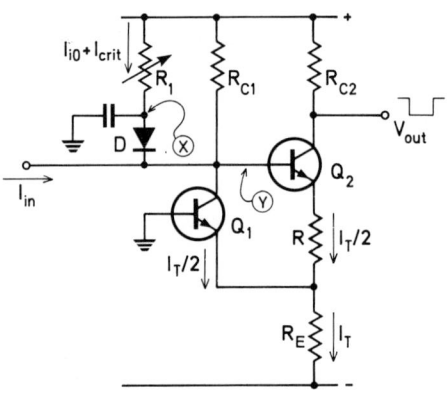

Fig. 4.16. Kandiah discriminator (KANDIAH [4.010])

The dynamic diode resistance

$$r_D = \frac{kT}{e} \frac{1}{I_D + I_{D0}} \qquad (4.16)$$

depends on the current I_D through the diode D ($I_{D0} \ll I_D$ denotes the saturation cutoff current of D). So long as I_D is high enough to yield $r_D < R$, the loop gain is $bA < 1$ and the circuit remains at rest with both transistors conducting. For a critical diode current I_{crit}

$$I_{crit} = \frac{kT}{e} \cdot \frac{1}{R} \qquad (4.17)$$

the diode resistance becomes $r_D = R$ and the loop gain reaches unity thus making the circuit unstable. The trigger flips over into the state

"Q_1 cut off / Q_2 conducting". With $R \approx 200\,\Omega$ the critical current is $I_{\text{crit}} \approx 125\,\mu\text{A}$.

With the aid of a potentiometer R_1 the standing diode current is adjusted to $I_D = I_{i0} + I_{\text{crit}}$. Input current pulses I_{in} of the indicated polarity cause the diode current I_D to decrease by I_{in}. Whilst $I_{\text{in}} < I_{i0}$ the diode current remains higher than I_{crit}. For $I_{\text{in}} \geq I_{i0}$, $I_D \leq I_{\text{crit}}$ is valid and the trigger flips over. The current I_{i0} thus defines the discriminating threshold for the input pulses.

From (4.17) and with $I_D \gg I_{D0}$ we get

$$\frac{dI_{\text{crit}}}{I_{\text{crit}}} = \frac{dT}{T}. \tag{4.18}$$

Hence the threshold changes by about $0.4\,\mu\text{A}/^\circ\text{C}$ when using $I_{\text{crit}} = 125\,\mu\text{A}$. Other temperature-dependent parameters of the circuit influence the threshold stability to a much lower extent.

Using the relation $I_{D0} \sim e^{-E_g/kT}$, where E_g denotes the energy gap of the semiconductor, equation (4.16) becomes more exactly

$$\frac{dI_{\text{crit}}}{I_{\text{crit}}} = \frac{dT}{T}\left(1 - \frac{E_g \cdot I_{D0}}{kT(I_{D0} + I_{\text{crit}})}\right). \tag{4.19}$$

Hence the temperature dependence of the threshold can be substantially reduced by adjusting $I_{\text{crit}} \approx I_{D0} \cdot E_g/kT$. KANDIAH [4.010] reports for a germanium diode with $E_g/kT \approx 30$ and $I_{D0} \approx 5\,\mu\text{A}$ at room temperature a threshold shift of only $0.05\,\mu\text{A}/^\circ\text{C}$ in the temperature interval $25\ldots40\,^\circ\text{C}$.

The sensitivity of the KANDIAH circuit is limited almost solely by the described temperature effects. In the form shown in Fig. 4.16 the circuit is bistable and must be reset by an external pulse fed into the base of Q_1, for example. A monostable behaviour is achieved by introducing a differentiator between Q_1 and Q_2 (point (Y)). The buffer condenser stabilizes the potential V_X of the diode anode. V_X is slightly pulse-rate-dependent, though it can easily be stabilized by a difference amplifier. For examples of practical circuits see KANDIAH [4.010].

The most important condition for correct operation of the KANDIAH discriminator is the validity of the relationship given in (4.16). According to GIANNELLI and STANCHI [4.015], and STRAUSS and BRENNER [4.016], this relationship is more properly applied to the dynamical emitter resistance r_E, as a function of the collector current I_C of a transistor, than to the semiconductor diodes (cf. also Chapter 3.23). Hence the use of the base emitter junction of a transistor instead of the diode D in the circuit Fig. 4.16 might offer advantages.

4.1.3. Differential Discriminators

Differential discriminators, often called single channel analyzers, are circuits producing a normalized digital output pulse for every input pulse, the amplitude V_{in} of which satisfies the condition $V_c < V_{in} < V_c + \Delta V_c$. Pulses with amplitudes lower than V_c or higher than $V_c + \Delta V_c$ are suppressed. Depending on the purpose of the discriminator, the limit V_c is called the lower threshold or channel position, the limit $V_c + \Delta V_c$ is called the upper threshold and the difference between both limits ΔV_c is called the channel width.

Essentially a single channel analyzer consists of two integral discriminators fixing the lower and the upper thresholds respectively, and of a simple digital logic circuit selecting events which trigger the "lower" Schmitt trigger ST_1 but not the "upper" one ST_2 (Fig. 4.17).

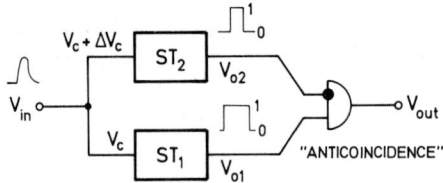

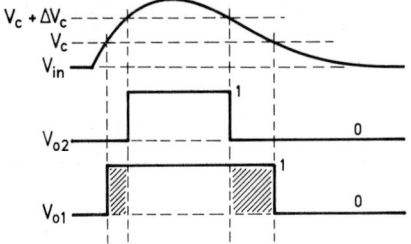

Fig. 4.17. Operating principle of a differential discriminator

Since the selection concerns operations on digital signals, the functional block notation for digital circuits is used, despite the fact that it will be introduced first in Chapter 6. This inconsistency in the presentation systematics can be accepted since the function of the circuits is self-explanatory.

The gate system denoted as "anticoincidence" should pass through the output pulse of the lower Schmitt trigger ST_1 unless it is blocked by the output pulse of the upper one ST_2. However, the circuit in the simple form shown in Fig. 4.17 would not operate satisfactorily. Since every physically significant pulse is wider at its base than at its top, the pulse V_{o1} always lasts longer than the inhibiting pulse V_{o2}. The gate is certainly blocked during the duration of V_{o2}, but the hatched parts of

Differential Discriminators

V_{o1} are passed through, resulting in output pulses. To avoid this effect the pulses V_{o1} and V_{o2} must be shaped: V_{o2} must always be longer than V_{o1} in order to have a total overlap in all possible cases.

Fig. 4.18 shows three frequently used methods of doing this conveniently. In the circuit (A) the output pulse is derived from the trailing edge of the lower Schmitt trigger signal ST_1 by a differentiator, the positive pulse corresponding to the leading edge being suppressed by means of a diode. The resulting signal is negative, i.e. a negative voltage corresponds to the logic 1 and a positive one to the logic 0. The inhibiting signal is stretched in a simple stretcher circuit, and its end is initiated by the output signal delayed by t_D, which activates the switch

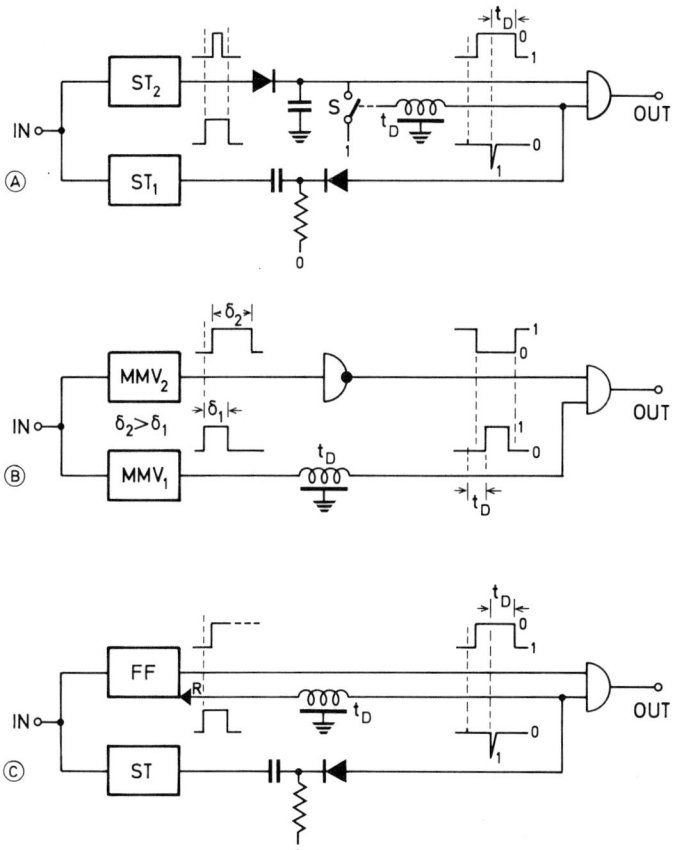

Fig. 4.18 A–C. Three frequently used methods to assure the pulse overlap in the anticoincidence gate of a differential discriminator

S and discharges the storage capacitor. In the negative logic used, the positive voltage corresponds to 0 (gate blocked) and the negative one to 1 (gate open). Obviously the inhibiting signal always overlaps the differentiated ST_1 signal if t_D is properly adjusted. Hence there is no output signal if both ST_1 and ST_2 are triggered. CHASE [4.007] describes two circuits of the type (A). Often no special delay is needed—the signal propagation times in the circuits employed are often sufficiently long. Because of its inherent slow response the circuit (A) is avoided in modern discriminators.

The analyzer (B) consists of two monostable multivibrators MMV_1 and MMV_2 with fixed output pulse lengths $\delta_2 > \delta_1$. The output pulse of MMV_1 is delayed by t_D, e.g. with the aid of a small delay line. If t_D is longer than the maximum possible delay of the inhibiting pulse with respect to the lower discriminator pulse, and if $t_D < \delta_2 - \delta_1$, the inhibiting signal always overlaps the MMV_1 output pulse. The output pulse of MMV_2 is converted in an inverter and controls the signal gate. The discriminators for the lower and the upper threshold must not themselves be connected as monostable multivibrators. The multivibrators MMV_1 and MMV_2 can be triggered by preceding Schmitt triggers etc. which perform the actual discrimination. The configuration (B) has been well known for a long time [4.017], and is still used often [4.018].

The upper discriminator in the analyzer (C) consists of a flip-flop FF. The lower threshold signal is gained as in (A) by means of a Schmitt trigger and the trailing edge differentiation. After a suitable t_D it resets the flip-flop in the starting position. Of course, the flip-flop FF itself can be set by a preconnected Schmitt trigger or monostable discriminator, or the negative differentiated pulse can be shaped by a monostable multivibrator prior to being fed into the output gate etc. The configuration (C) offers advantages if relatively high pulse rates must be dealt with [4.019, 4.127].

BRAFMAN [4.020] describes another analyzer system in which the delayed output pulse of the lower threshold discriminator must trigger an output pulse shaping circuit. It at the same time a signal from the upper threshold discriminator is present, the threshold of the output circuit is set up, thus inhibiting any triggering by the "lower" pulse.

Instead of using two monostable multivibrators (variant (B)), the pulse shaping can also be performed by differentiating the Schmitt trigger pulses by means of two different time constants [4.021].

WELTER [4.022] investigated various fast selection logic circuits for analyzers operating in the nanosecond range.

SATTLER [4.128] reports how to realize a simple single channel discriminator combining linear or logic integrated circuits.

Examples of practical selection gate circuits will be omitted here, since the digital circuit techniques will be discussed in more detail in Chapter 6.

Normally it is desirable that the channel width ΔV_c of a differential discriminator remains constant and independent of the actual setting of the channel position V_c. In such a case it is not admissible to set the bias for the upper and lower threshold discriminators by means of two independent potentiometers, since often $\Delta V_c \ll V_c$ and a small absolute change in the lower or upper threshold would cause a high relative change in the channel width. Hence, one potentiometer is used to adjust the channel position and another for the channel width. The related circuit diagram is shown in Fig. 4.19. The threshold $V_c = 0$ can be properly adjusted by means of the small auxiliary resistor R_3. Potentiometer POT 1 defines the channel position V_c. By means of the additional bias from POT 2 the upper discriminator threshold is shifted by ΔV_c.

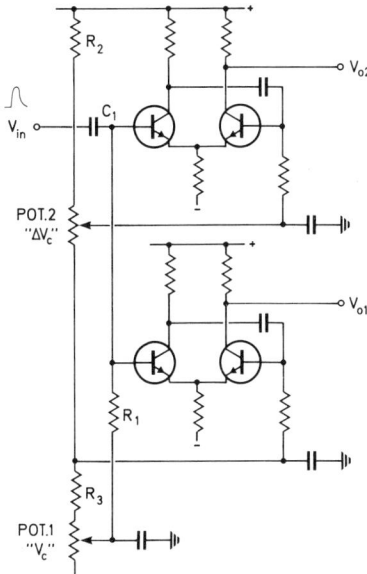

Fig. 4.19. Adjusting the channel position and the channel width by two independent potentiometers

This principle of threshold definition can be applied whenever discriminator circuits based on difference amplifiers, and thus having a threshold dependent on the difference of *two* adjustable potentials, are used. If it is impossible to adjust the two thresholds independently, a difference amplifier can be connected in front of an integral discriminator with fixed threshold, as shown in Fig. 4.20. The thresholds of the two discriminators are preferably fixed to signal amplitudes for which

$I_{C1} = I_{C2}$ and $I_{C3} = I_{C4}$ respectively. Then the temperature dependence of the preamplifiers is virtually zero.

Exactly speaking, the circuit Fig. 4.20 consists of two integral discriminators each with a window amplifier. The advantages of this configuration have been pointed out in Chapter 4.1.2. In this application

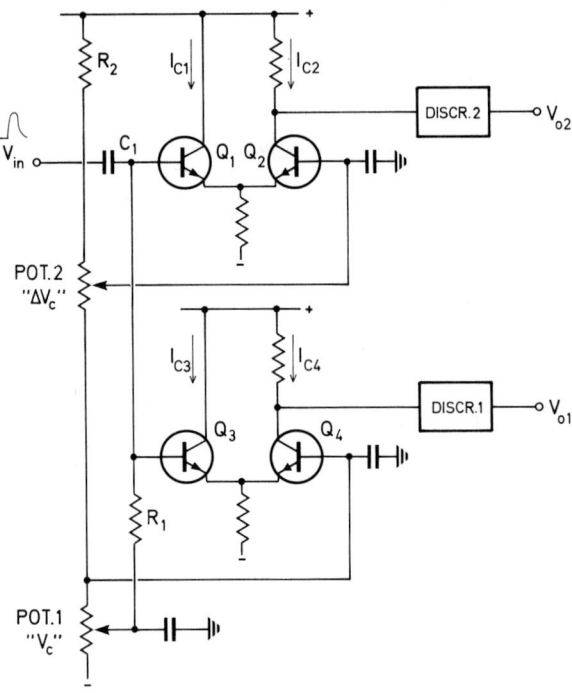

Fig. 4.20. Differential discriminator consisting of two window amplifiers and of two fixed-threshold integral discriminators

the window amplifier must not exhibit a pronounced "break-point" of the characteristic or a stable gain A, as aimed for in Chapter 3.3., if the discriminator thresholds lie approximately in the middle of the linear amplifying range (i.e. $I_{C1} \approx I_{C2}$ etc.). On the contrary the requirements on the window amplifier are higher if a single amplifier is used together with two discriminators with different thresholds in a configuration shown in Fig. 4.21. The thresholds of the lower and upper discriminators are denoted by V_{t1} and V_{t2}, respectively. $V_{out, WA}$ is the output voltage of the window amplifier with gain A and threshold V_w. For fixed V_{t1}, the channel position V_c is given mainly by the window amplifier threshold V_w, and the channel width by $\Delta V_c = (V_{t2} - V_{t1})/A$. The circuit offers the usual advantage of reducing the influence of the changes in V_{t1} and V_{t2} on V_c and ΔV_c. However, the channel position V_c and the channel width ΔV_c also depend on the gain A. The maximum channel width is limited to the linear range of the window amplifier. Hence for this application window amplifiers with reduced transition zone (i.e. pronounced break-point) between cutoff and the linear range and with stable gain A are desired.

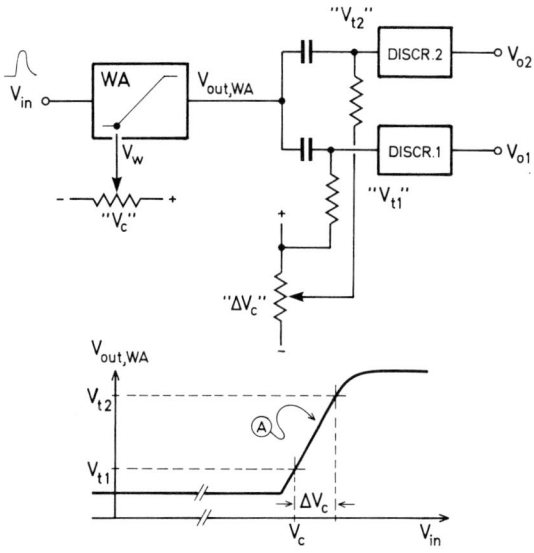

Fig. 4.21. Two discriminators with one common window amplifier

GATTI, PIVA, COLOMBO and COTTINI ([4.023] to [4.025]) worked out an alternative technique of pulse height analysis, the principle of which is shown in Fig. 4.22. A small voltage step ΔV is added to the input voltage pulse having the amplitude V (a rectangular pulse is assumed for the sake of simplicity). The step ΔV is delayed with regard to the leading edge of the pulse by Δt. The resulting staircase pulse is applied to an integral discriminator with the threshold V_t. There are three possible cases:

a) $V < V_t - \Delta V$: no triggering occurs,

b) $V_t - \Delta V < V < V_t$: the discriminator is triggered by the added step and thus delayed by Δt with regard to the pulse front,

c) $V_t < V$: the discriminator fires with no delay.

The output pulses of the discriminator are fed to a coincidence circuit which selects only those which are delayed by Δt. Accordingly only pulses with amplitudes satisfying

Analog-to-Digital Converters

the condition b) produce output pulses. Hence the circuit operates as a single-channel analyzer with the channel width ΔV and the channel position $V_t - \Delta V$. The constancy of the channel width is given merely by the stability of the step generator ΔV, the channel width a priori does not depend on the channel position V_t.

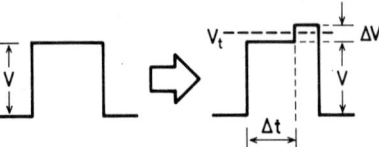

Fig. 4.22. Principle of the added-step technique

The staircase pulse can be formed by adding to the original input pulse a constant amplitude voltage pulse from a monostable multivibrator triggered by the delayed input signal. In this case ΔV is constant and the *absolute* channel width remains independent of V_t. Alternatively the staircase pulse is formed by the reflection of the rectangular input pulse at the end of a $\Delta t/2$ long delay line, terminated with $R_0 + \Delta R$ instead of its characteristic impedance R_0. As can easily be shown, in this case

$$\frac{\Delta V}{V} \approx \frac{\Delta R}{2 R_0} \qquad (4.20)$$

holds, assuming $\Delta R \ll R_0$. Hence the *relative* channel width remains constant and independent of V_t. The application of the added-step techniques to the multichannel systems will be discussed in the next chapter.

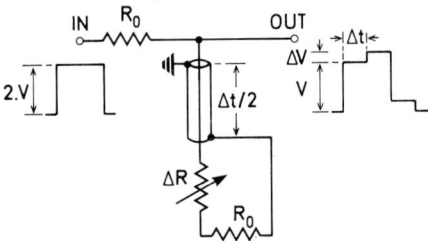

Fig. 4.23. Staircase pulse forming by means of an incorrectly terminated delay line

MORI [4.026] describes a single channel analyzer based on the conventional system of Fig. 4.17 but exhibiting a constant relative channel width $\Delta V_c/V_c$ independent of the actual setting of V_c.

4.1.4. Multiple Arrays of Differential Discriminators

The principle of the single channel analyzer (Fig. 4.17) can also be applied to the multichannel analysis of a given amplitude range, as has already been mentioned in the introduction to this chapter. The corresponding circuit diagram of Fig. 4.24, of course, must be modified (channel by channel) using one of the methods outlined in Fig. 4.18 in order to guarantee that the particular inhibiting pulses overlap the corresponding signal pulses.

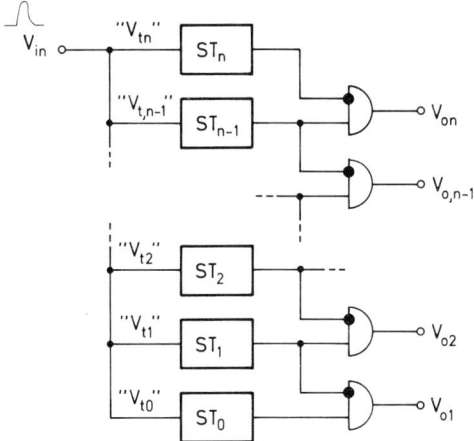

Fig. 4.24. Operating principle of a multi-channel discriminator

The multichannel array offers the following advantage: the amplitude range of interest is divided into n channels with the limits V_{c0}, $V_{c1} \ldots V_{cn}$. Hence the registration of the corresponding pulse rates at the outputs $V_{o1}, V_{o2} \ldots V_{on}$ in n independent scalers immediately yields the pulse height spectrum, which could be measured with the aid of a single channel analyzer only by progressively sampling the amplitude scale channel by channel. Thus for the same statistical accuracy the measuring time is reduced by a factor (n) corresponding to the number of channels. The analysis comprising the determination of the "address" of the pulse is performed very quickly; no more time is required than for the single channel analyzer. The disadvantage is the expensive circuitry of $n+1$ discriminators.

Before the more time-consuming but less expensive direct encoding of the pulse amplitude proved successful, multidiscriminator arrays were constructed, sometimes having ten channels or more (cf. the review by VAN RENNES [4.017], further [4.027] and [4.007]). At present such systems are made only for special purposes [4.028, 4.129], especially when an analyzer with a limited number of channels ($n \approx 10$) and with very high time resolutions in the nanosecond range is needed.

The crucial difficulty in the design of multichannel discriminators arises from the necessary equality and long term stability of the n channel widths, which are defined as respective differences of relatively high voltage values: $V_{c1} - V_{c0}$; $V_{c2} - V_{c1}$; ... $V_{cn} - V_{c,n-1}$. An especially elegant way of solving this problem consists in using the added step techniques according to GATTI et al. ([4.023] to [4.025], cf. also BONITZ and BERLOWITSCH [4.029] and WAUGH [4.130]), discussed at the end of the preceding chapter. Fig. 4.25 shows the block diagram of an added-step multichannel discriminator. In the pulse stretcher PS the input V_{in}

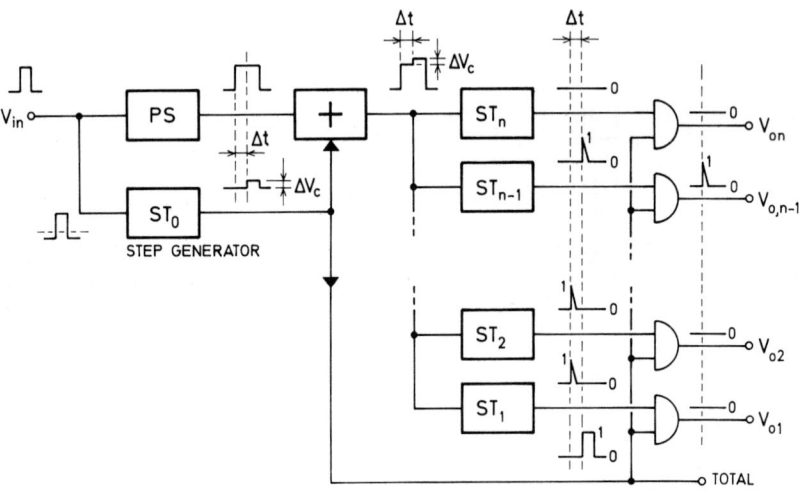

Fig. 4.25. A multi-channel discriminator using the added-step techniques

is first of all stretched and shaped to a rectangular pulse. A Schmitt trigger ST_0 of low threshold, which by the way defines the sensitivity of the device, actuates a step generator not shown explicitly. The generated square pulse with the well-defined and constant amplitude ΔV_c is added with the delay Δt to the stretched original pulse in an adder (+). Of the n threshold discriminators $ST_1, ST_2 ... ST_n$, only the one with the threshold lying within the voltage interval V_{in}; $V_{in} + \Delta V_c$ (e.g. the discriminator ST_{n-1}) is triggered delayed by Δt. A gate system, which is opened synchronously with the step pulse, thus passes only the output pulse $V_{o,n-1}$ to the connected register scalers. As can be seen immediately, the channel width depends solely on the step pulse amplitude ΔV_c, and the actual threshold voltages $V_{c1}, V_{c2} \ldots V_{cn}$ influence only the *position* of the corresponding channels. Hence changes in V_c distort the scale of the pulse height spectrum alone without affecting the differential spectrum shape. Normally ΔV_c is adjusted equal to the mean difference of the adjacent thresholds V_c in order to minimize the number of pulses lost or counted in two channels. In any case by counting the total number of analyzed pulses (output TOTAL) the integrated area under the spectrum curve can easily be brought in relation to the total count number.

4.1.5. Conservation of the Time Information in a Discriminator

The output pulse of an amplifier is correlated in time with the releasing event. Due to the signal path detector→preamplifier→amplifier, there is some signal delay, which, however, is constant on the average. If there are no non-linear operations on the pulse amplitudes (e.g. limiters, pulse shapers, window amplifiers), this delay too is independent of the pulse amplitude. When the amplitude-digital discriminator pulses are to be used later on for the time analysis in coincidence circuits or in time-to-pulse-height converters, the time information must be conserved in the discriminator.

In Fig. 4.26 the output pulses of an integral discriminator are shown which result from two input pulses with different amplitudes. The pulse origin ($t=0$) defines the instant of the releasing event (when neglecting

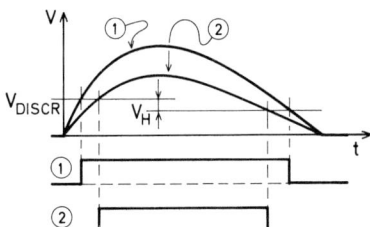

Fig. 4.26. Pulse shapes at the discriminator

the above-mentioned constant delay). As can be seen from this figure both parts of the discriminator pulse, which could be used for time definition, namely the leading and the trailing edges, are delayed by time intervals depending on the relative pulse height and the discriminator threshold V_{DISCR}. In differential discriminators too, the output pulse is derived from the leading or trailing edges of the lower threshold discriminator pulse, and hence is delayed by time intervals depending on the initial pulse amplitude (Fig. 4.18). Thus the shape of the amplifier pulse, especially its rise time or the slope of its trailing edge, influences the accuracy of the time definition by the discriminator output pulse. In addition to this time jitter, the pulses which only slightly exceed the threshold may trigger the discriminator with even more delay (cf. the discussion of the transition time, Fig. 4.05).

The amplitude-dependent delay of discriminator pulses is often denoted as "walking". The best way to overcome this difficulty is to divide the processing of the time and amplitude information related to one event into two separate channels according to the so called fast-slow coincidence techniques, which will be discussed in Chapter 5. Here only those modifications will be discussed, which are needed in order to obtain an amplitude-independent timing of the discriminator output pulse.

In principle any amplitude invariant point of the amplifier pulse can be used for correct amplitude-independent timing. WEINZIERL [4.030] and JOHANSSON [4.031] subtract an amplitude proportional pulse from the initial unipolar pulse. Due to the subtraction, the pulse shape is modified and its point of intersection with a fixed threshold becomes amplitude-independent. FAIRSTEIN [4.032] and GRUHLE [4.033] pointed out that the zero crossover time point of a bipolar pulse is independent of the amplitude.

For zero crossover timing an integral discriminator can be used, if the critical voltage V_{i2} (cf. the definition in Fig. 4.06) is made equal to the bias V_{i0}. Hence the discriminator reverts to the initial state at the input voltage $V_{in} = V_{i2} - V_{i0} = 0$. Thus the trailing edge of the discriminator output pulse corresponds to the zero crossing instant and it can be used for timing purposes. The accuracy is better than the initial pulse width, by about two orders of magnitude. The situation is shown in Fig. 4.27. The threshold $V_{DISCR} = V_{i1} - V_{i0}$ in this case corresponds to the hysteresis $V_H = V_{i1} - V_{i2}$.

When changing the threshold V_{DISCR} in the circuit of Fig. 4.27 the hysteresis V_H also must be changed. This is done most effectively by means of a diode limiter according to Fig. 4.08. The voltage V''_{B2} and the bias of the base of Q_1 are adjusted for $V_{i2} = V_{i0}$, and the triggering threshold $V_{i1} \approx V'_{B2}$ is changed by the variation of the bias of the diode D_1. Despite the very limited range within which V'_{B2} can be varied, every change in the hysteresis V_H also influences the transition time of the trigger, which itself affects the obtainable timing accuracy.

CHASE ([4.034], or [4.007] p. 167) describes a Schmitt trigger circuit, in which the triggering threshold V_{DISCR} can be varied over a wide range.

The principle is shown in Fig. 4.28 (however, in the original papers of CHASE the operational principle is described in a slightly different manner). The bias of a Schmitt trigger with a relatively constant and

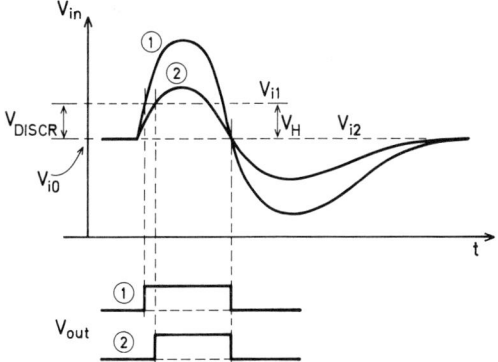

Fig. 4.27. Simple zero crossing trigger with threshold equal to hysteresis

small hysteresis is adjusted to yield the threshold V_{DISCR}. After the discriminator has been triggered an auxiliary circuit changes the reference potential V_{B2} of the base of T_2 to the value $V_{B2} = V_{i0}$. Hence the return to the initial state occurs when the input voltage crosses the zero line. Such circuits have been built occasionally [4.035], and a variant by EMMER [4.036] is shown in Fig. 4.29. The Schmitt trigger used in this discriminator consists of the transistors Q_1 and Q_2. The loop coupling is performed by a condensor C_3 and by an emitter-follower Q_4. The threshold is adjusted by means of the potentiometer POT.

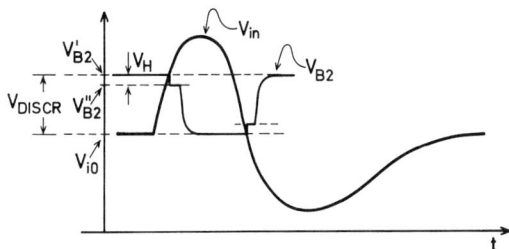

Fig. 4.28. Pulse shapes at a zero crossing discriminator with constant hysteresis

Analog-to-Digital Converters

An input voltage exceeding the threshold cuts off Q_2. The positive Q_2 collector pulse itself cuts off Q_7 and opens Q_5. The base voltage of Q_5 is limited by means of Q_6 to the value V_{i0} of the Q_1 bias, set by the potentiometer POT. Because of Q_5 the base bias of Q_2 also becomes V_{i0}. Hence the threshold for return to the initial state is given by V_{i0}. After return to the initial state, Q_7 opens and its negative differentiated emitter pulse saturates Q_9, producing a positive standard pulse about 0.3 µsec long. The discriminator threshold can be adjusted for positive bipolar pulses in the range of 0.05 ... 10 V. For input pulses with rise time of 0.1 µsec the walking of the output pulse is less than 10 nsec when varying the input amplitude over the whole range 1:20.

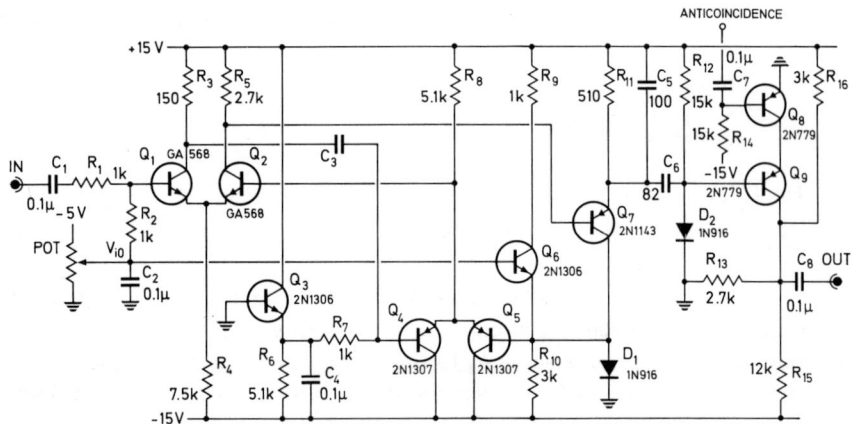

Fig. 4.29. Zero crossing discriminator according to EMMER [4.036]

Differential discriminators with amplitude-independent delay of the output pulse can also be realized. Preferably the configuration (C) of Fig. 4.18 is used, i.e. an analyser using a flip-flop as the upper discriminator. Since the output pulse is derived from the trailing edge of the lower Schmitt trigger signal, it is only necessary to use a zero-crossover sensitive trigger for the lower discriminator to obtain exactly timed output pulses. Of course, the circuit then operates with bipolar input pulses only.

Some more practical circuits are described in [4.131] and [4.132]. GEDCKE and MCDONALD [4.133] designed a "constant-fraction-of-pulse-height-trigger", which is a generalized case of the zero-crossover trigger.

STRAUSS [4.037] described a discriminator based on a monostable multivibrator with output pulse length dependent on the input pulse amplitude. With small input pulses the multivibrator fires later than with large pulses (cf. Fig. 4.26), however the pulse length is correspondingly shorter. Hence, independent of the input amplitude, the trailing edge of the pulse always occurs at the same instant and it can serve well for timing purposes.

The limit of the possible timing accuracy is given partly by the accuracy and stability of the zero-crossing-point detection, and partly by the error in the actual definition of the instant of the releasing event by the zero crossing point of a bipolar pulse. As has already been pointed out, the pulse is delayed with regard to the primary event by time intervals exhibiting some statistical deviation about the mean value (i.e. the fluorescence decay time and the electron propagation time in photomultipliers). Therefore the zero crossing point a priori is subject to variations in time. It can be shown that the mean square deviation of the timing with zero crossover is always higher than that of the leading edge timing (cf. [4.038] and the discussion in Chapter 5). Hence in coincidence experiments requiring the highest time resolution, zero crossing discriminators are not applicable. However, in measurements over large amplitude ranges, especially when the time information must be extracted from the integrated "slow" pulse, walking is the dominant effect and the use of zero crossing techniques offers advantages. For the highest time resolution over large amplitude ranges special precautions must be taken, such as the amplitude compensation of the leading edge timing (cf. [4.039] and Chapter 5.4.2.).

4.1.6. Fast Tunnel Diode Discriminators

Negative resistance components like tunnel or ESAKI diodes [4.040] offer an alternative possibility of realizing multivibrator and hence also discriminator circuits. Because of their short response times in the sub-nanosecond range, the tunnel diode discriminators are used especially in fast circuits.

Fig. 4.30 shows the basic circuit diagram of a tunnel diode (TD) multivibrator and its corresponding characteristics. The load resistor R and the unavoidable parasitic capacity C of the diode and of the related circuitry are connected in parallel to TD. The current I from a current source is divided between the diode (I_D) and the other two components (I_C, I_R). On the one hand, I_D depends on V_D in conformity with the diode characteristic $I_D = f(V_D)$ exhibiting a negative resistance R_n voltage range, on the other hand, the load R defines a load line $I_D = I - V_D/R$. Of the three points of intersection of these two curves, (assuming of course $|R| > |R_n|$) only the points (A) and (B) are stable. In these operating points $I = I_R + I_D$ holds and $I_C = 0$. If the input current I of the circuit exceeds the value I_1, the state (A) disappears; if I is decreased under I_2, the state (B) disappears. The plot of V_D (which is at the same time the output voltage of the circuit) versus the input current I exhibits a shape characteristic of multivibrators (cf. Fig. 4.02) with a hysteresis $I_H = I_1 - I_2$. During the transition, a current $I_C = I - I_R - I_D$ is available for recharging the capacity C. As can be seen from the characteristics, the transition I_1 is better defined and faster than the transition I_2. The smaller the difference between $|R|$ and $|R_n|$, the smaller is the hysteresis I_H.

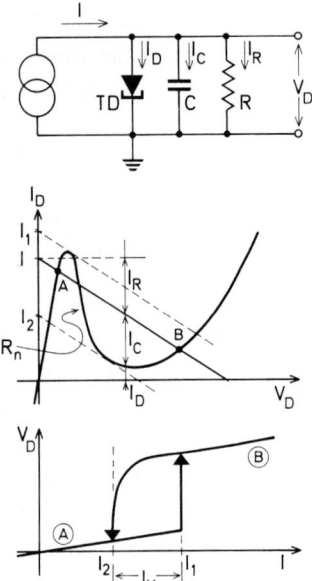

Fig. 4.30. Basic tunnel diode circuit

The tunnel diode TD in parallel with the resistor R can also be considered as a positive feedback amplifier with the (voltage) loop gain $bA = |R|/|R_n|$. The positive feedback is realized in a particularly simple manner, since the tunnel diode is an active dipole and the output as well as the input lie at the same terminal. The considerations stated in Chapter 4.1.1 concerning the magnitude of the hysteresis and the transition times as functions of the loop gain bA apply accordingly.

As can be seen from Fig. 4.30, a TD multivibrator, like the Schmitt trigger, can be biased outside of the hysteresis range (i.e. $I < I_2$ or $I > I_1$). In this case the operating range is limited only by the maximum ratings of the diode current. More sensitive discriminators are obtained when monostable or bistable circuits biased between I_1 and I_2 are used. Of course, bistable circuits must be reset by some external means [4.041,

4.049]. Two variants of a monostable multivibrator are shown in Fig. 4.31. A small resistor R ($|R| < |R_n|$) is used, thus giving only one intersection (A) of the static load line with the diode characteristic. The operating point (A) is adjusted by means of the standing current I_0.

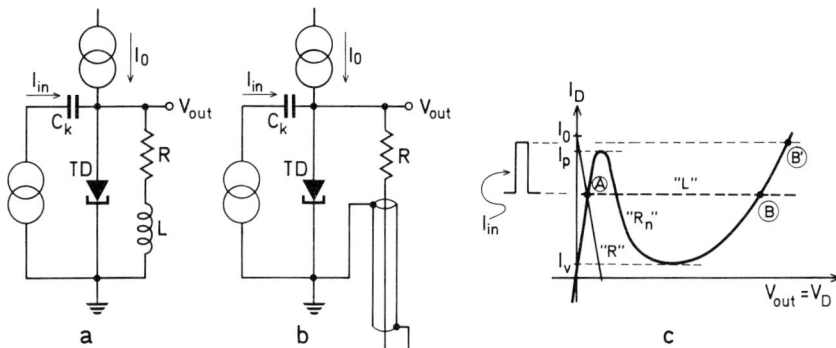

Fig. 4.31 a–c. Tunnel diode monostable multivibrator with coil L (a) or cable (b), and the related characteristics

In the circuit a) a coil with inductivity $L \gg |R_n| \cdot t_t$ is connected in series with R. Here t_t denotes the transition time of the circuit. Hence, at least during the transition, the coil L represents an infinite resistance and the dynamic load line "L" becomes horizontal. Any input current pulse I_{in} raising the dynamic load line over the tunnel diode peak current I_p causes the circuit to flip over into the state (B') and later (B). However, the point (B) is not stable and the circuit returns to the initial state (A) as soon as L passes enough current to reduce I_D below the valley current I_v. The resulting pulse length is approximately equal to $L/(|R|+|R_n|)$ [4.042].

In the circuit b) a time-dependent load is obtained by connecting in series a resistor R and a coaxial cable with the characteristic impedance $Z = R$. Obviously the circuit operates if $R < |R_n| < 2R$ holds. The pulse length is given by twice the delay time of the cable.

For a more detailed discussion of the tunnel diode discriminator, some considerations concerning the transition time t_t are necessary. In Fig. 4.32a the basic circuit diagram is shown once more with the standing current generator I_0 and a load $R = \infty$ (i.e. I_R is negligibly small). The input current pulse has an amplitude I_{in} and a length δ_{in}. The difference $(I_0 + I_{in}) - I_p$ is denoted by ΔI_{in}. Obviously only $\Delta I_{in} > 0$ triggers the circuit. Due to C the transition occurs only as fast as the para-

Analog-to-Digital Converters

sitic capacity is recharged. In Fig. 4.32 the current I_C available for charging C is plotted separately as a function of V_D. The curve is passed in the direction indicated by arrows, and the corresponding time scale $t(V_D)$ can be obtained by integrating C/I_C:

$$t(V_D) = C \int_{V_0}^{V_D} \frac{dV_D}{I_C(V_D)}. \tag{4.21}$$

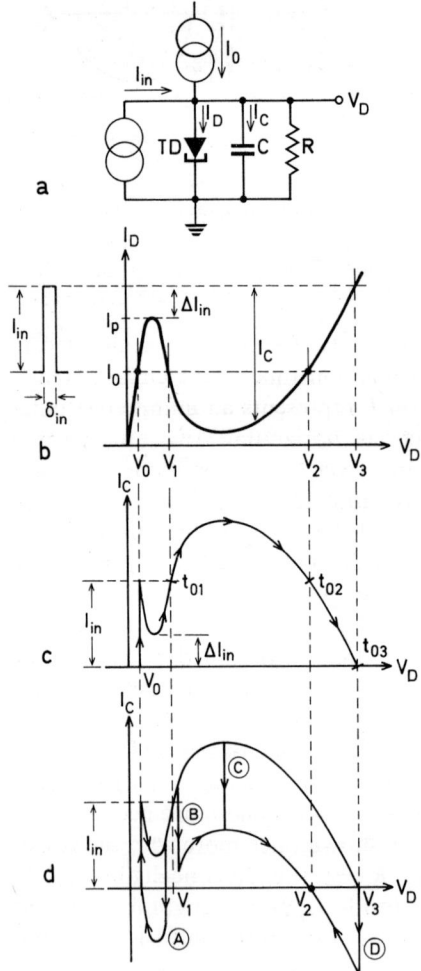

Fig. 4.32 a–d. Switching characteristics of a TD trigger

Fig. 4.33a shows some pulse shapes for $\delta_{in} = \infty$ (i.e. current step at the input) with ΔI_{in} as parameter. The small current I_C in the output voltage range $V_0 \ldots V_1$ yields a very slow pulse beginning. Hence in practice the pulse appears delayed by a time t_{01} defined by $V_D(t_{01}) = V_1$. The exact value of t_{01}, of course, depends on the actual shape of the tunnel diode characteristic. If I_0 lies in close proximity to I_p, $I_C(V_D)$ can be approximated by a parabola in the range $V_0 \ldots V_1$, with corresponding t_{01} of

$$t_{01} \approx \frac{\pi}{2} \frac{(V_1 - V_0)C}{\sqrt{I_{in} \cdot \Delta I_{in}}}, \tag{4.22}$$

assuming $\Delta I_{in} \ll I_{in}$. With $V_1 - V_0 = 50\,\text{mV}$, $C = 10\,\text{pF}$, $I_{in} = 1\,\text{mA}$ and $\Delta I_{in} = 100\,\mu\text{A}$ (4.22) comes to $t_{01} \approx 2.5$ nsec.

If the input pulse length δ_{in} is finite, two cases must be distinguished. If $\delta_{in} < t_{01}$ holds, the input current I_{in} is cut off at the instant at which $I_C < I_{in}$ is still valid (case (A) in Fig. 4.32d and 4.33b). Hence $I_C - I_{in}$ becomes negative, C will be discharged again, and the circuit returns to the initial state V_0. First if $\delta_{in} > t_{01}$, the disconnection of I_{in} results only in a diminution of I_C, but not in its polarity reversal, and the transition process is continued (cases (B) and (C) in Fig. 4.32d) and 4.33b)). Finally, if δ_{in} is longer than the whole transition time (t_{03}), a small overshoot results (case (D)).

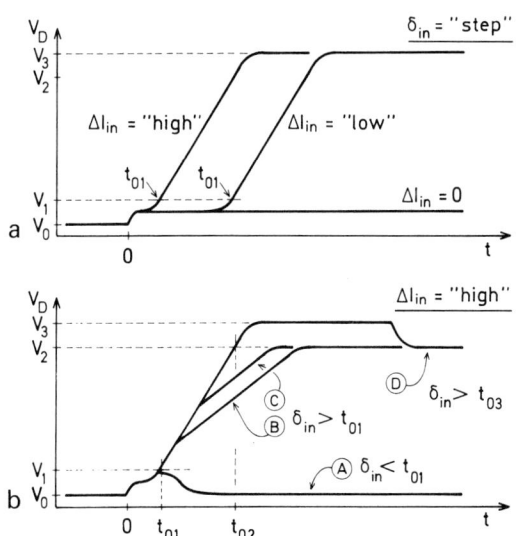

Fig. 4.33a and b. Tunnel diode multivibrator pulse shapes

Analog-to-Digital Converters

In order to trigger a tunnel diode discriminator therefore, two conditions must be fulfilled:

1. $\Delta I_{in} > 0$ must hold; the threshold given by I_p must clearly be exceeded.
2. $\delta_{in} > t_{01}$ must hold.

Since t_{01} itself depends on I_{in}, the second condition means that actually the whole integrated charge Q_{in} of the input current pulse must be higher than about $(V_1 - V_0) \cdot C$. Hence the tunnel diode discriminator is charge-sensitive for short pulses, and current-sensitive for long pulses (cf. [4.134]).

The shape of the output voltage pulse $V_D(t)$ is obtained by inversion of the relation (4.21). Unfortunately this inversion gives a non-linear differential equation which cannot be solved explicitly. Various graphical, numerical or approximative solutions are discussed in the papers [4.041] to [4.043] and [4.135] to [4.137].

In practical tunnel diode discriminator circuits, of course, it must also be taken into account that the output signal, too, runs into the input circuitry, since the input and the output are not separated. If for instance a TD discriminator is coupled by means of a coaxial cable to the photomultiplier, the cable must be terminated correctly at the photomultiplier output in order to avoid reflexions [4.044]. Alternatively the input of the discriminator can be separated from the detector output by means of a common base stage, as in Fig. 4.34 [4.045]. Of course, the collector of Q_1 can be coupled directly into TD, without the capacitor C_k.

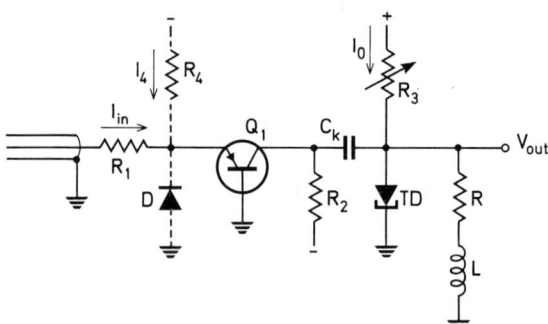

Fig. 4.34. Decoupling of the TD trigger input by means of a common-base transistor stage

The dependence of the discriminator threshold on the pulse shape and duration can be reduced considerably if a window amplifier is connected in front of the tunnel diode. The circuit of Fig. 4.34 can be easily modified by the introduction of a diode D (dotted line connections) to act as a window amplifier. Due to R_4, the diode D is forward-biased leading a current I_4. Only the difference $I_{in} - I_4$ is applied to the tunnel diode through Q_1. RIGHINI [4.046] described tunnel diode discriminators with biased emitter-followers as window amplifiers. WARD and YORK [4.047] used simple backward-biased semiconductor diodes for this purpose. In their circuit the discriminator threshold can be varied between 0.5 V and 10 V. Variation of the input pulse rise time between 0.5 and 20 nsec and of the pulse length between 5 and 100 nsec shifts the threshold by about only 10 mV. COLI [4.048] uses an additional input pulse shaping with the aid of a shortened cable at the discriminator input.

A voltage signal, of course, must be converted into a current signal, e. g. by means of a resistor (cf. R_1 in Fig. 4.34), prior to triggering a tunnel diode discriminator. HVAM and SMEDSDAL [4.050] investigated a voltage-sensitive tunnel diode discriminator with TD in the high impedance collector circuit of a normal emitter-coupled and biased (→POT!) difference amplifier (Fig. 4.35). The output signal across the tunnel diode is picked up using a second difference amplifier. PANDARESE and VILLA [4.051] used tunnel diodes in bridge configuration for discriminators. Some other TD discriminator circuits are described in [4.138] and [4.139].

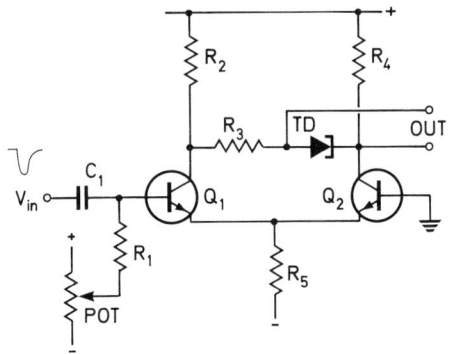

Fig. 4.35. Principle of the TD voltage discriminator circuit given by [4.050]

With fast TD threshold discriminators, the fast single-channel analyzer can also be realized, mostly by using one of the alternatives shown in Fig. 4.18. For the details of the related circuitry we refer to the original literature [4.018, 4.020, 4.140] and [4.022]. Because of the dependence of the output pulse length of a TD discriminator on the input pulse properties, an additional pulse shaping — commonly using another TD monostable multivibrator — is needed prior to the anticoincidence circuitry.

Analog-to-Digital Converters

Due to the dependence of the delay t_{01} between the input and output pulses on the input pulse height, the time information is not well conserved in a tunnel diode discriminator. ORMAN [4.052, 4.053] described a TD discriminator for bipolar pulses exhibiting a highly accurate zero-crossover timing. Fig. 4.36 shows the simplified circuit of such a discriminator reported by ALSTON and DRAPER [4.054]. The operating point (A) lies above the peak current I_p. The discriminator fires (Point (1)) when the input current $I_{in}(t)$ passes the threshold $I_v - I_0$. Because of $I_0 \approx I_p$, the return in the initial state (Point (2)) occurs approximately at the zero crossing point of the input current. Since I_p is greatly exceeded, the transition time is very short. Hence the steep trailing edge of the output pulse can be used for timing purposes. Any-

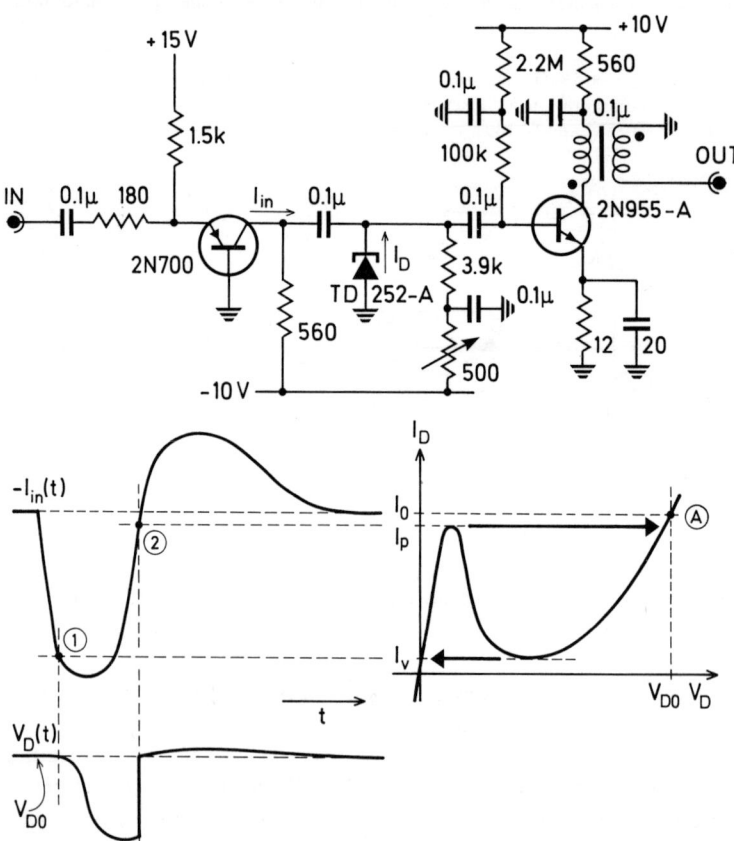

Fig. 4.36. Orman-type TD zero crossing discriminator according to ALSTON and DRAPER [4.054]

way, the discriminator has a fixed threshold $I_v - I_0 \approx I_v - I_p$, thus making a preconnected variable gain amplifier necessary. Other zero crossing discriminators with tunnel diodes are described by WIEGAND [4.055] and GARVEY [4.056].

4.2. Digital Encoding of the Pulse Height

The high energy resolution of modern detector systems necessitates a correspondingly high number n of analyzer channels. For instance, with semiconductor detectors with a relative FWHM of 0.1 % we need $n > 1000$ in order to display a spectrum peak in more than one channel. Despite the difficulties connected with the necessary stabilization of the particular channel widths and positions, an analog-to-digital converter according to the multi-discriminator principle (Chapter 4.14) would require such a number of discriminators and gating systems, which is not feasible for economic reasons. Hence a more economic solution for encoding must be found.

In industrial electronics analog values (voltage, position, angle etc.) must often be digitized and various techniques have been developed for this purpose (cf. e.g. SUSSKIND [4.057] or BORUCKI and DITTMANN [4.058]). Some of these techniques, such as the principle of the "voltage balance", can also be used in the encoding of pulse amplitudes — the particular circuits will be described briefly in Chapter 4.2.2. However, in nuclear pulse techniques in general the encoding method according to WILKINSON [4.059] has succeeded, based upon a first conversion of the amplitude into a pulse length, which can then be measured digitally by means of a frequency standard. The realization and the advantages of the WILKINSON type converters are discussed in Chapter 4.2.1.

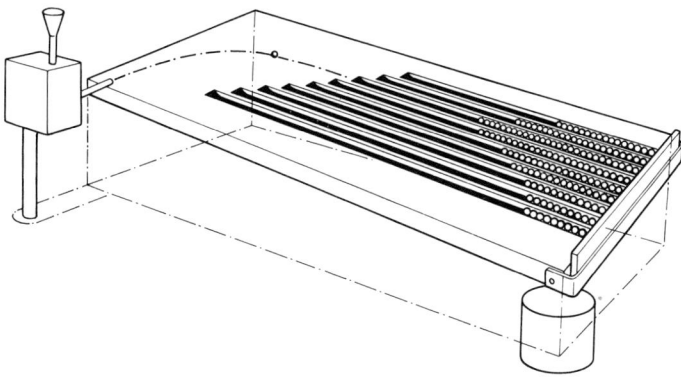

Fig. 4.37. The "kicksorter" pulse analyzer built by FRANK, FRISCH and SCARROTT [4.060]

Analog-to-Digital Converters

Since there is no well-defined boundary between the analog and the digital part of electronic systems, the treatment of some auxiliary digital circuits cannot be avoided in connection with the discussion of analog-to-digital converters, although the systematic discussion of digital circuits is reserved for Chapter 6.

The electro-mechanical pulse height analyzer by FRANK, FRISCH and SCARROTT [4.060] in 1951 is mentioned (Fig. 4.37) as a curiosity. This was an analog-to-digital converter and a 30-channel memory at the same time. The operating principle can easily be understood from the drawing: an electro-mechanical cue applies a kick to a small ball, the magnitude of the kick being proportional to the pulse amplitude (→ "address"). Corresponding to the magnitude of the kick, the ball alights in a more or less remote channel. The channel capacity reached 100 pulses (i.e. 100 balls), and the time resolution was about 0.2 sec. The name of this assembly — kick-sorter — was later on also used for purely electronic amplitude analyzers.

4.2.1. Converters of the Wilkinson Type

Fig. 4.38 shows the operating principle of a Wilkinson type converter. In the variant type a) the input pulse (A) of the amplitude V_{in} is at first stretched to the pulse (B) without any amplitude loss. Shortly after the input pulse (A) reaches its maximum, a linear ramp generator is triggered producing a voltage ramp (C) of constant slope and high linearity. At the same time a gate signal (D) is produced. A comparator circuit compares (B) and (C) and cuts off the gate (D) in the instant when (B)=(C). Hence the length of the gate signal (D) is proportional to the input amplitude: $\Delta t \sim V_{in}$. During Δt, in an electronic scaler pulses (E) of constant repetitive rate (e.g. from a quartz-stabilized oscillator) are counted. The number of pulses n is proportional to Δt and thus also to V_{in}. Hence the number n represents the desired digital equivalent of the pulse height and can be used directly as the address of the memory cell in which the event is to be stored.

The accuracy of this encoding is determined by three factors: By the accuracy of reproducing the input amplitude V_{in} in the stretched pulse (B), by the slope stability and linearity of the ramp, and by the stability of the pulse generator (E). However, since the proportionality $\Delta t \sim V_{in}$ holds, any variation of the pulse repetition frequency (E) or of the slope of (C) affects all channels in the same manner, and the channel width δV — otherwise the Achilles heel of amplitude analysis — remains relatively constant. Obviously δV is given by the time interval between two pulses (E), independent of the channel position.

Since it is relatively difficult to conserve the pulse height V_{in} during the desired time interval Δt of (often) more than 10 μsec in transistorized pulse stretchers (B), recent circuits use the converter variant type shown in Fig. 4.38b. In these circuits the converter capacitor is charged to the full input voltage V_{in}, and thereafter linearly discharged (F). The time Δt is defined by the duration of the condenser discharge up to the voltage zero (or any other suitable reference voltage). This circuit type will be discussed in more detail in what follows.

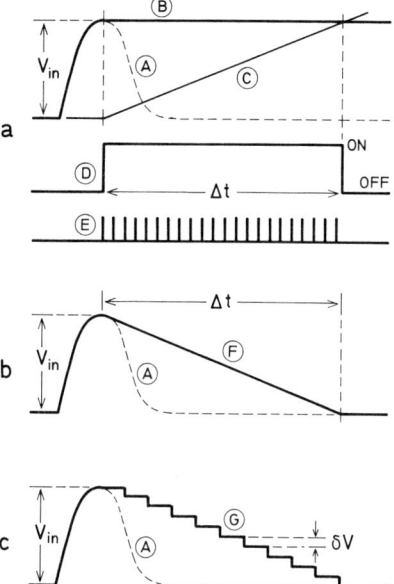

Fig. 4.38 a—c. Pulse shapes at various points of the Wilkinson analog-to-digital converter

A third variant (e. g. GUILLON [4.061], KANDIAH [4.062, 4.063]) is illustrated in Fig. 4.38 c, and the corresponding circuit diagram is shown in Fig. 4.39. Again a condenser (C_1) is charged to the full pulse amplitude V_{in}, though here the discharge is performed in a stepwise manner: with each step the condenser charge is diminished by $\delta V \cdot C_1$. The corresponding circuit operates as follows: the input pulse charges the condenser C_1 through the diode D_1. At the same time $C_2 \ll C_1$ is also charged. The point (X) is held by means of an impedance converter at the same potential as C_1. C_2, D_2 and D_3 form a so-called diode pump (cf. [4.003], p. 346). The condenser C_2 is charged via D_3 by a positive clock pulse of amplitude V_T. At the trailing edge of the clock pulse D_2 starts conducting and D_3 cuts off, and the negative charge $C_2 \cdot V_T$ flows through D_2 to the condenser $C_1 \gg C_2$. Hence, the voltage across C_1 is diminished by $\delta V \approx V_T \dfrac{C_2}{C_1}$ per clock pulse. A zero-crossing discriminator watches the instant of zero crossover. The number of steps required to bring the C_1 voltage to zero (which is the same as the corresponding clock pulse number) represent the digital equivalent of V_{in}. Obviously the amplitude V_T, and not the clock pulse frequency must remain constant in this circuit.

A converter according to Fig. 4.38 b with continuous linear condenser discharge needs only a single analog switching (starting the ramp). On the contrary, the diode

Analog-to-Digital Converters

pump converter requires two switchings (D_2, D_3) per step. Thus, for an n-channel conversion $2n$ switching processes are needed. Since the speed of an analog switching process is limited, obviously the diode pump converter is slower than the continuous one. It is used only exceptionally in recent analyzers [4.064].

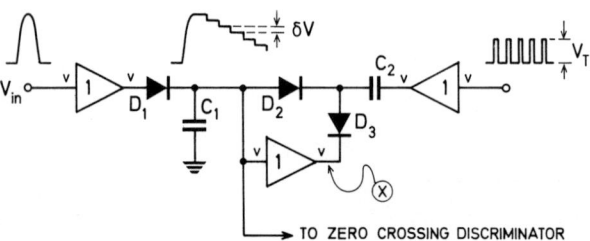

Fig. 4.39. Diode pump Wilkinson-type converter

At the input of an analog-to-digital converter a linear gate is always used, which is cut off during the conversion process and thus prevents this process from being disturbed by following input pulses. The gate control signal is longer than Δt by the time interval which is required for the processing of the digital output signal in the following digital equipment.

For the discussion of the non-linearity of the conversion process (cf. [4.063, 4.065, 4.066]), three possible pulse stretcher circuits are shown in Fig. 4.40. As already pointed out in Chapter 3.5, the simple diode circuit a) can be improved by a voltage feedback b), which is conceived most conveniently with the aid of a difference amplifier (+ denotes the non-inverting input, − the inverting one). Additionally, the diode D can be separated from the storage condenser C by means of a current amplifier B (Fig. 4.40c). Because of a possible pulse pedestal of the linear gate, the standing voltage V^* with closed gate in general differs from zero.

The condenser C charging process is interrupted at the time δ after the input pulse start. At this time the voltage across C is $V'_{in} \approx V_{in}$. Assuming an ideal diode characteristic (3.48) the circuit variant type a) results in

$$\frac{dV'_{in}}{dt} = \frac{I_0}{C}\left(\epsilon^{\frac{e(V_{in}-V'_{in})}{pkT}} - 1\right). \tag{4.23}$$

The factor p allows for differences in the conduction mechanism of different semiconductor diodes ($p \approx 1$ for Ge; $p \approx 2$ for Si). I_0 denotes the constant discharge current, which flows through the diode even during the charging process. The differential equation (4.23) can be integrated for a rectangular input pulse of the duration δ [4.063]:

$$V'_{in} = V_{in} - \frac{I_0 \cdot \delta}{C} + V_0 \cdot \log\left(\epsilon^{\frac{I_0 \cdot \delta}{C \cdot V_0}} + \epsilon^{-\frac{V_{in}}{V_0}} - 1\right), \quad (4.24)$$

where $V_0 = pkT/e$.

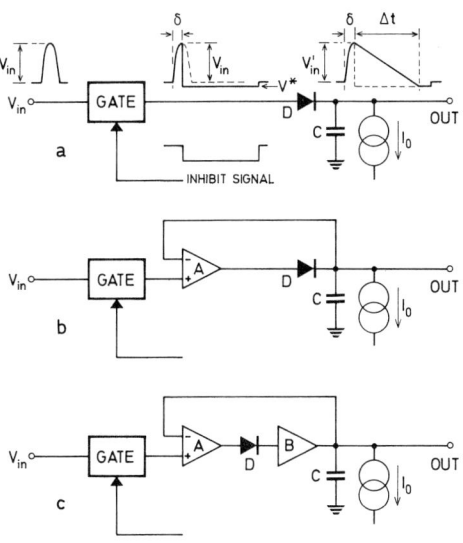

Fig. 4.40a–c. Operating principle of a pulse stretcher

Hence, a "channel width" dV'_{in} corresponds to a given pulse height interval dV_{in}, and the differential non-linearity $\varepsilon = (dV_{in} - dV'_{in})/dV'_{in}$ becomes from (4.24)

$$\varepsilon = \frac{1}{\left(\epsilon^{\frac{I_0 \delta}{CV_0}} - 1\right)\epsilon^{\frac{V_{in}}{V_0}}}. \quad (4.25)$$

The non-linearity ε hence depends on the input amplitude and becomes high when $V_{in} \approx V_0$. Therefore the converter circuit a) operates satisfactorily only with input pulses substantially higher than 25...50 mV.

Regarding the circuit variants in Fig. 4.40b and c, we see immediately that the only influence of the amplifier A on the circuit operation as compared with a) consists in changing the diode voltage scale by the gain $A: V_0 \to V_0/A$. Hence

$$\varepsilon = \frac{1}{\left(\epsilon^{\frac{AI_0\delta}{CV_0}} - 1\right)\epsilon^{\frac{AV_{in}}{V_0}}} . \qquad (4.26)$$

The gain B of the current amplifier does not appear in (4.26). As can easily be seen, the current amplifier merely disconnects D and C independent of the value of B.

An additional differential non-linearity is caused by the charge stored in D, which must be neutralized when the diode is cut off (cf. discussion in Chapter 3.5). Since this charge is approximately constant, the relative error caused is again highest for small pulses V_{in}. This error does not arise with the separating amplifier B between D and C.

After the diode D is cut off its junction capacity C_D lies in parallel with C. Due to the dependence of C_D on the reverse voltage across the diode, the discharge process is non-linear even if I_0 is absolutely constant, unless $C_D \ll C$ holds. The interconnection of a current amplifier between D and C scarcely diminishes this non-linearity, since C_D is only replaced by another voltage-dependent capacity (e.g. the collector-base junction capacity).

All errors dominate at small amplitudes. Considering this, different proposals have been made as to how the pulse height could be increased by adding a constant charge amount to every input pulse. Then the pulse height scale is shifted by a fixed amount to the higher values, which can be allowed for later on, e. g. by digital subtraction of a corresponding number of clock pulses (cf. [4.067] and comments at the end of Chapter 3.3).

For linearizing the discharge process of C, one of the well-known feedback techniques is used (bootstrap, Miller integrator). Alternatively a transistor in a constant-collector-current circuit is used for delivering I_0. Fig. 4.41 shows the circuit diagram of an analog-to-digital converter (type Fig. 4.40b) described by MANFREDI and RIMINI [4.065]. The input pulse is applied via an emitter follower Q_1 to the difference amplifier Q_2, Q_3, which charges the storage condenser $C=2000$ pF via Q_5 and D. The charging current is limited by a resistor $r=55$ in order to avoid overshoots. The feedback loop is closed by the double emitter follower Q_6, Q_7 whose load is formed by the constant current generator Q_{10}, Q_9. The collector base voltage V_{CB} of Q_6 and Q_7 is kept constant by means of the bootstrap Q_8, hence the collector-base junction capacity of Q_6 does not depend on the input pulse amplitude. After the linear gate is cut off, the diode D too cuts off, the feedback loop is interrupted, the collector potential of Q_3 drops to $+12$ V (limiter diode), and the linear discharging starts. The discharge current is defined by R_C and by the (constant) potential difference between the base of Q_6 and the collector of Q_{10}. When the voltage across C reaches V^*, the feedback loop is closed suddenly and the collector potential of Q_3 rises again. The negative collector pulse of Q_3 can be used as the oscillator gating signal ((D) in Fig. 4.38) after suitable shaping and limiting. In the circuit described the feedback difference amplifier itself serves as the comparator or zero crossing discriminator. For a discharging current corresponding to 20 mV/μsec ramp slope, the differential non-linearity is better then $\pm 1\%$ for amplitudes higher than 50 mV.

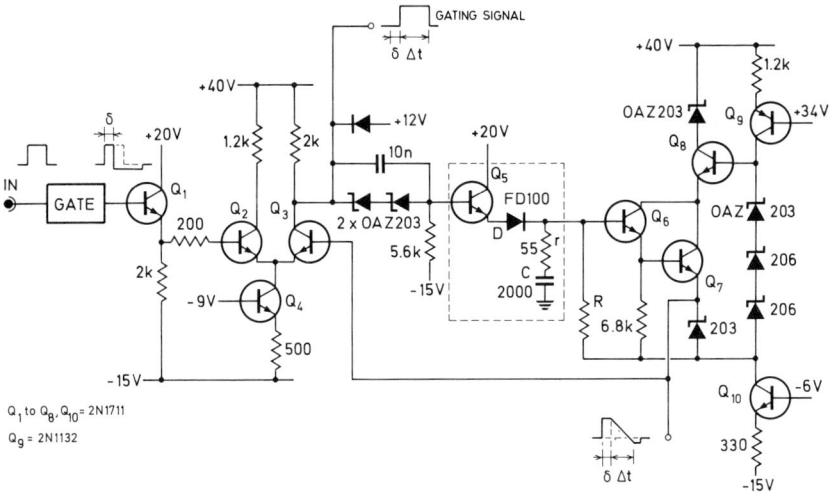

Fig. 4.41. Circuit diagram of an analog-to-digital converter according to MANFREDI and RIMINI [4.065]

Instead of discharging C to the end voltage V^*, the discharging process can be stopped as soon as the voltage over C reaches another preselected reference value V^{**}. Hence only the difference $V_{in} - V^{**}$ is digitized and the converter acts as its own window amplifier with the threshold V^{**} [4.068].

In order to avoid an additional inaccuracy of about one channel width, the pulse generator (E) and the gating signal (D) must be synchronized (Fig. 4.38). This can be achieved in two ways. Either an LC oscillator is triggered by the gating signal, thus locking the phase of the first oscillator pulse (i.e. the first oscillator wave) to the gating pulse, or the discharging process is started by the first pulse of a free-running oscillator which follows after the time instant δ, thus locking the gating pulse to the oscillator waves. The first variant is easier to realize, especially with frequencies below 10 Mcps. COMISKEY et al. [4.069] describe how to realize a 55 Mcps oscillator which can be switched on by a gating pulse. TURKO [4.141] investigated fast gated clock oscillators and gave a circuit diagram of a 100 Mcps gated oscillator. If highly constant (e.g. quartz stabilized) oscillators must be used, then the gating pulses must be synchronized [4.070].

Normally the vertex of the input pulse is used for the definition of the time point δ in which the charging process of the storage capacitor is finished. If a feedback difference amplifier according to the block diagrams of Fig. 4.40b) or c) is used, the pulse maximum can easily be determined by the fact that the feedback loop suddenly opens after the input voltage has passed its maximum [4.071]. Anywhere in the amplifier a time-correlated signal can be found and used for gating purposes (cf. the circuit in Fig. 4.41).

If the top of the input pulse is too flat, the determination of the maximum voltage and hence of the instant δ is too inaccurate, δ deviates too much from its mean value, and the differential linearity is adversely affected according to (4.25) and (4.26). In such a case the charging process is terminated preferably a fixed time interval after its leading edge, or before its zero crossing point (EMMER [4.036, 4.070]).

After this discussion of the particular aspects of a Wilkinson type converter the general idea of the logical build-up of an analog to digital converter will be discussed, using the block diagram of Fig. 4.42. The input signal V_{in} delayed by $T_D (\approx 1 \,\mu\text{sec})$ arrives at the linear GATE and thereafter at the pulse-height-to-time converter ATC. The gating signal (D) releases the oscillator OSC, and at the same time a STOP

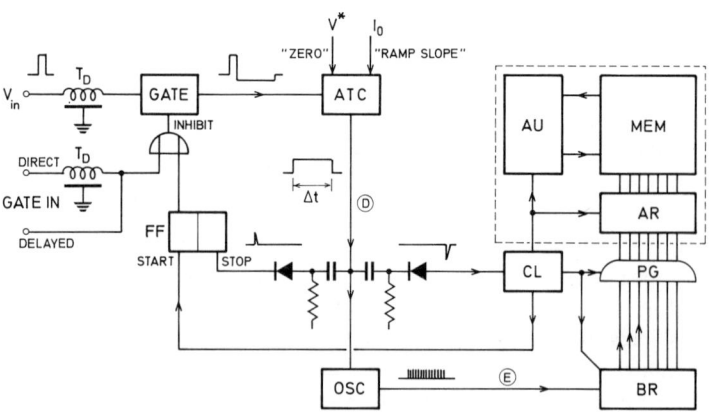

Fig. 4.42. Internal organization of an analog-to-digital converter with a buffer register

signal is derived from its leading edge, which sets the flip-flop FF. The output signal of FF cuts off the linear gate via a digital OR gate. The clock pulses (E) from the oscillator OSC are counted in a scaler BR (buffer register). After Δt, the trailing edge of the gating signal (D) triggers the control logic CL, initiating a control program: the number of counts stored in the buffer register BR is transferred by means of a parallel gate system PG to the address register AR of the memory MEM, the BR is cleared to zero, and the desired arithmetical operation on the content of the memory cell specified by AR (commonly the simple addition of 1) is performed in the arithmetic unit AU. The processing of the digital information after BR will be discussed in more detail in Chapters 6 and 8; here we limit ourselves to some general remarks. Immediately after the buffer register BR has been cleared the linear GATE can be opened (signal "START"). Means are often provided for blocking the linear gate by external signals e.g. when recording coincidence spectra etc. Depending on whether or not the external control signal is delayed against the measuring signal V_{in}, the control input "DELAYED" or "DIRECT" respectively can be used.

Instead of serving as a buffer register, the scaler BR can also be used as the address register, directly controlling the memory MEM. In this case the parallel gate transfer system PG and the address register AR are superfluous. Of course, the control logic CL may first clear BR to zero and open the linear GATE after the arithmetic operations in AU are terminated. Obviously this second variant is a slower one.

The time demand t_n for digitizing and storing a pulse of amplitude corresponding to n channels is

$$t_n = n \cdot \frac{1}{f} + t_c, \qquad (4.27)$$

where f denotes the oscillator frequency and t_c denotes the fixed time interval needed for the various switching operations, especially for the transfer BR→AR or for the arithmetic operations in AU. A common value of the frequency f is 2...10 Mcps, and in fast analyzers up to 100 Mcps. Hence for the hundredth channel $\Delta t = 100/f$ becomes 10...50 µsec, and in fast analyzers ~1 µsec. If the access time t_c of the memory is not too high in relation to Δt the buffer register can be omitted (i.e. BR can directly be used as the address scaler). In devices with very high channel numbers ($n > 1000$) or in very fast analyzers a buffer register offers advantages.

More examples of practical converter circuits can be found in papers [4.072] to [4.076]. The whole complex of the multi-channel analysis, including the analog-to-digital conversion, has been reviewed recently by CHASE [4.077], STANFORD [4.078] and GUILLON [4.064].

In multi-channel analyzing assemblies with very high channel numbers (i.e. $n \gtrsim 1000$) the stabilization of the system detector-amplifier-analyzer becomes unavoidable. The principle of the control loop stabilization is described in Chapter 2.57. However, according to DUDLEY and SCARPATETTI [4.079], when dealing with analog digital converters both the zero-line drift (i.e. reference voltage V^*) and the scale drift (i.e. the change in the overall gain and in the discharging current I_0 of the storage condenser) must be taken into account. Therefore the position of the channel "zero" ($=V^*$) must be stabilized using very small reference pulses, and at the same time also the gain and I_0 must be stabilized using reference pulses of a high amplitude corresponding to one of the upper channels. Stabilized analog-to-digital converters are described e.g. by CHASE [4.080] and GERE and MILLER [4.068].

The resolving time of an analog-to-digital converter can be improved by a pre-connected analog pulse memory. Due to the statistical distribution of the events to be analyzed, the time interval between two subsequent pulses is often much smaller than that which would correspond to the mean spacing (=reciprocal pulse rate). Of a pulse pair with time interval smaller than t_n (4.27), only the first is analyzed. The second pulse is suppressed, though the analyzer might thereafter wait "out of work" for a considerable time before the next event takes place. Even more extreme is the situation in experiments on pulsed particle accelerators. Fig. 4.43 shows a block diagram of an analog buffer memory serving as a derandomizer. The device consists of a number of pulse stretchers, which can be connected to the pulse input by means of the switches $S_{i1}, S_{i2}, \ldots$, and to the analog-to-digital converter ADC by means of the switches $S_{o1}, S_{o2}, \ldots$ A control logic controls the switches in the following manner: the input pulse charges one of the storage condensers which is just free (hence the input pulses start to be suppressed only after all memory positions are occupied). ADC starts the conversion of a further pulse stored in the memory immediately after the conversion of the preceding one has been terminated (hence no time losses occur through waiting for the next event: ADC is out of operation only if the whole memory is empty). This circuit was already realized in vacuum tube techniques [4.081] in 1955. Paradoxically analog buffer pulse memories were not used in transistorized circuits until recently, due obviously to the difficulties caused by the high cutoff currents of the transistors. ARQUE ALMARAZ [4.082] described a transistorized memory circuit using the principle shown in Fig. 4.43, in which the amplitude loss is less than one channel width for storage times up to 10 msec. Despite the analog buffer register, the overall energy resolution remained below 0.1 %! Other authors used cathode ray tubes for analog pulse height storage (cf. e.g. COSTRELL and BRUECKMANN [4.083]). Fast pulses can also be stored during a time of up to some hundred μsec in circulation memories using delay lines (e.g. RUMPHORST et al. [4.084]).

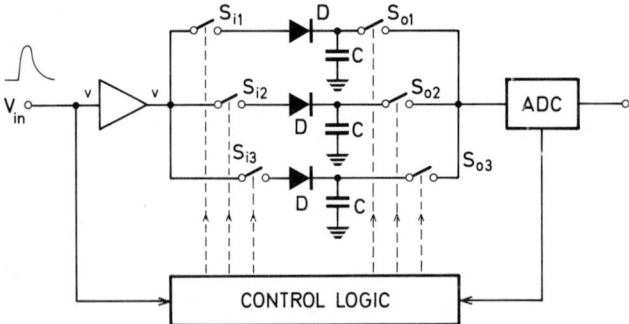

Fig. 4.43. Operating principle of an analog buffer register (derandomizer)

Test pulse generators [4.007, 4.063] are used to check the integral and the differential non-linearity of the analog-to-digital converter systems. These test pulse generators generate pulses with uniform amplitude distribution over a given pulse height range. The principle (Fig. 4.44) of such circuits is discussed in detail by DRAPER and ALSTON [4.085]. A saw-tooth voltage (B) of extreme linearity is sampled at statistically distributed instants (A), thus yielding pulses of variable amplitude (C). The probability for every pulse amplitude between zero and the maximum value is the same. Instead of a statistical pulse sequence (A) e. g. from an auxiliary radiation detector with suitable source, a periodic sequence from an oscillator can also be used. However, in the latter case the saw-tooth wave and the sampling pulses must not be synchronized. Using (C) as the input pulses of the ADC, the differential non-linearity can easily be estimated. Because of the uniform pulse height distribution, an ideal multichannel analyzer would store equal pulse numbers in each channel (besides the statistical fluctuations). Hence the relative deviations of the channel content immediately indicate the differential non-linearity.

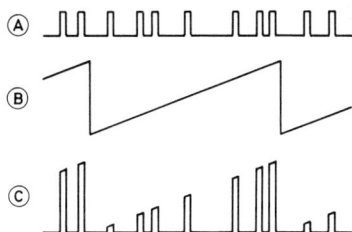

Fig. 4.44 A–C. Principle of a test pulse generator for testing the differential non-linearity of analog-to-digital converters

ADC linearity measurement techniques and related equipment (e. g. digitally controlled precision reference pulsers) have been discussed in papers [4.142] to [4.145].

4.2.2. Other Converter Systems

The conversion of a high pulse corresponding to one of the upper channels in a Wilkinson type converter is very time-consuming. This can be improved by using the principle of the voltage balance known from digital voltmeters. Fig. 4.45 shows the block diagram of an amplitude balance converter (also known as successive binary approximation type converter) operating as follows. The input pulse is stretched in the pulse stretcher PS and compared with a reference voltage V_{ref} in the comparator COMP. The digital output signal of the comparator (e.g. "0" for $V_{in} > V_{ref}$ and "1" for $V_{in} < V_{ref}$) controls a system of say five flip-flops FF 1 to 5 by means of a suitable logic CL. The reference voltage is given by the state of the flip-flops: the diode current gates switching binary graded currents into the operational amplifier $-A$ form a digital-to-analog converter. If all the flip-flops FF are set to "zero" (i.e. positive voltage at the outputs 1 to 5) the currents through R, $2R$, $4R$

Analog-to-Digital Converters

etc. flow into the flip-flop circuits and $V_{ref}=0$. If one or more flip-flops are set to "one" (negative voltage) the right-hand diodes of the corresponding gates are cut off and the currents flow to the input of the operational amplifier. Due to the binary gradation of the resistors, the reference voltage V_{ref}, which is proportional to the current sum, can amount to any multiple of $V_s/32$ in the interval $0 \ldots V_s$.

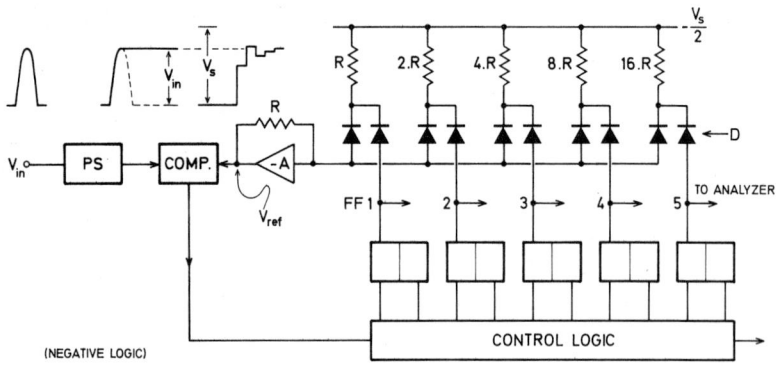

Fig. 4.45. Successive binary approximation type converter (amplitude balance)

The conversion of V_{in} into a digital code is performed by successive setting and — if necessary — resetting of FF 1, FF 2,..., FF 5. By this procedure V_{in} is compared at first with $\frac{1}{2}V_s$, then with $\frac{1}{4}V_s$ (if in the first step $V_{in}<\frac{1}{2}V_s$) or with $\frac{3}{4}V_s$ (if in the first step $V_{in}>\frac{1}{2}V_s$), then with $\frac{1}{8}V_s$ (if in the second step $V_{in}<\frac{1}{4}V_s$) or with $\frac{3}{8}V_s$ (if in the second step $V_{in}>\frac{1}{4}V_s$) or with $\frac{5}{8}V_s$ (if in the second step $V_{in}<\frac{3}{4}V_s$), etc. The control logic CL controls the decision tree in order to diminish the difference $V_{in}-V_{ref}$ by each step. At the end of the comparison $|V_{in}-V_{ref}| < V_s/32$. The corresponding value of V_{ref} can be picked up from the FF outputs in a binary code.

The conversion with n channels requires $\log_2 n$ steps. In spite of the fact that a definite time is necessary for performing each step, the reduction of the duration of the whole conversion process in relation to the WILKINSON type converter may be significant for very high channel numbers ($n > 1000$).

The disadvantage of the amplitude balance conversion techniques is that it is difficult to achieve the necessary differential linearity. The channel width is affected more than proportionally by errors in the small resistors of the digital-to-analog converter ($R, 2R$). If, for instance, R deviates from the nominal value by only -0.1%, and if all other resistors $2R, 4R, 8R$ and $16R$ are absolutely precise, all channels have the nominal width except for channel 16, which is wider by $+1.6\%$. The channel error is proportional to the channel number, so that even an amplitude balance with 256 channels ($=8$ comparison steps) having differential non-linearity better than 1% is obviously not within practical possibility.

Various measures have been proposed for improving the linearity. FRANZ and SCHULZ [4.086] combined the added-step technique acording to GATTI and coworkers ([4.023] to [4.025], cf. Chapter 4.1.3) with the binary amplitude balance. The reference voltage V_{ref} serves only for the definition of the channel position, and the channel width is given by the height of an auxiliary pulse which remains the same independent of the actual value of V_{ref} (Fig. 4.46). At first (with the aid of a circuit similar to that in Fig. 4.45) the binary coded reference voltage smaller than V_{in} is estimated, for which the difference $V_{in} - V_{ref}$ is minimum. Then an auxiliary pulse of the amplitude δV is added to V_{ref}. By means of a coincidence stage it can be decided whether or not $V_{ref} + \delta V > V_{in} > V_{ref}$. An analog-to-digital converter operating on this principle with 256 channels exhibited an integral and differential linearity of 1% and an amplitude-independent dead time of $t_n = 10\,\mu sec$. SCHUSTER [4.146] and HRISOHO [4.147] discussed some other aspects of improving the differential linearity of the successive step approximation type ADC.

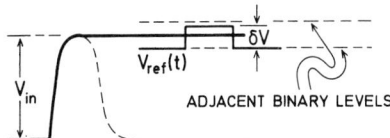

Fig. 4.46. Stabilizing the channel width of an amplitude balance type converter by the added-step technique (FRANZ and SCHULZ [4.086])

LENG and PATWARDHAN [4.087] proposed a decimal coded amplitude balance with only two decades. 9 equal resistors are used within one decade, the desired number (0 to 9) being switched in parallel when necessary. The only higher discontinuity then exists between the "10" decade (resistor R) and the "1" decade (resistors $10 \cdot R$). Since the most extreme ratio of the resistors is 10, the resistor inaccuracy influences the channel width error magnified only by about a factor of 10, and a differential linearity of 1% can be achieved using resistors with tolerances of 0.1%.

COTTINI, GATTI and SVELTO [4.088, 4.089, 4.148], developed a very interesting method of smoothing channel widths, which is of interest not only for the amplitude balance ADC and which shall be therefore discussed in some detail (Fig. 4.47). The technique is similar to the measurement of the length of a bar using a rule with well defined total length but with inaccurate divisions. Despite the false divisions, the bar length can be measured precisely if the bar is successively measured starting from the zero, first, second etc. graduation mark, and if the arithmetical mean of all measurements is taken (so called sliding scale principle). In Fig. 4.47 a voltage interval is divided into 20 channels with relative deviations from the nominal value of up to $\pm 50\%$. In each successive measurement the origin of the measuring range is shifted to the right by about one channel width, so that channel 5 (say) successively becomes channel 4, 3, 2, 1 etc. The deviation of the scale shift from the nominal value of one channel width is indicated at the right of the diagram. From Fig. 4.47 the reduction of the channel width fluctuation can easily be seen. The electronic realization of the sliding scale consists in completing e.g. a 256 channel converter of Fig. 4.45 with a 64 channel shift voltage generator. The shift voltage is analog added to V_{ref}, the address resulting from encoding V_{ref} is digitally corrected for the actual shift voltage value. After each input pulse the 64 channel generator shifts by one position, and after 64 pulses it is reset to the initial state. The influence of the smoothing on the pulse

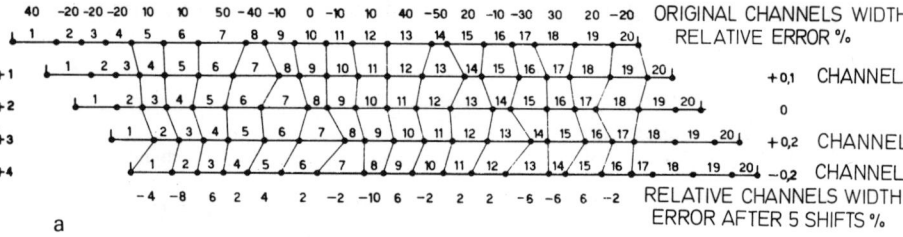

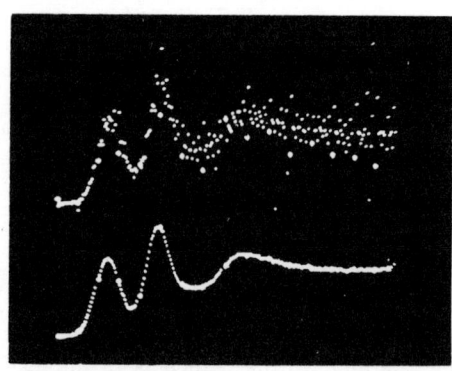

Fig. 4.47a and b. Channel width smooting by means of the sliding scale principle according to COTTINI et al. [4.089]. Left side, top: Spectrum display without smoothing; bottom: improved by smoothing

height spectrum shape is clearly demonstrated on the right side of Fig. 4.47. Using the sliding scale technique, ROBINSON et al. [4.149] constructed a 12 bit (4096 channel) ADC of amplitude balance type, having a dead time of $\leqslant 25\,\mu\text{sec}$.

Various other encoding techniques have been investigated. GÅSSTROM [4.090] reviewed the older papers (up to 1959). MACMAHON [4.091] modified the diode pump converter Fig. 4.39 in the following manner: Two diode pumps carry the charge from the storage condenser C_1. The coarse one with $C_2 = 10 \cdot C$ diminishes the charge in large steps, the fine one with $C_2 = C$ in steps ten times smaller. A discriminator switches over from the coarse charge reduction to the fine one as soon as the voltage across C_1 falls below the channel 10. The time resolution of the circuit is hence improved. However, it remains well over the dead time of modern fast WILKINSON type converters due to the large number of required analog switchings. COLOMBO and STANCHI [4.150] proposed a new ADC in which the signal pulse *area* is evaluated by means of an integrator, which compares it to the number of elementary charges corresponding to the total charge: this number is the digital equivalent. PIZER [4.092] triggers a Hartley oscillator linearly damped by means of a diode. The starting amplitude and thus also the total number of the oscillations (the amplitudes decrease linearly with the time) is proportional to the integrated amplitude of the input pulse. The oscillator frequency is about 10 Mcps. The converter is especially suitable for fast photomultiplier pulses. Its sensitivity is sufficient for direct coupling to the photomultiplier anode, obviating the use of an amplifier. ALBERIGI-QUARANTA and RIGHINI [4.093, 4.151], let the input pulse circulate in a circulation delay line memory with linear attenuation using a biased diode in the loop. The pulse amplitude is reduced by a fixed amount (= channel width) per cycle. Hence the number of cycles necessary for the reduction of the pulse amplitude to zero is proportional to the input pulse height. This number is counted in a digital scaler.

4.3. Pulse Shape Discriminators

If the detector signal depends not only on the energy loss ΔE but also on the ionization density of the particle trace, the pulse shape carries some information about the kind of releasing particle. Scintillation counters with organic or inorganic scintillators are very well suited for particle identification, as has already been pointed out in Chapter 2.5.2. The physical fundamentals of particle identification using scintillation counters are reviewed by OWEN [4.094]. Also semiconductor detectors ([4.095] to [4.097] and [4.103]) and proportional counters [4.098, 4.099] may yield pulses with particle-dependent shapes.

Specially prepared scintillators such as "phoswich" (phosphor sandwich) can be used for particle identification purposes ([4.100] to [4.102]). Two layers of scintillators exhibiting different fluorescence decay times are placed together. Heavy particles are absorbed in the slow front scintillator for example, thus producing slow pulses. On the other hand the electrons predominantly excite the fast scintillator layer behind the slow one, and thus produce fast pulses.

The difference in shape between the signal and noise pulses of a scintillation counter can be used for noise reduction. This special application of pulse shape discrimination techniques has been described in Chapter 2.5.4.

Analog-to-Digital Converters

In general, pulse shape discrimination consists in distinguishing between two or more types of signals with different characteristic times, e.g. $I_1(t)$ and $I_2(t)$ in Fig. 4.48. The distinction must be made independent of the actual integrated voltage amplitudes of the particular pulses. A very simple solution, which has been reported by BROOKS [4.104], consists in integrating the detector current by two integrators

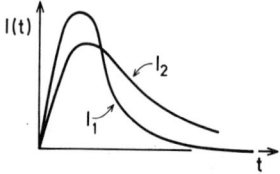

Fig. 4.48. Two different current pulses, suitable for pulse shape discrimination

with different time constants. An attached analog circuit forms the difference between the integrated signals. The realization of this principle with a scintillation counter is further simplified by the fact that the anode and one of the last dynodes can serve as two independent current sources (Fig. 4.49). The positive dynode current pulse is integrated by means of C_1 with a long time constant, the negative anode current pulse is integrated on C_2 with a short time constant. The diodes D_1 and D_2 merely serve for pulse stretching. With the aid of various adjusting resistors the circuit can be adjusted to produce a higher negative anode signal in the case of fast pulses. Hence no positive overshoot occurs for fast pulses (e.g. for γ-rays in an organic scintillator). On the contrary, a neutron pulse exhibits a slow component, which charges C_1,

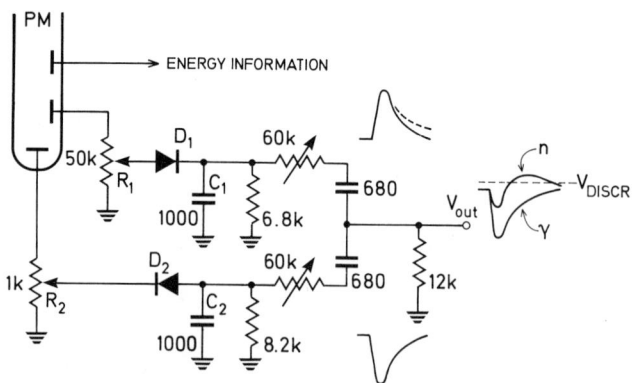

Fig. 4.49. Pulse shape discriminator according to BROOKS [4.104]

but not C_2. Thus for slow pulses the positive dynode signal predominates, yielding a positive overshoot. A simple voltage discriminator at the circuit output V_{out} can therefore discriminate between γ and neutron pulses. The energy information is obtained by integrating the current at the $(n-1)$th dynode.

This technique has been used by various authors [4.094]. SUHAMI and OPHIR [4.105] described a very simple method of realizing this discriminator circuit using the current mode modules according to ARBEL.

Another discrimination technique makes use of the fact that the integrated voltage pulses corresponding to $I_1(t)$ and $I_2(t)$ of Fig. 4.48 exhibit different rise times. Since the zero crossover point of a double differentiated pulse depends on the rise time of the original signal, after double differentiation the crossover instant of the pulses carries the information necessary for particle identification. Moreover, the shift of the crossover point does not depend on the actual pulse amplitude. Hence, for the purposes of pulse shape discrimination it is only necessary to estimate the zero crossover delay in regard to the pulse origin. Fig. 4.50 shows two possibilities of a corresponding circuit. In the leading edge discriminator LED the instant of the pulse origin is estimated, in the zero crossing discriminator ZCD the zero crossover point.

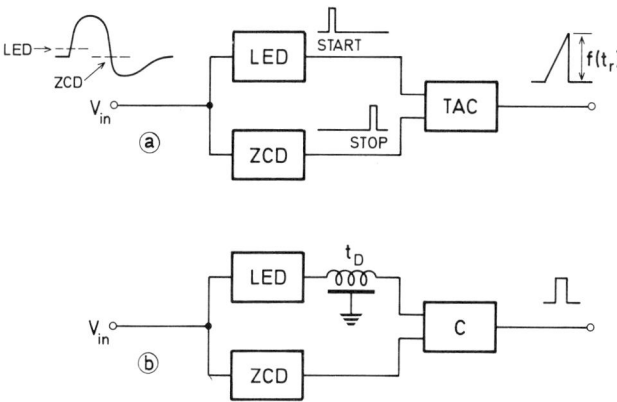

Fig. 4.50a and b. Pulse shape discrimination by comparison of the leading edge (LED) and the zero crossing (ZCD) timing

In the variant type a) the time delay between the two pulses is converted into a proportional voltage pulse in a time-to-pulse-height converter TAC. The resulting pulse amplitude is a function $f(t_r)$ of the rise time t_r of the original signal. A simple single channel discriminator can then select the events with a given t_r, which correspond to a given type of particle. In the circuit type b) the leading edge pulse, delayed by t_D, is checked in the coincidence stage C, as to whether or not it is coincident with the zero crossing pulse. The selection of the different events is performed by varying t_D.

The zero crossing technique offers a welcome opportunity of pulse shape discrimination using pulses which have already been shaped for other purposes, namely for pulse height analysis. Hence, the pulse shape discriminator can be added to already existing assemblies without any major modifications to the system. Recent papers (e.g. [4.036], [4.106] to [4.112], [4.152]) therefore show preference for this technique. Because of the non-zero threshold of the leading edge discriminator LED the output signal depends slightly on the pulse amplitude. BASS et al. [4.113, 4.114] avoid this difficulty by measuring the time distance between the leading edges of the positive and negative pulse parts respectively, instead of the time delay between the positive leading edge and the zero crossover. When the positive and negative thresholds are properly adjusted, this time difference is nearly amplitude-independent.

Both discriminators LED and ZCD can be realized by means of a single Schmitt trigger circuit, the hysteresis of which is made equal to the threshold (cf. Chapter 4.1.5, Fig. 4.27). The length of the output pulse of such a discriminator is a function of the rise time t_r. NADAV and KAUFMAN [4.115] used a tunnel diode trigger according to ORMAN (Fig. 4.36) for purposes of pulse shape discrimination. SCHWEIMER [4.116] described a similar circuit realized in a simple manner using only three transistor and one tunnel diode (Fig. 4.51). The discriminator is attached to a scintillation detector consisting of the liquid scintillator NE 213 and the phototube XP-1040. The current pulse from the 11th dynode is integrated by R_1 and C_1 and amplified by Q_1. The transistor Q_2 serves as an inverter and current amplifier, and C_2 together with the input resistance of the stage forms the necessary second differentiator. The tunnel diode TD is biased as in Fig. 4.36. Hence the duration of the output pulse corresponds to the length of the positive part of the bipolar pulse at the collector of Q_2. During the TD output pulse duration a constant current flows through $R = 470\,\Omega$ into $C = 1\,\text{nF}$, where it is integrated. The integrated pulse is further amplified by Q_3, and the amplitude of the Q_3 output pulse depends on t_r and hence upon the particle type, without being dependent on the particle energy.

From the V_{out}/E-diagram of Fig. 4.51 we see that a clear distinction between protons and electrons is possible down to about 500 keV particle energy. Below 500 keV the pulse height fluctuates to such an extent that an overlap between the proton and the electron region occurs. The lowest limit of distinction is given by the statistical fluctuations of the pulse shape which are highest for the smallest pulses consisting of only a few photocathode electrons ([4.117] to [4.119]).

In the above-mentioned discrimination techniques only a part of the available information about the pulse shape differences can be used. GATTI and DE MARTINI [4.118] reported a technique for which the information yield is an optimum. This technique will be described using a practical example.

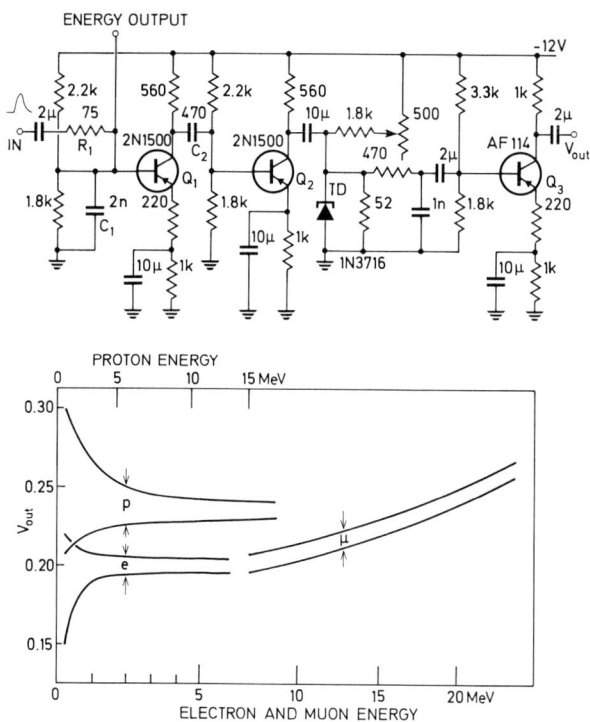

Fig. 4.51. Simple pulse shape discriminator by SCHWEIMER [4.116]

Fig. 4.52 shows the experimentally estimated shapes of current pulses of a scintillation counter with anthracene for α particles and γ rays. These curves will be simply denoted by $\alpha(t)$ and $\gamma(t)$. Both pulses are equally high, i.e.

$$\int_0^T \alpha(t)dt = \int_0^T \gamma(t)dt = N, \qquad (4.28)$$

Analog-to-Digital Converters

where N denotes the number of released photocathode electrons. T is an arbitrary time interval within which both $\alpha(t)$ and $\gamma(t)$ become practically zero. Different parts of the pulses are now attenuated in a time-dependent linear filter to different extents. Hence two signals S_α and S_γ result

$$S_\alpha = \int_0^T \alpha(t) P(t) dt \quad \text{and} \quad S_\gamma = \int_0^T \gamma(t) P(t) dt, \quad (4.29)$$

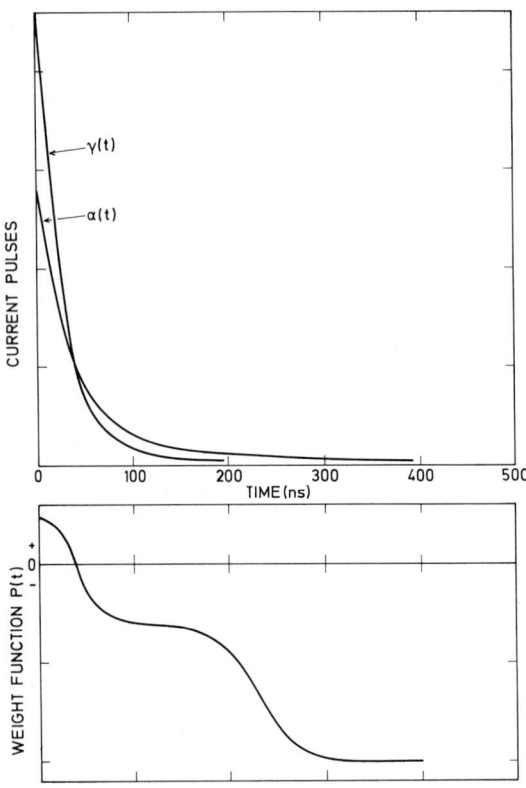

Fig. 4.52. Current pulse shapes for alpha and gamma rays in scintillation counter with anthracene crystal, and the weight function $P(t)$ according to GATTI and DE MARTINI [4.118]

where the weight function $P(t)$ represents the time-dependent attenuation factor of the linear filter. The optimum function $P(t)$ can be derived using a condition such that it yields the minimum relative mean-square deviation of the difference $S_\alpha - S_\gamma$. This condition gives

$$P(t) = \frac{\alpha(t) - \gamma(t)}{\alpha(t) + \gamma(t)}. \tag{4.30}$$

The relation (4.30) can easily be interpreted: those pulse parts where the relative difference is greatest receive the greatest weight $P(t)$. The function $P(t)$ is also plotted in Fig. 4.52 for the above example.

According to the rules of the Laplace transformation (Appendix 8.1), a network with the response function $\hat{Q}(p)$ produces an output signal $V_{out}(t)$

$$V_{out}(t) = \int_0^t I_{in}(t') Q(t - t') dt', \tag{4.31}$$

where $I_{in}(t)$ denotes the input signal and $Q(t)$ is the response of the network to $\delta(t)$ at the input; i.e. $Q(t) = \pounds^{-1}\{\hat{Q}(p)\}$. Obviously the integral in (4.31) becomes the one in (4.29) with $\alpha(t)$ or $\gamma(t)$ as the input current, if $Q(t) = P(T-t)$ is chosen and if the output signal is sampled at $t = T$. Fig. 4.53 shows the realization of a network having approximately $Q(t) = P(T-t)$ with the aid of two delay lines of the lengths $T_1/2$ and $T_2/2$, which are terminated by two small resistors $R_1 < Z_0$ and $R_2 < Z_0$ respectively (Z_0 denotes the characteristic line impedance).

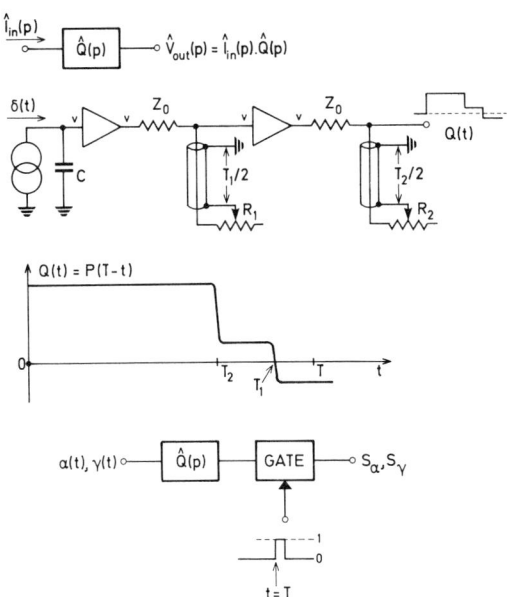

Fig. 4.53. Filter circuit related to the weight function $P(t)$ of Fig. 4.52

The network is self-explanatory. The output signal is sampled by means of a linear gate at time $t=T$. For the sake of simplicity the polarity of $Q(t)$ is reversed in the diagram.

Assuming a non-disturbed Poisson statistical distribution GATTI and DE MARTINI loc. cit. calculated a theoretical resolution limit of 13 keV. The limit achieved in experiments was 39 keV. The difference is mainly due to the fact that the fluctuations of the pulse amplitude of a scintillation counter do not obey the non-disturbed Poisson distribution.

A vast number of recent publications ([4.153] to [4.158]) dealt with the problem of pulse shape discrimination.

5. Evaluation of the Time Information

Analog information is supplied not only by the amplitude or shape of a pulse, but also by the time point of its appearance.

This information is either directly related to physical magnitudes having the dimension of time (as, for example, the decay time of an excited state of a nucleus), or to other magnitudes such as particle energy, velocity etc., by means of some intermediate process. We need only instance particle velocity determination by measuring flight time over a path of fixed length ("time-of-flight method"). As with pulse amplitudes, the time information too must be digitized prior to further processing. This chapter is devoted to the discussion of the related circuits.

The physical fundamentals of time analysis in nuclear metrology are reviewed in various text books and monographs. We need only to refer to the papers by BELL [5.001], DE BENEDETTI and FINDLEY [5.002] and SCHWARZSCHILD [5.021].

5.1. General Considerations, Resolution

Two separate problems must be distinguished in processing time information:

1. the distribution of the time intervals between two defined events should be measured, or

2. certain specified event pairs or event groups should be selected, which are correlated in time.

In general the first problem is solved by digitizing the time intervals in an appropriate converter and subsequently storing the digital information in a memory. The second problem is solved by means of various coincidence circuits.

The determination of the mean decay time of an excited nuclear state populated by a β-decay (Fig. 5.01) may serve as a typical example of the first type of problem. The β particle and the γ quantum are detected with the aid of two separate detectors. The β pulse signals the birth ("START") of the state, the γ pulse its death ("STOP"). By meas-

Evaluation of the Time Information

uring the distribution of the delay of stop signals with regard to the related start signals, the mean decay time τ_γ of the state is estimated.

The same figure can serve to illustrate the selection of events correlated in time (i.e. coincident), which is used, for example, in the determination of the source strength. N_0 denotes the decay rate of the β level, i.e. source strength of the corresponding probe of the β active isotope.

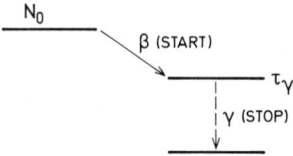

Fig. 5.01. A typical problem of time analysis

The count rates N_β and N_γ of the β and γ pulses, respectively, are given by the counting efficiencies ε_β and ε_γ of the detectors

$$N_\beta = N_0 \cdot \varepsilon_\beta, \quad N_\gamma = N_0 \cdot \varepsilon_\gamma. \tag{5.01}$$

The count rate N_c of the coincident $\beta-\gamma$ pulses is given by

$$N_c = N_0 \cdot \varepsilon_\beta \cdot \varepsilon_\gamma. \tag{5.02}$$

Since neither ε_β nor ε_γ is known accurately, it is not possible to determine N_0 from (5.01) by measuring N_β or N_γ. However, it suffices to measure N_β, N_γ and N_c simultaneously, since

$$N_0 = \frac{N_\beta \cdot N_\gamma}{N_c} \tag{5.03}$$

gives the source strength.

A coincidence circuit recognizes two pulses as "coincident" if the time interval between them is smaller than a certain given value τ_c. The time interval τ_c — or $2 \cdot \tau_c$, since one of the pulses may precede or follow the other — is called the resolution of the circuit. Due to the finite resolution $2\tau_c$, not only the true coincidences N_c, but also chance coincidences N_{ch} (caused by two pulses belonging to two different decay events occurring within $2 \cdot \tau_c$) are recorded. For weakly correlated count rates N_β and N_γ (i.e. for $\varepsilon_\beta \ll 1$ and $\varepsilon_\gamma \ll 1$) we get

$$N_{ch} = 2 \cdot N_\beta \cdot N_\gamma \cdot \tau_c. \tag{5.04}$$

Of course, N_{ch} should be small compared with N_c. The ratio

$$\frac{N_{ch}}{N_c} = 2 \cdot N_0 \cdot \tau_c \tag{5.05}$$

is proportional to τ_c. In order to be able to work with high decay rates N_0 — necessary for high statistical accuracy — very small resolution times τ_c are desirable.

On the other hand, if all time-correlated events are to be recorded without loss, τ_c must not be less than a certain lower limit. Therefore (5.02) is valid only if $\tau_c \gg \tau_\gamma$, i.e. if the emission of the β particle and of the γ quantum occurs "coincidently" when compared with τ_c. Besides this fundamental physical limitation, the imperfection of the detector system must also be taken into account: the detector signal is delayed with regard to the releasing event by statistically fluctuating time intervals, such as the collection times of electrons and ions or holes, the fluorescence decay times of the scintillators, the photomultiplier electron propagation times etc. The extent to which these fluctuations influence the timing accuracy depends upon the techniques used for the shaping of the detector signal. Another reason for the timing inaccuracy, the amplitude-dependent delay of the signal ("walking"), has already been mentioned in Chapter 4.1.5.

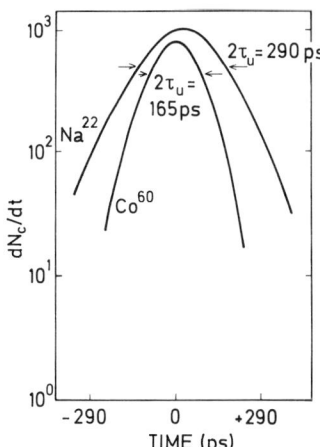

Fig. 5.02. Time interval spectra (resolution curves) of a fast coincidence assembly according to [5.003]

The term "resolution" is used for denoting two fundamentally different magnitudes. Besides the electronic resolution time $2\tau_c$ introduced above, the resolution often denotes also the width $2\tau_u$ of the distribution of the time intervals between two pulses corresponding to two absolutely coincident primary events, which is due to the imperfections of the electronic apparatus. As an illustration, Fig. 5.02 shows the time interval spectra dN_c/dt of a very fast assembly using two phototubes RCA C 70045 with plastic scintillators for the $\gamma-\gamma$ cascade of Co^{60} and for the annihilation radiation of Na^{22}.

The curves of Fig. 5.02 can be interpreted as follows. In coincidence measurements where all true coincidences should be registered, $2\tau_c \gg 2\tau_u$ must be chosen. In the absence of systematic delays the integration of dN_c/dt within the limits $-\tau_c \ldots +\tau_c$ yields the percentage of the registered coincidence events. On the other hand, if the spectrum dN_c/dt is to be measured point by point using a simple coincidence circuit and a delay line of variable length, $2\tau_c \ll 2\tau_u$ must be chosen. The greater of the two values τ_c and τ_u finally determines the accuracy with which a time interval distribution of the *primary events* can be measured in a coincidence assembly. Even if $\tau_c \approx 0$ could be realized, the mean decay time τ_γ of a nuclear state must be $\tau_\gamma \gtrsim \tau_u$ for a good accuracy of measurement. Hence the value of τ_u determines the ultimate "resolution" of the assembly in time measurement experiments.

There is an interdependence between the pulse height measurement and the timing. Mostly the time correlation between pulses of specified amplitudes should be estimated, so that the measurement of the pulse height must precede the time interval measurement. With the aid of the zero crossing techniques (Chapter 4.1.5), amplitude-digital pulses can be formed with simultaneous conservation of the time information to a few nsec. The output pulses of zero crossover discriminators may be processed in coincidence circuits having $2\tau_c \geq 10$ nsec. If shorter resolution is desired, separate processing of the time and amplitude information according to the "fast-slow" coincidence techniques proposed by BELL and PETCH [5.004] offers advantages (Fig. 5.03). The detector signals are led through two fast acting pulse shaping stages PS into the fast coincidence circuit with resolution τ_{cf}. Simultaneously the "slow" integrated signals (often derived from another detector output) are led through linear amplifiers to the amplitude discriminators 1 and 2. The slow triple coincidence stage with resolution τ_{cs} yields an output pulse if, and only if, the time and amplitude criteria are fulfilled at the same time. The resolution τ_{cs} must be large enough in order to compensate for any amplitude-dependent delay of the discriminator pulses. Systematic delays in the two slow channels must be compensated by a corresponding fixed delay in the fast coincidence channel. Of course, the separation of the time and pulse height information processing according to the fast-slow techniques can also be used when the time interval distribution of the detector signals is measured (Chapter 5.4.1).

An example of a recent fast-slow coincidence system has been given by IACI and LO SARIO [5.107].

Since no precautions for the amplitude conservation must be taken in the fast channels, they can be designed for optimum time-information conservation. The resolution τ_{cf} is limited only by the properties of the detector and fast pulse shaper systems (τ_u). If the delay of the pulse shaper signals depends on the signal amplitude, the actual resolution τ_u in a fast-slow assembly still depends on the setting of discriminator levels in the slow channels. When single channel discriminators with narrow windows are used, the effect of pulse "walking" is eliminated due to selection of pulses of almost always the same amplitude, and τ_u reaches its lower limit given solely by the detector properties. On the contrary, if working with integral discriminators, very wide height ranges are processed, and the amplitude-dependent delay fluctuation may dominate the resolution τ_u.

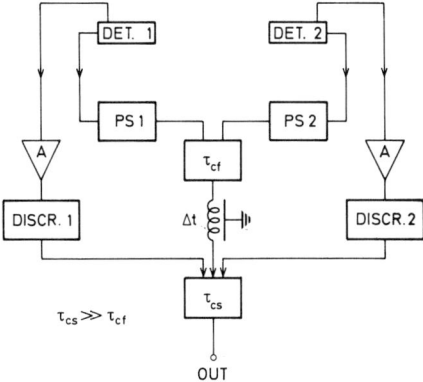

Fig. 5.03. Principle of a fast-slow assembly

5.2. Pulse Shapers for Coincidence Circuits and Time-to-Digital Converters

Most of the coincidence circuits and time converters operate with input pulses of standard shape and standard height, which must be derived from the detector signals by means of some non-linear operation conserving as much of the time information as possible. For slow and medium-fast coincidence circuits, the digital output pulses of the amplitude discriminators with or without zero crossover timing can be used, if necessary after an additional shaping to standard pulse length in a monostable multivibrator. For fast coincidence circuits with very short resolution time, special pulse shaping stages are necessary, which are connected to the detector output either directly or via a special very fast amplifier.

217

Evaluation of the Time Information

Although in principle any point of the detector pulse may serve for timing purposes, the shaping circuits can be divided into two categories. The circuits of the first category respond to the leading edge of the pulse, those of the second category respond to the zero crossover point of a suitably shaped detector signal.

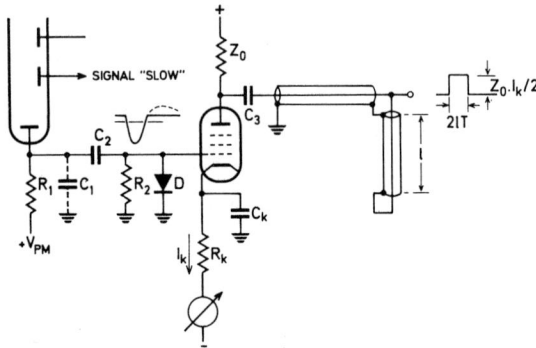

Fig. 5.04. Pentode limiter according to BELL, GRAHAM and PETCH [5.005]

The tube limiter according to BELL, GRAHAM and PETCH [5.005] is a classical example for the leading edge timing circuit. Though in existence for more than 15 years, it can still be found in some recent fast-slow assemblies. The operating principle can easily be understood from Fig. 5.04. In the quiescent state the penthode is biased to conduct a cathode current I_k. The load of the tube consists of the characteristic impedance Z_0 of a coaxial cable ($\sim 100\,\Omega$). Then pentode is cut off by the incoming anode pulse of the photomultiplier, yielding a voltage step of amplitude $I_k \cdot Z_0/2$ in the anode circuit, which is differentiated to the length $2l/T$ by means of a shorted delay line with the length l. By means of a careful layout of the circuit with fast tubes (e.g. E810F), rise times lower than 1 nsec can be realized. Depending upon the time constants $R_1 C_1$ (C_1 = parasitic anode capacity $\approx 20\,\text{pF}$) and $R_2 C_2$ (C_2 is a high voltage coupling capacitor), the anode current pulse is more or less integrated. If $R_2 C_2 \gg R_1 C_1$ cannot be chosen, the resulting undershoot must be suppressed by means of a diode clamp D. Of course, $R_1 C_1$ must be large in relation to the fluorescence decay time τ_{fl} of the scintillator used, or else the particular single electron pulses may become resolved and may produce multiple pulses in the anode circuit of the limiter. The photomultiplier gain is commonly adjusted for cutting of the tube by single electron pulses. Hence the first electron of a pulse actuates the limiter with a corresponding improvement in the system resolution τ_u.

Such single-electron-sensitive limiters of course respond also to the photomultiplier noise pulses, which occasionally may cause some difficulties. If only high signal pulses need be processed, a simple discrimination against (low) noise pulses can be performed by introducing a fast biased diode before the pentode control grid (ASPELUND [5.006]).

The negative pulses at the photomultiplier anode can be up to several tens of volts high. Hence special precautions are necessary in order not to exceed the maximum ratings in transistorized limiters. Fig. 5.05 shows a fast limiter according to SUGARMAN et al. [5.007] which operates on the photomultiplier current signal. The transistor 2N700 is used in the common base configuration and so has a high frequency limit f_α. The standing current I_C is adjusted with the aid of the variable resistor 250Ω. Negative pulses cut off the transistor, yielding a voltage pulse with amplitude $I_C Z_0/2$ across the terminating resistor Z_0. The pulse length is given by the length of the clipping cable. The diode 570-G protects the transistor against excessively high PM signals. Somewhat better rise times can be achieved ($\lesssim$1 nsec) as in tube limiters. Transistorized limiters of the type shown in Fig. 5.05 are now commonly used (cf. [5.008]). An improvement of the circuit, namely a better decoupling of the photomultiplier from the coincidence circuit, is performed by introducing an additional common base amplifier stage between the collector of the limiter transistor and the clipping cable [5.007].

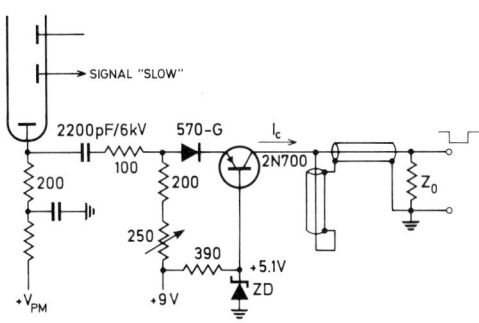

Fig. 5.05. Transistorized limiter with diode protection according to SUGARMAN et al. [5.007]

The SUGARMAN limiter cannot suppress small noise pulses. This feature can be achieved by connecting a biased emitter-follower before the common base limiter. Such a circuit (Fig. 5.06) is described by BARNA et al. [5.009]. The quiescent current I_C of Q_2 flows into the clipping

Evaluation of the Time Information

cable and Z_0. Transistor Q_1 is cut off by a small bias. A negative input pulse opens Q_1 and switches over the current I_C from Q_2 to Q_1, thus yielding an output standard pulse. Diode D_1 limits the positive voltage excursion of the input, diode D_2 together with the base-emitter junction of Q_1 limits the negative one. The output pulse rise time is about 1 nsec.

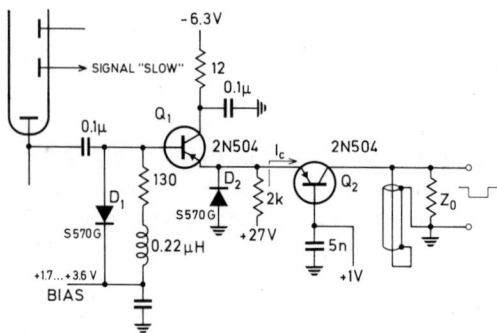

Fig. 5.06. Limiter circuit according to BARNA et al. [5.009]

Of course a transistorized limiter circuit employing the common emitter configuration (similar to the vacuum tube circuit of Fig. 5.04) can be constructed. However, due to the slower properties of the common emitter circuits in regard to the common base configuration, these limiters have rise times of a few nsec (SIDI and SOLD [5.010]).

The tunnel diode enabled trigger type pulse shapers to be used instead of limiters. The tunnel diode circuits have been discussed in detail in Chapter 4.1.6. Their high switching speed and the dependence of the output pulse delay on the input pulse height have been pointed out. The delay becomes especially high if the triggering pulse just reaches the trigger threshold. However, when by means of a fast-slow assembly only pulses of a given amplitude range — much higher than the threshold of the pulse shapers — are selected, TD multivibrators are equally well suited for coincidence purposes as fast limiters.

Fig. 5.07 shows a TD pulse shaper according to WHETSTONE and KOUNOSU [5.011]. The photomultiplier anode circuit is decoupled from the actual discriminator TD_1 by means of a transformer 1:3 (for decoupling purposes other authors often use a transistor stage in a common base configuration, as in Fig. 4.34). The sensitivity can be adjusted with the aid of the TD biasing current, and the maximum sensitivity comes to 50 mV from a 50 Ω cable. Except for the first 5% above the threshold, the delay of the multivibrator output pulse changes only by about 1 nsec if the input amplitude changes by a factor of 10. With $L \approx 0.5\,\mu H$ the pulse width is 4 nsec. Since the actual pulse width still depends upon the input pulse height, the first multivibrator TD_1 triggers another monostable multivibrator TD_2, which then delivers a pulse, standard in amplitude and length, for the attached coincidence circuit. Other circuits are described by VAN ZURK [5.108, 5.117] and MURN [5.109].

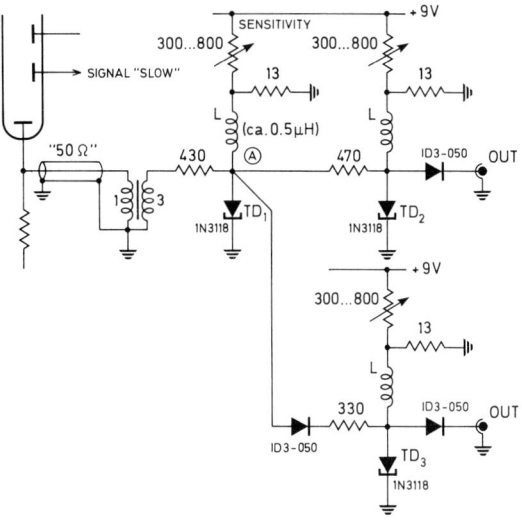

Fig. 5.07. Tunnel diode pulse shaper according to WHETSTONE and KOUNOSU [5.011]

In circuits like Fig. 5.07 the problem of a desirable signal splitting can easily be solved. Using decoupling diodes or resistors, more than one secondary multivibrator (say TD_2, TD_3...) can be connected to the output (A) of the discriminator TD_1, to deliver the desired number of output pulses to independent channels. Many coincidence circuits can therefore be supplied by pulses from one photomultiplier. High fan-out is often necessary in complex experiments (cf. the note at the end of this chapter).

However, the tunnel diode discriminators are mostly used for the zero crossing timing. BJERKE et al. [5.012] and WIEGAND [5.013] have described a circuit producing the zero crossing signal by triggering an undercritically damped 80 Mcps LC resonator (transformer + parasitic capacities), which is mounted directly on the photomultiplier socket (Fig. 5.08). The bipolar pulse is led via a resistor $27\,\Omega$ to a sensitive TD monostable multivibrator responding to the pulse undershoot. The

Evaluation of the Time Information

multivibrator output pulse is 3 nsec wide, its rise time is 150 psec. The pulse is differentiated by the 4 pF condensor and controls thereafter the transistor 2N700. In a circuit of the type shown in Fig. 5.08, BJERKE et al. [5.012] achieved a walking of less than 0.5 nsec over an input amplitude range of 1:20.

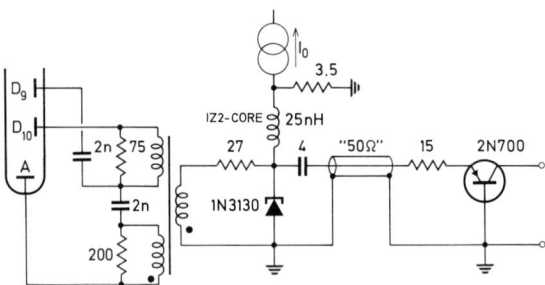

Fig. 5.08. Zero crossing pulse shaper according to WIEGAND [5.013]

Cable clipping offers another possibility of forming bipolar pulses [5.012]. ORMAN [5.014] uses the circuit of Fig. 5.09 for clipping purposes. The photomultiplier current pulse is fed through a common base transistor stage to a shorted cable, and the superposition of the original and reflected pulses is picked up with the aid of another common base stage. Of course, the cable length must correspond to the original photomultiplier current pulse length. The use of the base-emitter junction of transistor Q_3 as a "short circuit" to terminate the clipping cable offers interesting possibilities: at the collector of Q_3 the original signal is available for other purposes such as integration etc. For the detection of zero crossing point, ORMAN uses the trigger described in Chapter 4.1.6

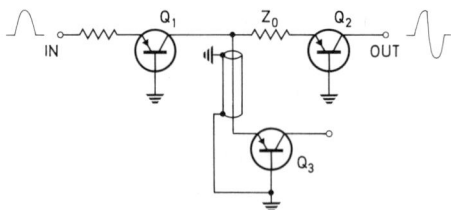

Fig. 5.09. Fast bipolar pulse shaping by means of delay cable differentiation according to ORMAN [5.014]

(Fig. 4.36), with hysteresis equal to the sensitivity, hence the crossover point is marked by the trailing edge of the output pulse. The walking of the output pulse is smaller than 1 nsec when varying the input amplitude between 8 and 60 mA. Another ORMAN type zero-crossing TD discriminator is described by BERNAOLA et al. [5.110]. ABBATISTA et al. [5.111] analyzed the timing purposes of TD monostable circuit and compared it to the ORMAN circuit. Other TD pulse shapers for double differentiated detector signals are described in [5.118] to [5.122].

GARVEY [5.015] at first limits the bipolar cable differentiated pulse in two fast common base stages and differentiates it thereafter with a time constant of about 350 psec. The resulting needle pulse corresponding to zero crossover triggers a conventional TD monostable multivibrator. By proper adjustment of the limiters the amplitude-dependent walking can almost be eliminated. As can be seen from Fig. 5.10, for amplitudes higher than 8 mA the delay change is not measurable.

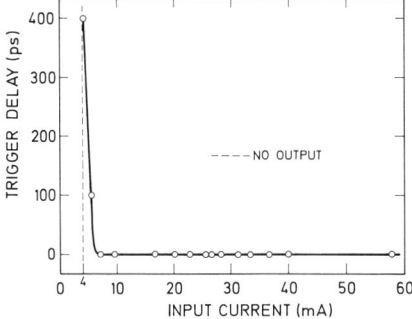

Fig. 5.10. Relative trigger delay dependence on the input amplitude according to GARVEY [5.015]

Since most of the fast coincidence systems work with scintillation counters, all examples described above relate to these detectors. Anyway, all the pulse shapers mentioned, and especially the TD trigger circuits, operate also with semiconductor detectors or proportional counters. Of course, the relatively small output signals must first be amplified in fast amplifiers, which degrade the time resolution.

Special pulse shapers for semiconductor detectors are described in [5.112] to [5.114]. When used with large Ge(Li) detectors, the time jitter due to slow and variable pulse rise times must be compensated by appropriate means (cf. also Chapter 4.2).

In addition to high current gain, the photomultiplier tubes also offer the additional advantage of two or more independent signal outputs. Hence, if e. g. a fast current pulse is picked up from the anode, a slow integrated voltage pulse can be delivered by one of the last dynodes. This simplifies the separation of the fast and slow signals. When using semiconductor detectors special artifices must be applied to separate the fast and slow signals. Two self-explanatory circuits reported by SCHEER [5.016] and WILLIAMS and BIGGERSTAFF [5.017] are shown in Fig. 5.11 (compare also the possibility of integration of the fast signal at Q_3 in Fig. 5.09).

Evaluation of the Time Information

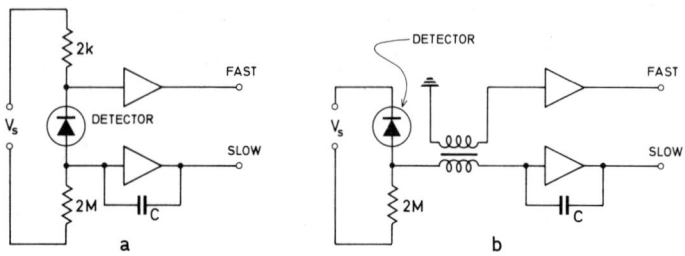

Fig. 5.11a and b. Separating fast and slow signals in semiconductor detectors according to SCHEER [5.016] (circuit a) and WILLIAMS and BIGGERSTAFF [5.017] (circuit b)

Under otherwise equal conditions, the zero crossover timing exhibits a more inferior statistical accuracy than does the leading edge timing. A rough estimate can be derived from Fig. 5.12. The curve (a) corresponds to the integrated current pulse of a photomultiplier, its rise time being given by the fluorescence decay time τ_{fl} and by the electron propagation time fluctuations in the photomultiplier (cf. Chapter 2.5.2). For the sake of simplicity, the bipolar pulse (b) is assumed to be formed by double delay line clipping of (a).

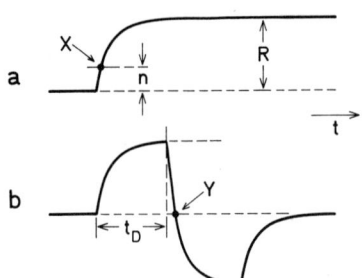

Fig. 5.12 a and b. Pulse shapes in leading edge (a) and zero crossing (b) timing

The amplitude of (a) corresponds to R photocathode electrons. A leading edge trigger with a threshold at n electrons fires at the point X, and the zero crossing trigger fires at the point Y. Obviously Y lies in the middle of a double-size copy of the leading edge, which has been delayed by a fixed amount of time t_D and inverted. Hence Y corresponds to a triggering threshold of $n = \dfrac{R}{2}$. The statistical fluctuation of the time interval, within which n out of the total R electrons appear, can be calculated. According to BELL [5.001, 5.018], the time resolution $2\tau_u$ of a symetrical pair of scintillation counters comes to

$$2\tau_u = 3.33[\sigma_p^2/n + n(\tau_{fl}/R)^2]^{\frac{1}{2}}, \qquad (5.06)$$

if n first electrons are used for timing purposes. σ_p denotes the mean square deviation of the electron propagation time in the photomultiplier (i.e. $\sigma_p = t_p/\sqrt{2}$ with t_p from (2.46)). The relation (5.06) holds good only for $n \ll R$. The resolution $2\tau_u$ exhibits an optimum for $n/R = \sigma_p/\tau_{fl}$. Because $\sigma_p/\tau_{fl} \ll 1$ is true for fast photomultiplier tubes, the leading edge timing $n \ll R$ must be preferred to zero crossing timing $n = R/2$. The later case is especially disadvantageous when slow inorganic scintillators are used. This fact is clearly demonstrated by the following table showing calculated resolutions $2\tau_u$ for inorganic and organic scintillators [5.018]:

Scintillator	τ_{fl} (nsec)	σ_p (nsec)	R (for 1 MeV)	$2\tau_u$ (nsec) "zero crossing" $n/R=0.5$	$2\tau_u$ (nsec) "leading edge" $n/R=\sigma_p/\tau_{fl}$	Ratio "zero crossing" to "leading edge"
NaI(Tl)	250	1	6000	7.6	0.55	13.7 (!)
Plastic	4	1	600	0.43	0.22	1.95

Hence, zero crossover timing is not suitable if high resolution coincidence assemblies are to be realized with slow inorganic scintillators. GATTI et al. [5.019] described a circuit altering the shape of a bipolar pulse in such a way that the zero crossing occurs not at $n/R = 0.5$, but at $n/R \ll 1$ (Fig. 5.13). The voltage pulse is converted into a current pulse by R_1 and — in addition — is differentiated by the capacitor C.

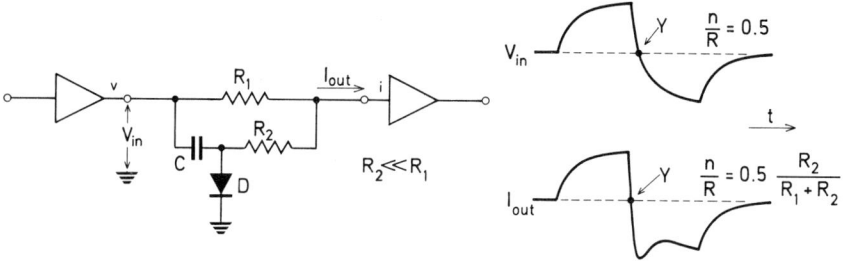

Fig. 5.13. Improving the bipolar pulse shape according to GATTI et al. [5.019]

Diode D cuts away the positive peaks, though the negative current peaks are led through $R_2 \ll R_1$ to the input of the current amplifier (or current sensitive trigger). Hence the zero crossing edge is made faster.

Evaluation of the Time Information

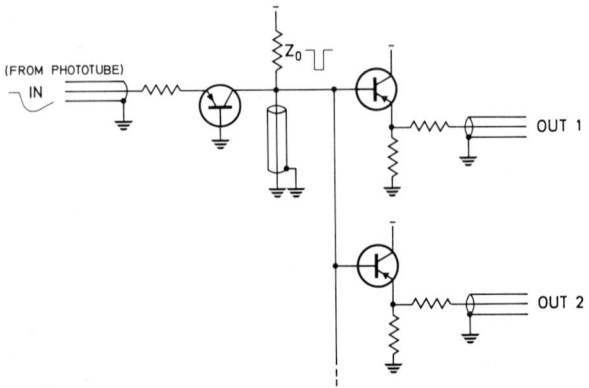

Fig. 5.14. Operating principle of a fast fan-out circuit

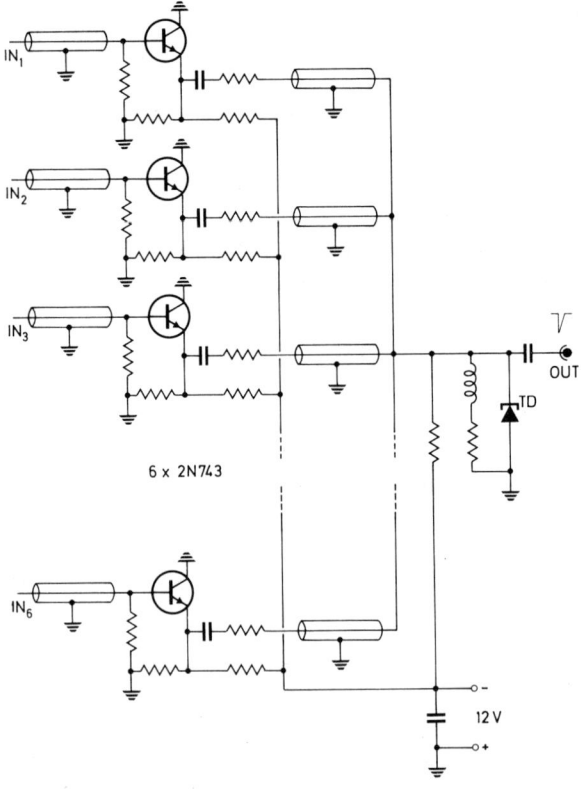

Fig. 5.15. Example of a fast signal mixer circuit (BENOIT et al. [5.022])

If $R_2 C$ is assumed to be smaller than the pulse length, but greater than the original rise time, the equivalent threshold of the point Y can easily by calculated to

$$\frac{n}{R} = \frac{1}{2} \cdot \frac{R_2}{R_1 + R_2}. \tag{5.07}$$

Due to the finite switching speed of the diode D the improvement of the resolution $2\tau_u$ is slightly lower than that corresponding to (5.07).

Various authors have treated the problem of calculating $2\tau_u$ for different pulse shapes and different triggering thresholds n/R. The general problem of estimating an optimum filter (or of the corresponding pulse shape) for the resolution $2\tau_u$ with given threshold n/R has been solved by GATTI and SVELTO [5.020].

Some other pulse shaper circuits, partly with "snap-off" diodes or avalanche transistors, are discussed in [5.123] to [5.127]. STANCHI [5.128] proposed the use of so-called bitripoles in pulse shapers: bitripoles are TD diodes driven into high frequency oscillations and exhibiting two distinct regions of negative resistance. ROTA et al. [5.129] improve the time resolution by double sampling the pulse: two time references are obtained from two different timing systems and the pulse is accepted for time analysis only if both time references are equal within predetermined limits.

In complex experimental assemblies often the pulse from one detector must feed several coincidence circuits, in general with different resolution times τ_c. On the other hand, one coincidence stage must often accept signals from several independent detectors. In order to avoid any mutual influence between the particular detectors or circuits, the particular signal paths must be carefully separated by means of signal distribution circuits (fan-out circuits) or signal mixers. For fan-out circuits mostly simple emitter-followers are used, fed by the output signal of a limiter or trigger type pulse shaper. Such a circuit is shown in Fig. 5.14. The timing accuracy depends on the types of transistors used. Another possibility of signal distribution consists in using the photomultiplier signal for cutting off several limiters in parallel. The current steps in each of the independent collector circuits can be shaped by separate clipping cables to pulses of desired lengths, which vary according to the outputs.

Strictly speaking, the signal mixers are fast OR gates. Commonly an additive mixing of outputs of different limiters is used with a subsequent pulse shaping in an additional shaper, preferably of the trigger type. The circuit of Fig. 5.15 is described by BENOIT et al. [5.022]. A TD monostable multivibrator serves for standardizing the output pulse shape. Other examples of fast fan-out and mixer circuits are reviewed in [5.023], p. 423.

5.3. Coincidence Circuits

5.3.1. Ideal Coincidence Stage

An ideal infinitely fast AND gate for pulses with standard (i.e. digitized) amplitudes represents the ideal coincidence circuit (Fig. 5.16). As can easily be seen, the resolution time is $2\tau_c = \delta_1 + \delta_2$, where δ_1 and δ_2 denote the respective pulse lengths of the input pulses. Almost always

Evaluation of the Time Information

the two pulse shapes are equal, hence $\delta_1 = \delta_2 = \delta$ and $2\tau_c = 2\delta$. Of course, AND gates with more than two inputs can also be used as (multiple) coincidence stages. The described idealized situation is approximated if the rise times of the input pulses and the time constants (response times) of the gate can be neglected in comparison to δ. Such conditions are necessary, especially when a high long-term stability of the resolution $2\tau_c$ is desired, e.g. for exact calculation of the chance coincidence pulse rate (5.04). All types of gates in a conventional layout or in integrated circuit techniques can be used as coincidence stages (e.g. RTL, DTL, DCTL etc.). Hence all of the circuit examples described later in this chapter exhibit similarities to basic gate circuits of one of the circuit logics, reviewed in Chapter 6.

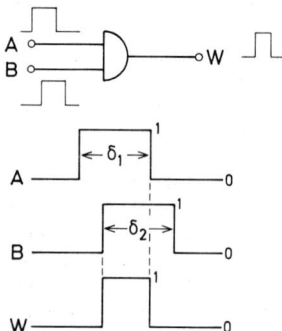

Fig. 5.16. An AND gate acting as ideal coincidence stage

The reason for treating coincidence stages separately and not as a part of digital circuits in Chapter 6 is the following: in a digital system the time too loses its analog meaning — two digital pulses occur either simultaneously, in the same clock interval, or not. Here, however, we are dealing with pulses, the time interval between which does exhibit an analog character. The coincidence circuit with the resolution $2\tau_c$ first has to digitize the analog time information. Hence two identical circuits may serve for two different purposes and must consequently be treated under different aspects.

Regarded in detail, the AND gate in Fig. 5.16 can be considered to consist of a pulse mixer with a subsequent amplitude discriminator (Fig. 5.17). The mixer either adds the input pulses (a) linearly or performs on them some non-linear operation (b). In a linear mixer the output during the overlap time interval is twice as high as for a single pulse. Thus the threshold of the discriminator must be higher than the single pulse amplitude, but smaller than twice the pulse amplitude. The linear mixer can be realized using any circuit allowing a linear or a quasi-linear addition, especially passive resistor networks, coaxial cables, addition of currents at low impedance terminals, etc. In a non-linear mixer (b) the coincidence signal is (much) higher than twice the single input pulse amplitude. Hence the range of usable discriminator threshold is correspondingly greater. Since only components with switching characteristics, such as diodes, transistors etc., can be used for performing non-linear operations, the non-linear mixers are mostly somewhat slower than the linear ones, which consists of passive components only.

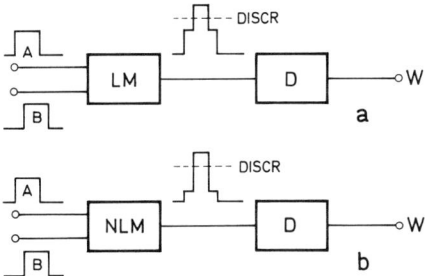

Fig. 5.17a and b. Coincidence stages with linear (a) and non-linear (b) pulse mixer

In the case of rectangular input pulses with negligible rise times and of an infinitely fast discriminator, the resolution time $2\tau_c$ of a coincidence stage shown in Fig. 5.17 remains as 2δ, independent of the actual discriminator threshold. However, if the rise time cannot be neglected, a different situation arises, as instanced in Fig. 5.18, for the case of a linear mixer and two input pulses with a trapezoidal shape. Now the voltage V_M by which the mixer output signal exceeds the single pulse level is a function of the mutual pulse distance t_D, and the resolution $2\tau_c$ becomes a function of the applied discriminator threshold.

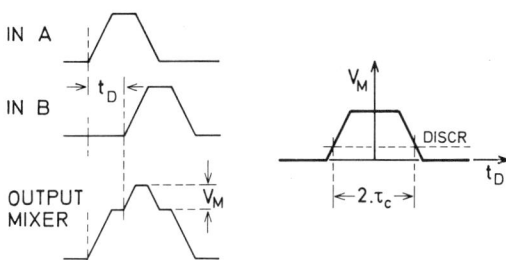

Fig. 5.18. Two trapezoidal pulses at the inputs of a linear mixer

Since the curve $V_M(t_D)$ is directly correlated to the original pulse shape, the shape of the input pulses can be estimated [5.024, 5.025], from a measurement of the mixer output pulse height as a function of applied delay t_D. This technique can occasionally be used for measurement of subnanosecond pulses if no sampling oscilloscope is available.

Obviously trapezoidal or triangular shapes of $V_M(t_D)$ result also for nonlinear mixers. Slow mixers (i.e. mixers with long integration time constants with respect to the input pulse length) deliver triangular characteristics $V_M(t_D)$, too, even if the input pulses are strongly rectangular. High discriminator thresholds (e.g. V_{int} in Fig. 5.19) yield very short resolving times $2\tau_c$, which are occasionally much shorter than the actual pulse length. Still shorter times τ_c can be achieved with the aid of differential discriminators, with the channel dV_s set somewhere at the edge of $V_M(t_D)$ curve (case (B) in Fig. 5.19). In this case the resolution $2\tau_c$ is given solely by the slope of the mixer characteristic and by the channel width dV_s, and can reach a few picoseconds. Of course, the delay t_D^* corresponding to the channel position must be compensated by a delay line of fixed value in one of the signal paths.

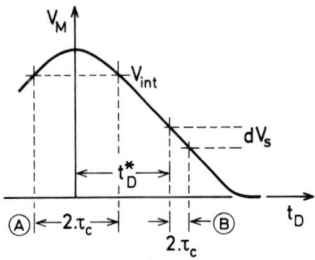

Fig. 5.19. Achieving short resolution times with integral (A) and differential (B) discriminators

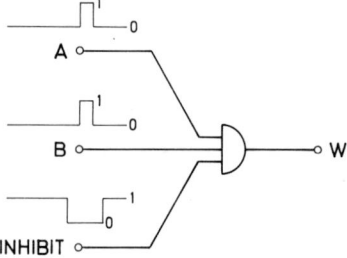

Fig. 5.20. Principle of the anticoincidence gate

For the sake of clarity in Fig. 5.19 only the right half of the $V_M(t_D)$ curve is shown. However, obviously the discriminator channel dV_s also cuts the left edge of $V_M(t_D)$. Hence both events will be recorded with respective positive and negative mutual delays $\pm t_D^*$ (assuming a symmetrical shape of $V_M(t_D)$). If t_D^* is compensated so as to count non-delayed coincidences in the right channel, coincidences delayed by $-2t_D^*$ are recorded in the left channel. If no true delayed coincidences can occur, the "left" count rate consists of chance coincidences only. If no precaution for elimination of the "left" coincidences (e. g. by a superimposed slower coincidence stage with a resolution of $<2t_D^*$) the chance coincidence pulse rate is doubled. On the other hand, the "left" count rate — if properly identified — can be subtracted from the "right" one, thus correcting automatically for chance coincidences (BLAUGRUND and VAGER [5.026], cf. also [5.132]).

It remains to discuss the concept of anticoincidence. In various experimental assemblies the coincidence signal must be suppressed when an inhibiting signal occurs simultaneously. In principle it suffices to connect one of the coincidence inputs to a permanent signal corresponding to logical 1, and to reduce it to logical 0 by the inhibiting pulses (Fig. 5.20). The coincidence signals and the inhibiting signal hence differ in their respective polarities. The inhibiting pulse must be made somewhat longer than the coincidence input pulses in order to overlap with certainty every true coincidence, which must be suppressed. Though any of the three AND gate inputs in Fig. 5.20 may serve as an anticoincidence input, in practice the anticoincidence input often differs from the coincidence inputs, e.g. by acting directly on the discriminator, or on the discriminator output. Very often an inverter precedes the anticoincidence input in order to allow all coincidence or inhibition pulses to have the same polarity. Various practical circuit examples are discussed in the following Chapter 5.3.2. POLLY [5.130] investigated the influence of different pulse shaping circuits on the efficiency of anticoincidence gates.

5.3.2. Practical Circuits

Only a few of the vast number of circuits described in the literature will be treated here. Fig. 5.21 shows a circuit diagram according to EMMER [5.027]. This circuit is based on the linear addition of currents in a clipping cable and is destined for medium fast purposes ($2\tau_c = 10 \ldots 100$ nsec).

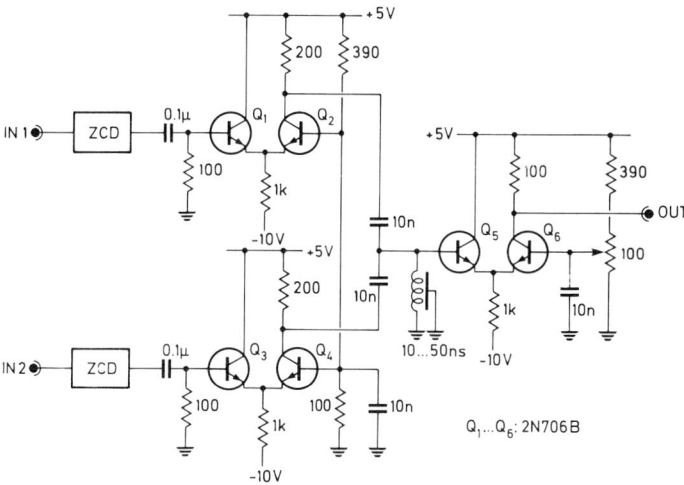

Fig. 5.21. Medium fast coincidence stage for use with zero crossing discriminators (ZCD) according to EMMER [5.027]

Evaluation of the Time Information

Output pulses of zero crossing discriminators from Fig. 4.29 are applied to the inputs of this coincidence stage. In the initial (quiescent) state the transistors Q_2, Q_4 and Q_6 conduct. A positive input pulse at any of the two inputs cuts off Q_2 or Q_4, thus switching the constant standing current into the clipping cable. The pulse length is, of course, given by twice the cable length. The biased amplifier Q_5/Q_6 serves as a discriminator responding only to pulses of double amplitude. The load resistors of Q_2 and Q_4 are properly chosen to terminate the cable ($Z_0 = 100\,\Omega$). The trailing edge of the input pulse produces a negative pulse at the clipping cable, which however is cut away by the discriminator and does not influence the output.

A tunnel diode monostable multivibrator can be used instead of a transistorized difference amplifier. Since a TD multivibrator is faster than the described amplifier, shorter clipping cables can be used. A circuit of this type is described by SIDI and SOLD [5.010]. BARNA, MARSHAL and SANDS [5.009] use the current addition in a clipping cable connected to a diode discriminator D_1/D_2 (Fig. 5.22). Current steps at both inputs are differentiated and yield at (X) a negative voltage pulse with an amplitude of -0.6 V in the case of full overlap. Through the biased diode D_2 a critically damped oscillator is triggered, producing a current pulse having an approximate duration of $0.25\,\mu\text{sec}$, which is further amplified in a current amplifier. The bias voltage is applied via the diode D_1. This biasing circuit ensures that the cable is always terminated properly either by $D_1 + 33\,\Omega$ or by $D_2 + 39\,\Omega$, independent of the state of D_1 and D_2. Resolution times $2\tau_c$ of about 2 nsec can be realized.

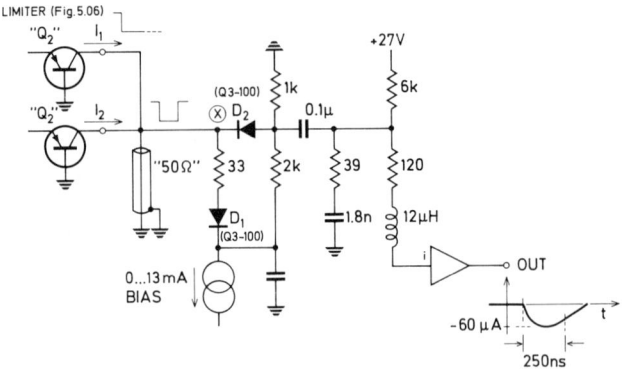

Fig. 5.22. Fast coincidence stage described by BARNA et al. [5.009]

If the limiter stages cannot be connected directly to the coincidence circuit, a separate pulse shaping at each input is preferably performed. In this case all input steps are cable differentiated and the short pulses are led to the mixer through coaxial cables terminated properly and

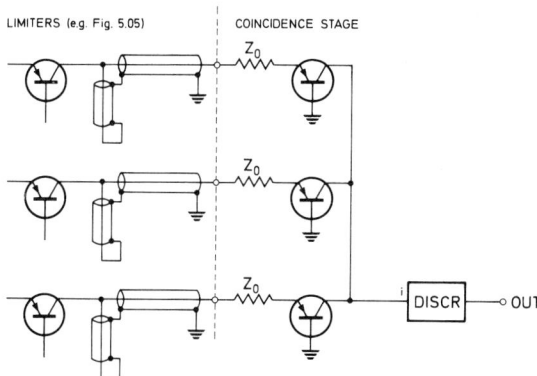

Fig. 5.23. Operating principle of the coincidence circuit according to SUGARMAN [5.007]

decoupled from the mixer by common base stages. First the collector currents of these stages are added at a point of low impedance (e.g. current discriminator). The circuit diagram shown in Fig. 5.23 has been described by SUGARMAN et al. [5.007]. A diode circuit from Fig. 5.22, or still better a tunnel diode trigger, can be used as a discriminator.

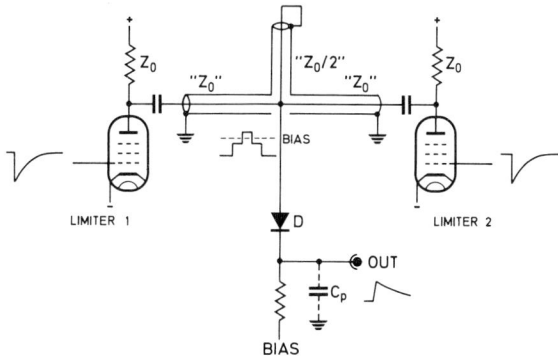

Fig. 5.24. Coincidence circuit by BELL, GRAHAM and PETCH [5.005]

Evaluation of the Time Information

Fig. 5.24 shows a coincidence circuit with linear signal mixing according to BELL, GRAHAM and PETCH [5.005], one of the few "classical" circuits of nuclear electronics. The circuit described in the original paper is designed for pentode limiters, but any other limiter can be used if it is terminated by the characteristic cable impedance Z_0. Obviously the clipping cable must exhibit half the characteristic impedance of the connecting cables Z_0 (instead of a clipping cable $Z_0/2$ two clipping cables Z_0 in parallel can also be used). The reflected signals are absorbed in the limiter terminating resistors. Due to the parasitic capacity C_p, the output signal of the simple diode discriminator is stretched. This facilitates further signal processing.

Finally there follows an example of linear additive voltage signal mixing at tunnel diode discriminators. The circuit Fig. 5.25 has been described by WHETSTONE and KOUNOSU [5.011]; it accepts pulses from TD pulse shaping circuits according to Fig. 5.07. The voltage pulses of standard height and length are converted into currents by means of the 200 Ω resistors, and the currents are added at the input of the first TD trigger. The triggering threshold is adjusted so as to respond only to

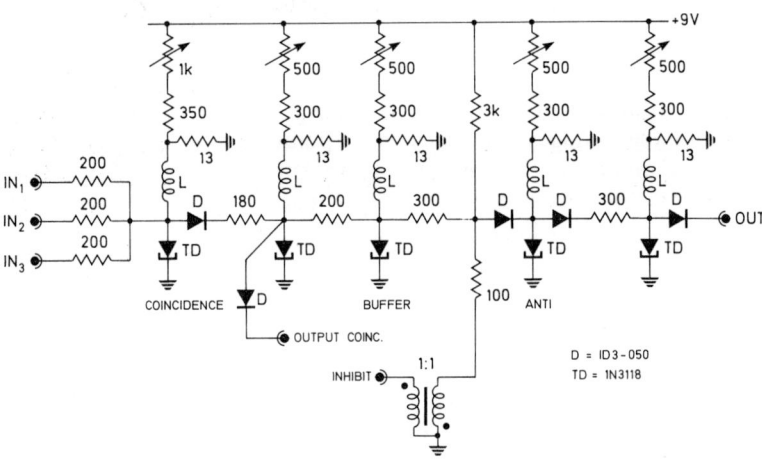

Fig. 5.25. Tunnel diode coincidence circuit by WHETSTONE and KOUNOSU [5.011]

three times the single input current. Hence only triple coincidences are registered. (However, the threshold can be adjusted by means of the potentiometer POT below the single current value. In this case the stage responds to each single input pulse at any of the three inputs and acts as a fast OR gate. By proper adjustment the threshold logic "2 out of 3" can also be realized — cf. also [5.131].) The output signal of the coincidence stages is then shaped in another TD multivibrator. In an additional stage the signal can be mixed with an inhibiting external pulse, hence allowing the anticoincidence operation. Since the inhibiting pulse must have opposite polarity, it is first inverted in a pulse transformer. Depending on the length of the input pulses, resolving times $2\tau_c$ of a few nanoseconds are achieved in the circuit described.

Non-linear mixers can be divided into two categories, namely series and parallel type circuits. The use of two (or more) electronic switches connected in series for coincidence purposes was proposed in 1929 by BOTHE [5.028] — the original circuit with a double-grid tube exhibited a resolution of 1.4 msec. Two examples of series type mixers are shown in Fig. 5.26. In the tube variant, both control grids of an appropriate tube (e.g. 6 BN 6, E 91 H) are biased to cut-off. Anode current flows only if *both* grids are opened simultaneously by positive signals, so that the output signal from the anode load resistor indicates a coincidence. FISCHER and MARSHALL [5.029] achieved sub-nanosecond resolution times in circuits with beam pentodes type 6 BN 6. However, since this circuit does not offer advantages over other known fast semiconductor circuits, which moreover are much smaller, it is of historical interest only. The transistorized version of the BOTHE circuit in Fig. 5.26 is used too, but only in simple slow ($2\tau_c \approx 1\,\mu$sec) coincidence and anticoincidence stages.

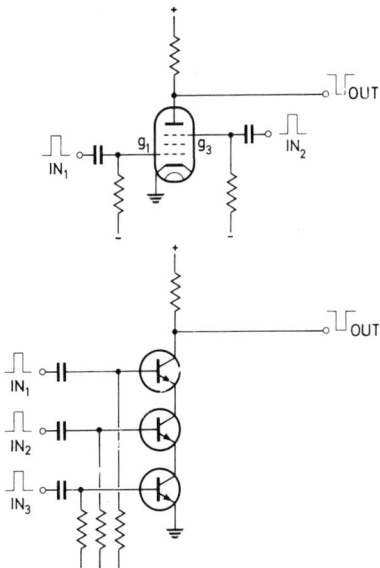

Fig. 5.26. The BOTHE coincidence stage with vacuum tube and with transistors, respectively

Rossi [5.030] described in 1930 a coincidence circuit with two or more triodes in parallel. The Rossi circuit — at first with vacuum tubes, especially pentodes, vacuum or semiconductor diodes, and later with transistors — has become the most widely used coincidence stage. Some of the circuit types are shown in Fig. 5.27. In circuit (A) all base resistors R_B are small ($<\beta \cdot R_l$), hence all transistors are saturated. The current through the load R_l is divided more or less uniformly between the particular transistors. If one or two transistors are cut off by negative input pulses, the remaining transistor takes over the whole current. If, however, all three transistors are cut off, a positive output pulse arises (or the available current is switched over into a low-resistance load connected to the output). The resolution is limited partly by the time constant $R_l C_p$ (C_p=parasitic output capacities), partly by the switching speed of the transistors. Parasitic effects, such as the finite saturation resistance of the transistors, or capacitive voltage division between base-collector capacity and C_p, of course produce a small output pulse even in the case of incomplete coincidence, thus making a subsequent discriminator still necessary. The limit of resolution is reached when the input pulses are so short, that the available current even for complete coincidence cannot charge C_p higher than the parasitic effects do — in this case the discriminator cannot distinguish between complete and incomplete coincidences.

The transistors used as switches in the circuit (A) can be replaced by diodes as shown in circuit diagram (B). The situation is somewhat less favourable than with transistors, since the saturation resistance of the transistor is replaced by the forward resistance r_D of the diodes and by R_d (= the output resistance of the source of input pulses). The parasitic pulses in the case of incomplete coincidences are correspondingly higher. On the other hand there exist diodes which are faster than transistors, and a version of the Rossi circuit with such diodes may be of interest for fast applications. DE BENEDETTI and RICHINGS [5.031] described a fast double-diode coincidence stage where the diode cathodes are connected directly to the clipping cables of limiter-type pulse shaper, hence making R_d equal to half the characteristic cable impedance. Another (third) diode serves as the discriminator. This diode discriminator with parasitic output capacity acts at the same time as a pulse stretcher. The stretched output pulse is amplified in a slow ($\sim\mu$sec) amplifier and digitized in an integral discriminator. DE VRIES [5.032] placed the three diodes in a properly laid-out coaxial arrangement. This circuit exhibited a resolution $2\tau_c$ of a few tens of picoseconds. His theoretical investigations show a lower limit of about 10 psec for the resolution, due to the parasitic effects and to the amplifier noise.

BRUNNER [5.033] also investigated the theoretical resolution limits of a coincidence assembly.

Finally the diodes can be replaced by the base-emitter junctions of transistors in emitter-follower configuration, as shown in version (C) of Fig. 5.27. The high current through the load resistor R_l is led to the

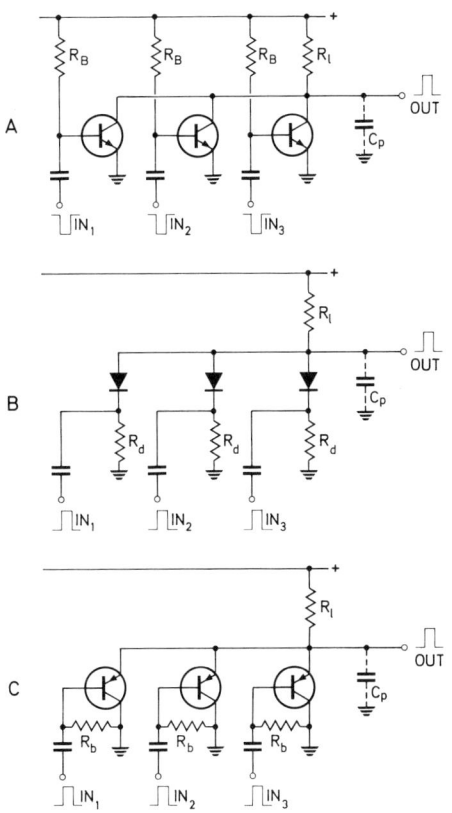

Fig. 5.27 A–C. The ROSSI coincidence stage with transistors (A), with diodes (B), and with transistors in emitter-follower configuration (C)

ground without loading the input, so that relatively high values of R_b can be used. This version too is suitable for fast applications: DUMAS, AUBRET and BENOIT [5.034], for example, realized coincidence modules with a resolution of a few nanoseconds using this circuit principle.

237

Evaluation of the Time Information

Fig. 5.28 shows a complete diagram of a ROSSI circuit (A) according to GOULDING and MCNAUGHT [5.035]; this is used as the slow coincidence stage with a resolution in the submicrosecond range in "fast-slow" coincidence assemblies. The transistors Q_2 to Q_4 operate in a common load resistor R_6 (383 Ω). The output pulse is differentiated by R_{10} and C_6 possibly in parallel with an external capacitor, and opens the biased amplifier Q_5 which acts as a discriminator. There is an emitter coupling between Q_5 and Q_1. These two transistors (via R_6 and C_6) form an amplifier with positive feedback and thus a monostable multivibrator circuit. Hence, after the discriminator fires, the multivibrator delivers a standard output pulse of 0.3 μsec length, independent of the shape and duration of the input pulses. The anticoincidence (inhibiting) pulse increases via Q_1 the potential of the common emitters of the remaining transistors, so that Q_2 to Q_4 cannot be cut off by the coincidence pulses. (The polarities in Fig. 5.28 and 5.27 (A) are just the opposite.)

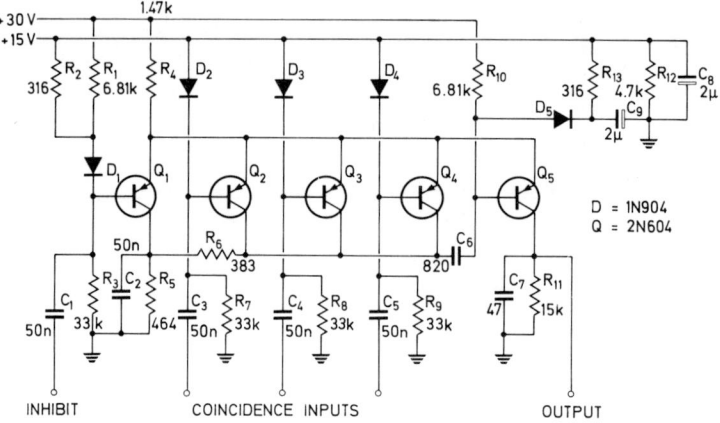

Fig. 5.28. Slow three-input Rossi type coincidence circuit according to GOULDING and MCNAUGHT [5.035]. The circuit accepts input pulses of $+1\,\text{V} \pm 10\%$

GARWIN [5.036] modified the ROSSI circuit by using a non-linear component as the load. A transistorized version of the GARWIN coincidence circuit has been described for example by BAKER [5.037]. The operating principle of the circuit is shown in Fig. 5.29. In the initial state the three (or more/or less) transistors deliver a current $I_1 + I_2$ flowing partly through R_1 and partly through R_2 and the diode D. The anode of the diode is blocked by a large condenser, hence the dynamic load is equal to the forward resistance r_D of the diode. The base resistors R_B deliver base currents which do not saturate the transistors. For the sake of simplicity all collector currents I_C are assumed to be equal: $3 \cdot I_C = I_1 + I_2$. Further, I_2 should be somewhat higher than $2 \cdot I_1$. In general for n inputs the relations

$$I_2 \gtrsim (n-1) \cdot I_1 \quad \text{or} \quad I_2 \gtrsim (n-1) I_C \tag{5.08}$$

should hold.

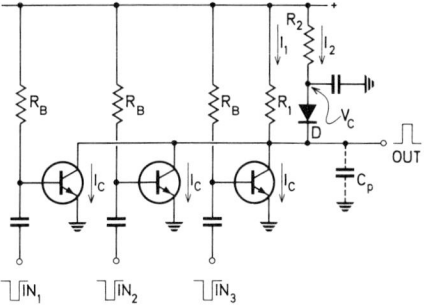

Fig. 5.29. Operating principle of the GARWIN coincidence circuit [5.036]

When one of the transistors is cut off, the current is diminished by I_C. Because $R_1 \gg r_D$, the current I_1 remains constant and only the diode current I_2 is diminished to $I_2 - I_C$, yielding a small output pulse of amplitude $r_D \cdot I_C$. When two transistors are cut off, the current is diminished by $2 \cdot I_C$. However, since $2I_C \lesssim I_2$, the diode still remains conducting and a small output pulse of amplitude $r_D \cdot 2I_C$ results. Not until all three transistors are cut off at the same time does the current loss becomes $3I_C > I_2$, the diode cuts off too, and a high output pulse $R_1 \cdot I_1$ results.

Since the transistors are not saturated, the GARWIN circuit is very fast, and suitable for applications in the nanosecond range. However, only the small part I_1 of the total current $I_1 + I_2$ is used for output signal generation. The diode D increases the parasitic capacity and — in addition — the stored charge must also be neutralized by I_1 after cutting off the diode. For short resolution times, therefore, very fast diodes must be used. When working with high pulse rates the changes of the anode potential V_C of the diode must be taken into consideration ($C \neq \infty$).

Various other coincidence circuits will be mentioned only briefly. BALDINGER, HUBER and MEYER [5.038] developed a coincidence stage operating with non-shaped photomultiplier pulses. In principle, the circuit consists of a balanced diode bridge which does

Evaluation of the Time Information

not deliver an output signal for single pulses, but which is disadjusted by two coincident pulses. Even for high resolution (~ 10 nsec) the bridge coincidence circuit needs only small input pulses of about 0.1 V amplitude.

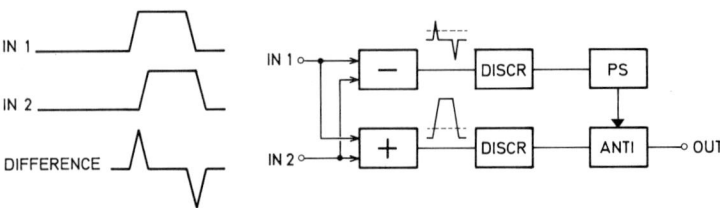

Fig. 5.30. Principle of the differential coincidence circuit

All coincidence stages dealt with in this chapter are based on the addition (or on some non-linear mixing of more-or-less additive character) of the input pulses. BAY [5.039] proposed the so-called differential coincidence circuits [5.040, 5.041], based upon subtraction of the input pulses. A transistorized version of this circuit is described by KULLANDER [5.042], and the operating principle is shown in Fig. 5.30. Obviously the difference between two equally-shaped pulses disappears when the pulses are absolutely coincident, but reaches its maximum possible value when one of the pulses is delayed by more than the pulse rise time. The disappearance of the difference signal can thus be used as the criterion for coincidence. The resolution $2\tau_c$ is given merely by the pulse rise time, it does not depend on the pulse length and can be much shorter than the last one. Fig. 5.30 also shows the block diagram of the circuit. In a normal "additive" coincidence stage (mixer + and the discriminator DISCR) the rough coincidence of the two input pulses is first checked. The difference ($-$) is formed, say, by inverting one of the input pulses and mixing it in an additive mixer. The discriminator DISCR responds only to the positive signal peak. The discriminator output pulse is stretched in the pulse stretcher PS and cuts off the anticoincidence stage ANTI. Hence the output pulse indicates the absence of a positive difference signal. With difference circuits resolutions in the sub-nanosecond range can be obtained.

5.3.3. The Chronotron Principle

Even as the pulse height spectrum can be measured with the aid of a single channel analyzer by varying the channel position, so the distribution of the time intervals between two correlated events can be measured with the aid of a simple fast coincidence stage by varying the signal delay in one of the signal paths. Hence, as with multi-channel amplitude analyzers (Chapter 4.1.4), multiple arrays of coincidence stages with particular delays raised step by step can also be realized. Two possible alternatives are shown in Fig. 5.31. In the first circuit each signal is delayed in a separate cable. The signals are picked up at intervals of Δt and fed to fast coincidence stages. In the second circuit the signals run in opposite directions along a single cable which also mixes them linearly. Hence only discriminators *D* are needed for the detection of a coincidence. The single cable alternative is more sensitive to incorrectly terminated cable ends than is the double-cable version.

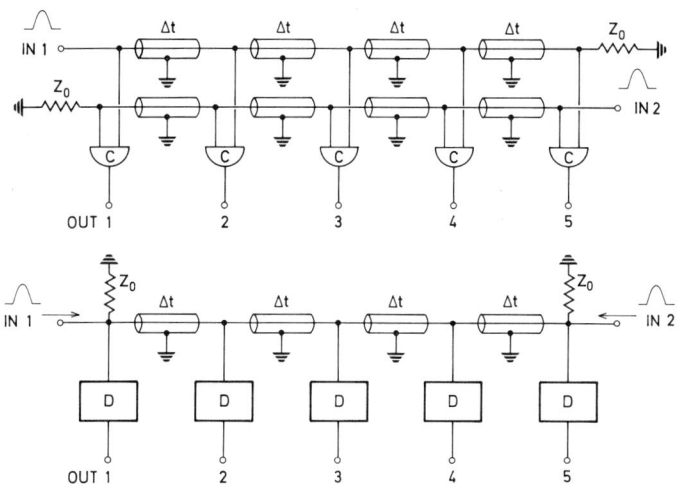

Fig. 5.31. Two versions of a multi-channel coincidence array

The channel distance is given by $2 \cdot \Delta t$, and the channel widths by the resolutions $2\tau_c$ of the particular coincidence stages. It is very difficult to obtain the same resolution for all stages, or even to make the resolution $2\tau_c$ equal to the channel distance $2\Delta t$. Since a coincidence stage with high long-term resolution stability can only be realized if the resolution is determined by the pulse width δ, $\delta = 2\tau_c$ should be aimed for. However, in almost all measurements only the case $2 \cdot \Delta t < \delta$ is interesting. Thus the channel width $2\tau_c = \delta$ would be greater than the channel distance $2\Delta t$ and one event would be registered in more than in one channel. This effect can be avoided by the so-called chronotron array. In chronotron circuits the channel width is always equal to the channel distance, and one event is always registered once and only once. Fig. 5.32 illustrates the operating principle of a chronotron (to be exact, the single-cable chronotron). The overlapping pulses are discriminated by biased diodes *D* and stretched by means of the condensors *C*. At each condenser a voltage proportional to the actual pulse overlap at the corresponding cable position is stored and amplified by the slow amplifiers *A*. A START pulse, picked-up (e.g.) from the ex-

Evaluation of the Time Information

treme left diode, triggers a monostable multivibrator MMV. Its output pulse releases the beam deflection of a cathode ray oscilloscope, and subsequently opens the gates G in intervals of ΔT. Hence the outputs of the amplifiers are sampled at the rate $1/\Delta T$. ΔT is chosen such (e.g. 0.5 μsec) that a conventional slow oscilloscope can be used. From the plot of the condenser voltages at the oscilloscope screen the highest voltage — and thus also the position of the coincidence in the cable — can easily be determined. Exactly speaking, the chronotron technique consists of a time transformation $\Delta t \rightarrow \Delta T$. The transformed fast pulse is plotted on the screen of a slow oscilloscope and its delay in relation to another pulse (releasing the time base) is measured.

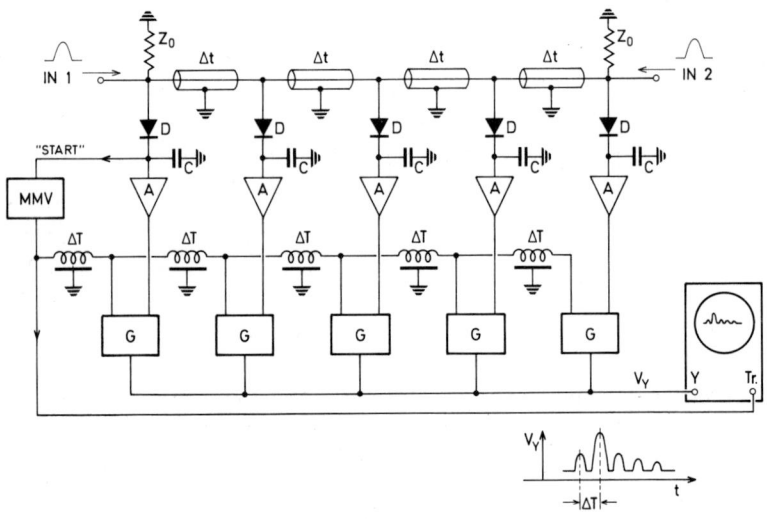

Fig. 5.32. A single-cable chronotron (coincidence at the second diode)

The detection of the highest pulse can also be performed by electronic means, e.g. by connecting slow discriminators with variable thresholds to the amplifier outputs instead of the gates G. In this case the multivibrator MMV generates a saw-tooth voltage, decreasing the discriminator thresholds slowly from a high initial value. Suitable auxiliary circuits ensure that the discriminator which responds first blocks all the others. Hence the discriminator output pulse localizes the position of the heighest overlap and thus the channel of registration.

Recently the relatively clumsy chronotron circuits have been replaced by the handier time-to-pulse-height converters. The chronotrons are used for special purposes only. The chronotron principle was discovered by NEDDERMEYER et al. [5.043]. Other chronotron circuits are reviewed in [5.044, 5.045]. BJERKE, KERNS and NUNAMAKER [5.046] described a transistorized seven channel chronotron with channel widths variable between 3 and 7 nsec.

5.4. Digital Encoding of the Time Interval

As has been discussed early in this Chapter (cf. Fig. 5.01), the time information analysis normally consists in measuring the distribution of time intervals between corresponding START and STOP pulses. However, the origin of the start and stop signals and the measuring range of interest are different according to the actual experiment. When the decay times of excited nuclear states are measured, both the start (= birth of the state) and the stop (= death of the state) pulses are randomly distributed in time. To each start pulse there can be only one corresponding stop pulse but, depending on the particular detectors, the count rates of the start and stop pulses can be different. In time-of-flight measurements and in other experiments on pulsed particle accelerators, one of the signals which has a periodic character can be delivered by the machine. The "counting efficiency" for this start signal is unity. The stop signals are delivered by one detector or by a system of detectors. Although the detection efficiency of these detectors is less than unity, for large particle bursts there may be more than one stop pulse per start pulse. Due to the periodicity of the start pulses it is irrelevant whether two pulses corresponding physically to each other are used for the time interval definition, or whether the stop pulse is analyzed in terms of the *following* machine start pulse (cf. for instance [5.133]). Hence the denomination of start and stop pulses may not have a causal significance.

In time analyzers the start pulse commonly releases some action which is stopped by the following stop pulse. In the absence of a stop pulse, this action can be interrupted by means of some auxiliary circuits only after a wait corresponding at least to the measuring range. During this period the analyzer is blocked for the analysis of new events. Therefore, to reduce dead-time, pulses with the lowest count rate are normally chosen as start pulses, in order to have a stop signal for as many start signals as possible.

The measuring range, i.e. the upper level of delay between the start and stop pulses, varies from a few nanoseconds in the measurement of short-lived states up to some tens of microseconds in time-of-flight measurements on slow neutrons. The corresponding channel widths vary between a few picoseconds and some tenths of a microsecond. The techniques used in short-time-interval analysis have been reviewed by several authors, e.g. by GATTI and SVELTO [5.047] and AMRAM [5.097]. BONITZ [5.048] published a comprehensive review, quoting the relevant literature up to 1962. The most recent (1968) review on these topics is that by OGATA et al. [5.134].

There is no difficulty in the digital encoding of sufficiently long time intervals. It is only necessary to count periodic pulses of a constant repetition frequency during the time interval to be measured; the state of the scaler at the instant of the stop pulse indicates the digital equivalent of the analog information (this technique has already been used in the pulse height encoding described in Chapter 4.2.1). However, the direct encoding is limited by the scaler counting speed. Since reliable pulse counting is limited at present to frequencies below about 200 Mcps, the lower limit of the channel width which can be reached in type of encoders is about 5 nsec.

Shorter time intervals must first be stretched prior to their encoding according to the principle described above. Although interval stretching, described later, would be the most direct way of digital encoding, an indirect technique based on a time-to-pulse-height converter is commonly used. The time interval is first transformed into another analog quantity, the pulse height, which is subsequently analyzed by an amplitude analyzer. The reason for this roundabout procedure lies above all in the fact that multichannel amplitude analyzers became standard laboratory equipment. Fortunately, this double conversion does not adversely affect the measuring accuracy.

5.4.1. Direct Digital Encoding

Fig. 5.33 shows the principle of direct digital encoding. The START pulse sets a flip-flop FF which opens an AND gate. Through this gate suitably shaped pulses from a free-running oscillator OSC with constant frequency enter the scaler (of course, the scaler must be cleared to zero prior to counting the oscillator pulses). The STOP pulse resets the flip-flop and blocks the AND gate. Hence the number of counts registered in the scaler is proportional to the time interval between the start and stop pulses. The stop pulse also releases the information transfer from the scaler to a buffer or address register (READ-OUT). In the absence of a stop pulse, the process of counting is stopped by the overflow of the scaler, and in this case no read-out operation is performed. The whole device consists of standard digital basic circuits which will be described in more detail in Chapter 6.

In the case of a free-running clock oscillator, both the start and the stop pulses introduce a timing error of up to one channel width ($=$ interval between two clock pulses). Hence the total inaccuracy may exceed the channel width. This error is reduced to less than one channel width by the use of a gated oscillator which first starts to oscillate when the flip-flop is set. The phase of the oscillations of such a oscillator is locked to the releasing start pulse.

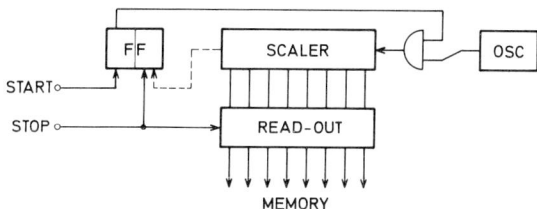

Fig. 5.33. Direct digital encoding of time intervals

The first time analyzer using direct digital encoding has been described by SCHUMANN [5.049]. The capacity was limited to 1024 channels by the ferrite-core memory, and the highest oscillator frequency was 2 Mcps, corresponding to the lowest channel width of 500 nsec. With faster components, channel widths of 50 nsec [5.050], 10 nsec ([5.051] to [5.054], [5.135]), and recently 5 nsec [5.055] can be reached. Two or more stop pulses per single start pulse can be analyzed if the scaler count state can be transferred to one or more buffer registers without stopping or interrupting the counting process (cf. Chapter 6) [5.056, 5.057, 5.136]. THENARD and VICTOR [5.058, 5.059] described a time analyzer with programmed clock-pulse frequency, thus allowing different channel widths to be chosen for different time-interval ranges.

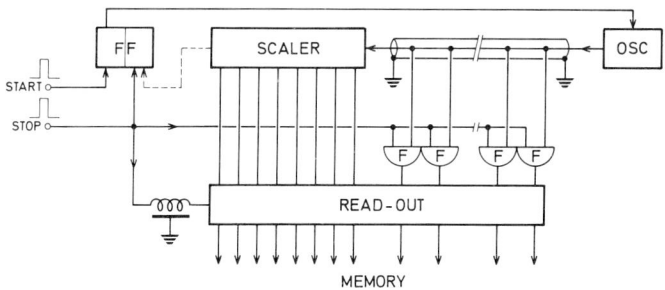

Fig. 5.34. Interpolation of the clock pulse interval

Evaluation of the Time Information

The resolution can be improved by dividing the interval between two clock pulses by some interpolation technique similar to the chronotron. Fig. 5.34 shows a block diagram of such a time analyzer using a triggered oscillator OSC. The release of the counting process does not differ from that shown in the assembly of Fig. 5.33. However, the clock pulses are led in the scaler through a delay line, the length of which corresponds to the clock-pulse interval. The delay line is tapped at n equally spaced points. The stop pulse switches off the oscillator in the normal way. Moreover, in n fast coincidence stages F its delay is measured against the last clock pulse. The coincidence indicates the position of the clock pulse in the delay line at the instant of the stop pulse. Hence the n coincidence outputs form the "last digits" of the digital equivalent of the whole time interval, and are read out into a buffer or address register at the same time as the scaler count state. The channel width is obviously $1/n$-th of the clock-pulse interval.

Meyer [5.060] achieved a channel width of 1 nsec using a 100 Mcps scaler and a corresponding oscillator with ten-fold interpolation. Because of their better long term stability, free-running oscillators are superior to triggered ones. When a free-running oscillator is used, the instants of both the start and the stop pulses must be estimated by an interpolation. Prior to read-out, the two interpolation values must be added digitally. A corresponding circuit also having a channel width of 1 nsec has been described by Durand et al. [5.061].

5.4.2. Principle of a Time-to-Pulse-Height Converter

The time-to-pulse-height converter circuits can be divided in two categories according to their operating principle, viz. start-stop converter or overlap converter. In the *start-stop converter* (Fig. 5.35a) a storage element — almost always a condenser C — is connected to a constant current generator I_0 by the START signal, and disconnected by the STOP signal. The resulting pulse amplitude is proportional to the time interval T between the start and stop pulses. The switching can be achieved by using fast tubes or semiconductor components. In the *overlap converter* (Fig. 5.35b) the START pulse must exhibit a well-defined length L, and the STOP pulse must be somewhat longer. Both pulses are added in a mixer M, and their overlap is detected by means of a simple threshold discriminator D (the whole assembly is familiar to the reader — it corresponds to the coincidence stage Fig. 5.17). The length of the discriminator output signal indicates the extent of overlap. The output signal is integrated in a linear integrator INT, and the amplitude of the resulting pulse is proportional to $(L-T)$. An overlap converter does not produce an output signal for a single start pulse as

Principle of a Time-to-Pulse-Height Converter

a start-stop converter does. This may be advantageous if high count rates of start pulses, which are not always followed by stop pulses, must be processed. However, the overlap converter cannot check the correct sequence of the input pulses — a stop pulse followed by a start pulse also yields an output pulse. Such events with "negative" time intervals must be eliminated by means of additional auxiliary circuits.

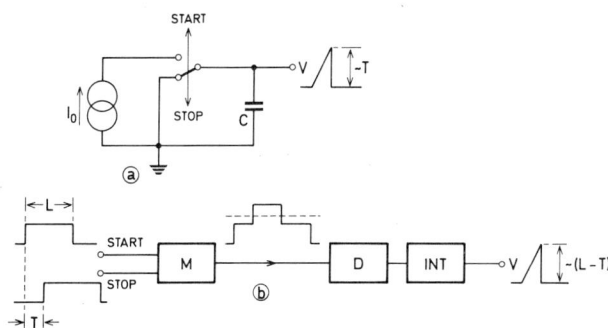

Fig. 5.35a and b. A start-stop converter (a) and an overlap converter (b)

Practical converter circuits of both categories will be discussed in the following two chapters 5.4.3 and 5.4.4.

When a pulse height selection must be made in addition to the time analysis, or when the pulse-height-dependent "walking" of the limiter

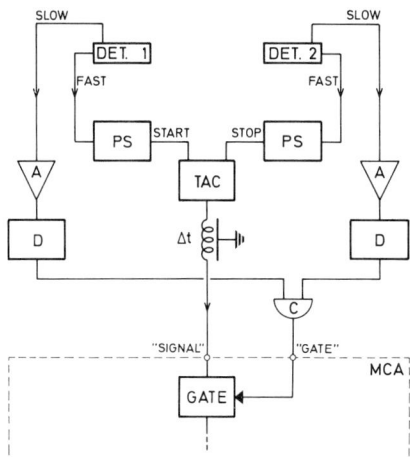

Fig. 5.36. A fast-slow assembly with time-to-pulse-height converter

247

Evaluation of the Time Information

pulses are to be eliminated by limiting the accepted pulse heights to narrow ranges, a "fast-slow" arrangement shown in Fig. 5.36 is used. Fast signals are shaped in the pulse shaper circuits PS and treated in the time-to-pulse-height converter TAC; slow signals are amplified (A) and digitized in discriminators D. The discriminator output pulses are fed to the slow coincidence stage C, the output of which controls the linear GATE of the multichannel analyzer MCA. Hence only those events for which both amplitude criteria are fulfilled are registered in MCA.

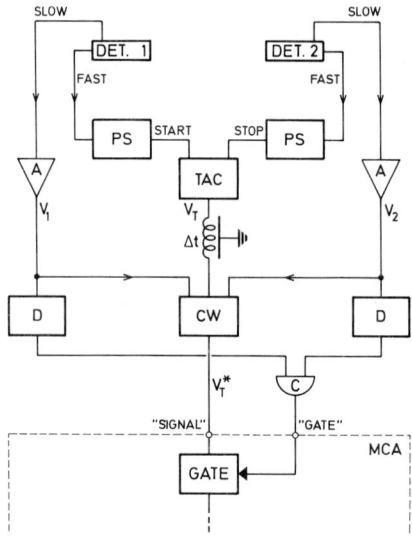

Fig. 5.37. A fast-slow assembly with correction network CW allowing for amplitude-dependent walking of the fast pulse shaper output pulses

Due to the amplitude-dependent walking of the fast pulse-shaper output pulses, a very narrow amplitude range must be selected by the discriminators for the highest time resolution. If the discriminator windows cannot be chosen small enough (e.g. because of a limited measuring time) the output signal of the time-to-pulse-height converter must be corrected as a function of the actual amplitudes in the slow channels [5.021, 5.062]. The principle is illustrated in Fig. 5.37. The converter pulses having amplitude V_T are fed, together with the slow pulses V_1 and V_2, into the correction network CW which produces output pulses V_T^* with the amplitude

$$V_T^* = V_T + f(V_1, V_2). \tag{5.09}$$

The correction function $f(V_1, V_2)$ is often very simple. Assuming a start-stop converter with V_T proportional to the time interval T, and leading-edge timing in limiters, the limiter output pulse delay with regard to the releasing event decreases if the slow pulse amplitude increases. Hence an amplitude increase in the start channel 1 increases the time interval T, and an amplitude increase in the stop channel 2 diminishes T. If the discriminator D windows are not too wide, a linear approximation is sufficient and (5.09) becomes

$$V_T^* = V_T - \varkappa_1 V_1 + \varkappa_2 V_2. \qquad (5.10)$$

The correction thus consists in the simple addition and subtraction of small fractions $\varkappa_1 \approx \varkappa_2$ of the slow pulse amplitudes, respectively.

RODDA, GRIFFIN and STEWART [5.063] described a correction network for the realization of the relation (5.10). The circuit consists of passive components only and is therefore very stable (Fig. 5.38). With the aid of the selector switches and the 100 Ω potentiometers, the fractions $\varkappa_1$ and $\varkappa_2$ are experimentally estimated for optimum resolution. One of the amplitudes $(\varkappa_2 \cdot V_2)$ is added to the converter signal V_T, the other $(\varkappa_1 \cdot V_1)$ is inverted in the transformer and subtracted from V_T. In any case, the optimum adjustment of $\varkappa_1$ and $\varkappa_2$ requires some experimental practice. Using this assembly the authors were able to estimate the lifetimes of positrons in different materials to an accuracy of ±3 psec. Another linear approximation (5.10) connection network has been described by SEN and PATRO [5.137].

THIEBERGER [5.064] discussed the realization of correction circuits for more general correction functions $f(V_1, V_2)$ from (5.09).

HAUSER et al. [5.138] described the application of the internal electronics of a commercial sampling oscilloscope as a very fast time-to-pulse-height converter (resolution FWHM ≈ 3·10⁻¹¹ s).

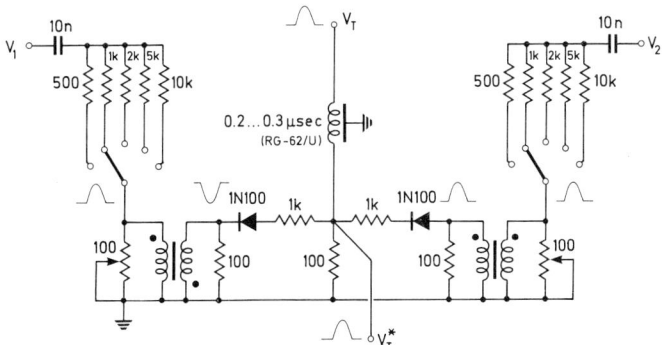

Fig. 5.38. Correction network according to RODDA, GRIFFIN and STEWART [5.063]

5.4.3. Start-Stop Converter

Fig. 5.39 shows a start-stop converter according to GATTI, VAGHI and ZAGLIO [5.019] which requires voltage steps at the inputs. In the quiescent state both Q_9 and Q_{10} conduct. The transistor Q_9 serves as a constant current generator, the exact value of I_0 is adjusted by means of the potentiometer $500\,\Omega$. The condenser C is discharged down to the saturation voltage of Q_{10}. A start step cuts off Q_{10} through Q_8, thus starting a linear charging process of C. The charging process is first interrupted by the stop step, which cuts off Q_9 with the aid of the emitter follower Q_6/Q_7. Hence the amplitude of the output pulse $T \cdot I_0/C$ is proportional to the start-stop time interval T. The component values indicated in the circuit diagram are chosen for $T = 0\ldots 100$ nsec. In counting practice, of course, the input steps have a finite duration δ_{start} and δ_{stop}. Obviously $\delta_{\text{stop}} > \delta_{\text{start}} > T_{\max}$ must hold ($T_{\max}$ denotes the upper limit of the measuring range). Contrary to the overlap converter, the actual lengths δ_{stop} and δ_{start} of the input pulses do not affect the output signal.

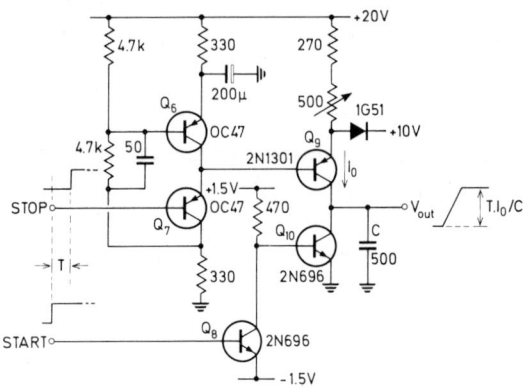

Fig. 5.39. Start-stop converter described by GATTI et al. [5.019] for the time range $0\ldots 100$ nsec

Before transistors were available, a double-grid vacuum tube such as 6BN6 was used instead of the transistor combination Q_9/Q_{10} (e.g. [5.065]). MEILING et al. [5.066] investigated a converter based on a difference amplifier (Fig. 5.40). The start step disturbs the current equilibrium between the two tubes, the stop step being of exactly the same amplitude as the start step, restores the equilibrium. The current difference is integrated by means of the parasitic capacity C_p. Since neither of the two active components has to be cut off, this circuit is very sensitive and operates even with input amplitudes of a few millivolts. However, the output pulse height is directly proportional to the amplitude of the start step, which therefore must be extremely constant.

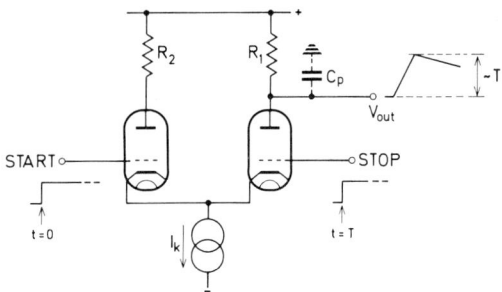

Fig. 5.40. Start-stop converter according to MEILING et al. [5.066]

CULLIGAN and LIPMAN [5.067] store the current released between the start and the step signals in a choke prior to integrating it in a condenser. The operating principle of this converter and a circuit diagram by OPHIR [5.068] are shown in Fig. 5.41. In the LC-converter (case b) the inductance L is connected to a constant-voltage source V_0 during the start-stop interval T. The current through L is a linear function of the time, hence its value I_L at the end of the time interval T is

$$I_L = V_0 T/L. \tag{5.11}$$

After the voltage source V_0 has been disconnected by the stop signal, I_L charges the condenser C via the diode D_1. Assuming an ideal diode, the resulting amplitude V_{LC} of the voltage step across the condenser C can easily be calculated from the condition of equality of the respective choke and condenser energies:

$$V_{LC} = T \cdot V_0 / \sqrt{LC}. \tag{5.12}$$

The output amplitude V_C of a conventional C-converter (case a) comes to $V_C = T \cdot I_0/C$. V_C is limited by the fact that an electronic switch can conduct only a limited current I_0. Since the choke current I_L in the LC-converter, even in the case of the maximum start-stop interval T_{max}, also must not exceed I_0, the voltage V_0 is limited by the relation

$$I_{L,max} = V_0 \cdot T_{max}/L = I_0. \tag{5.13}$$

Introducing V_0 from (5.13) into equation (5.12), the ratio of V_{LC}/V_C can be calculated to

$$\frac{V_{LC}}{V_C} = \frac{\sqrt{LC}}{T_{max}}. \tag{5.14}$$

Obviously CULLIGAN's LC-converter is superior to the C-converter for fast applications, where T_{max} is small, since it produces higher output pulse amplitudes. For example, with $I_0 = 10$ mA, $C = 1000$ pF and $L = 25$ µH V_{LC} is higher than V_C by one order of magnitude if $T_{max} \leqslant 25$ nsec [5.068].

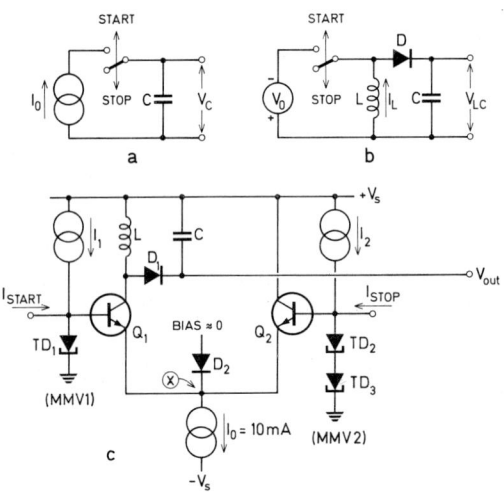

Fig. 5.41 a–c. Operating principle of the LC converter according to CULLIGAN and LIPMAN [5.067] (b) as compared with the conventional C converter (a). Basic diagram of a practical circuit described by OPHIR [5.068] (c)

The LC-converter, described by OPHIR loc. cit. and shown in Fig. 5.41 c, operates as follows: in the quiescent state the current $I_0 = 10$ mA flows through the diode D_2. The impedance of the point (X) is very low (diode forward resistance) as long as the transistors Q_1, Q_2 together draw less than 10 mA current. At first both Q_1 and Q_2 are cut off. The start signal saturates Q_1, hence the whole potential difference between $+V_s$ and (X) appears at L. The current I_L increases as a linear function of the time. The higher stop signal saturates Q_2, which takes over the whole available current I_0 and cuts off Q_1. Due to the low impedance emitter coupling the circuit operates extremely fast.

Contrary to the circuits discussed above, the converter shown in Fig. 5.41c does not require input steps, but only short current pulses ($\ll T_{max}$). The necessary holding pulses of the respective lengths δ_{start} and δ_{stop} are produced by the monostable tunnel diode multivibrators MMV1 and MMV2 (which are not shown in full detail in the circuit diagram).

WIEBER [5.070] described a C-converter using two very flexible TD monostable multivibrators (Fig. 5.42). The standing current I_0 flows at first through both diodes D_1 and D_2. The start pulse cuts off the diode D_1, hence directing I_0 to the condenser C, which begins to discharge linearly. The stop pulse cuts off the diode D_2, too, and the current I_0 flows into TD2. Simply changing the capacities in the multivibrators TD1 and TD2 and changing the integrating capacitor C, makes it possible to measure time intervals ranging from a few to about 200 µsec. The electronic system resolution (FWHM of the distribution of exactly coincident test pulses) amounts to 6 psec.

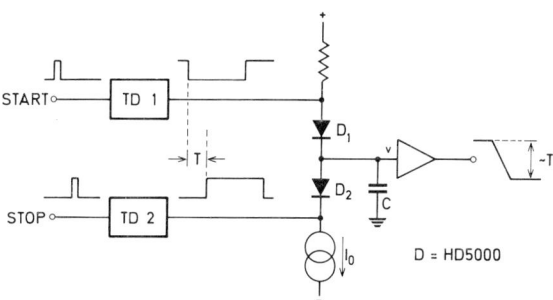

Fig. 5.42. Start-stop converter by WIEBER [5.070]

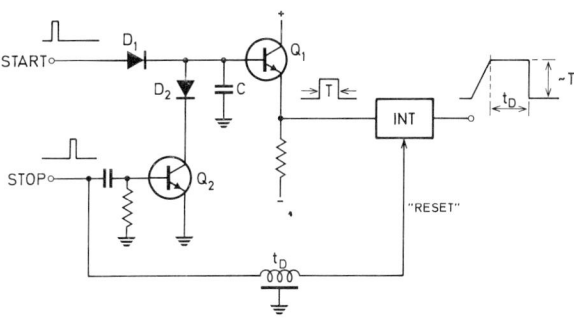

Fig. 5.43. Start-stop converter proposed by BRUN et al. [5.069]

Evaluation of the Time Information

A pulse stretcher, which holds the start signal until the stop pulse clears it, can also be used for the definition of the start-stop time interval T. Fig. 5.43 shows the conception of this idea by BRUN et al. [5.069]. The start pulse charges C through the diode D_1, the stop pulse discharges C by means of Q_2 and D_2. The output pulse is integrated after its amplitudes have been standardized, the integrator is reset by the stop pulse delayed by t_D.

Since the advent of fast tunnel diodes, a TD bistable multivibrator set by the start and reset by the stop pulses is preferred for the definition of the interval T ([5.071] to [5.073]). The resulting flip-flop output pulse with the duration T is further integrated by means of a suitable circuit. BRAFMAN [5.074] uses for the start and stop channels two separate flip-flops which are both reset by an additional dead-time circuit. DARDINI et al. [5.139] described another two-flip-flop time-to-pulse-height converter.

Fig. 5.44 shows another type of start-stop converter. The start pulse triggers a sawtooth voltage generator SG. The voltage ramp is sampled

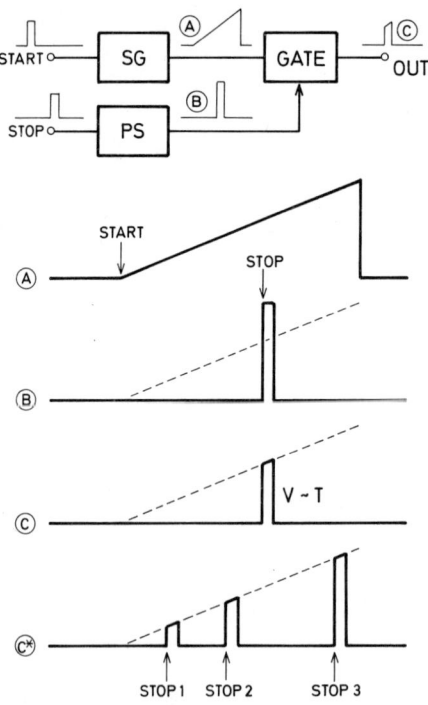

Fig. 5.44 A–C. Voltage ramp type start-stop converter

in the instant of the stop pulse by a suitably shaped narrow pulse (pulse shaper PS) with the aid of a linear GATE. Obviously the output pulse height is proportional to the time interval T. CHRISTIANSEN [5.075] investigated this operating principle using a circuit with a beam deflection tube E80T, BLOESS and MUENNICH [5.076] described a transistorized ramp converter. The ramp generator used by the latter authors delivered a 70 nsec long linear sawtooth voltage pulse. The measuring range $T_{max} = 70$ nsec is given by the sawtooth pulse length, and the system resolution for test pulses is claimed to be 0.1 nsec.

The ramp-type converter can be adapted to accept more than one stop pulse for each start pulse, if the used amplitude analyzer is equipped with a fast analog pulse buffer memory (derandomizer) such as shown in Fig. 4.43 [5.077]. The plot C^* in Fig. 5.44 illustrates the situation for three stop pulses. The amplitudes of the three output pulses are proportional to the particular start-stop intervals. The pulses are stored in the derandomizer and subsequently analyzed. Another multi-stop TAC for 4 pulses has been described by WHITE [5.140].

In all the converters described above the start pulse starts an action which must be terminated by means of an external control circuit if there is no stop pulse within T_{max}. Such a control circuit consists normally of a simple discriminator connected to the converter output, which indicates when the output voltage exceeds the voltage corresponding to T_{max}. If necessary, converter pulses resulting from sole start events can be prevented from entering the multichannel analyzer by means of an additional gate system (e. g. [5.078]).

With the aid of a fast coincidence stage before the start input of the converter a sole start pulse can be suppressed even prior to starting a conversion action. This has been proposed by BALLINI and POMELAS [5.079]. The principle of such an arrangement is shown in Fig. 5.45.

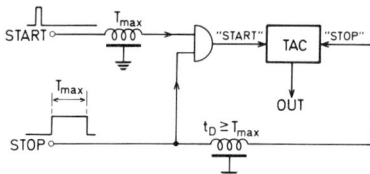

Fig. 5.45. Principle of a start-stop converter not responding to isolated start pulses

The length of the stop pulse corresponds to the range T_{max} of the converter TAC. The narrow start pulse is delayed by T_{max} and fed to a fast AND gate controlled by the stop pulse. Obviously the start pulse

Evaluation of the Time Information

can pass the gate and enter the converter only if it is followed by a stop pulse within T_{max}. The stop pulse itself must also be delayed by $t_D \gtrsim T_{max}$ before entering the converter. Sole stop pulses do not produce any output pulse.

WEISBERG and BERKO [5.080, 5.081] used the principle illustrated in Fig. 5.45 in a very compact circuit, the block diagram of which is shown in Fig. 5.46. The start pulse delayed by T_{max} triggers a monostable tunnel diode multivibrator MMV1, which produces a narrow current pulse ($\delta_{start} \ll T_{max}$). The non-delayed stop pulse is shaped in MMV2 to a current pulse of the duration $\delta_{stop} = T_{max}$. The output currents of both multivibrators add at the input of a tunnel diode Schmitt trigger circuit. The triggering threshold and the hysteresis of the TD trigger circuit are adjusted so as to set the circuit at the double-pulse amplitude (Point A), but to reset it below the single-pulse amplitude (Point B).

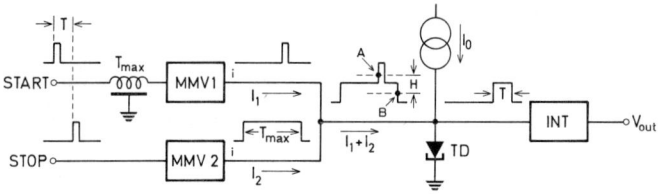

Fig. 5.46. Start-stop converter according to WEISBERG [5.081]

Hence the trigger circuit remains set between the delayed start pulse and the end of the stop pulse. Since the start pulse delay corresponds exactly to the length of the stop pulse, this time interval amounts exactly to T. The output pulse of the TD Schmitt trigger circuit must still be integrated in a suitable circuit INT. The resolution of this converter is tens of psec.

5.4.4. Overlap Converter

As has been shown in Chapter 5.4.2 (cf. Fig. 5.35), the overlap converter is derived from a fast coincidence stage with integrated output. Hence almost all of the coincidence circuits described in Chapter 5.3.2 can also be used as overlap converters. A selection of popular circuits is shown in Figs. 5.47 and 5.49.

Since the output voltage is proportional to $(L-T)$, the pulse length L must be very stable. For this reason cable pulse shaping is used almost without exception. In the circuit Fig. 5.47 (A) according to GREEN and

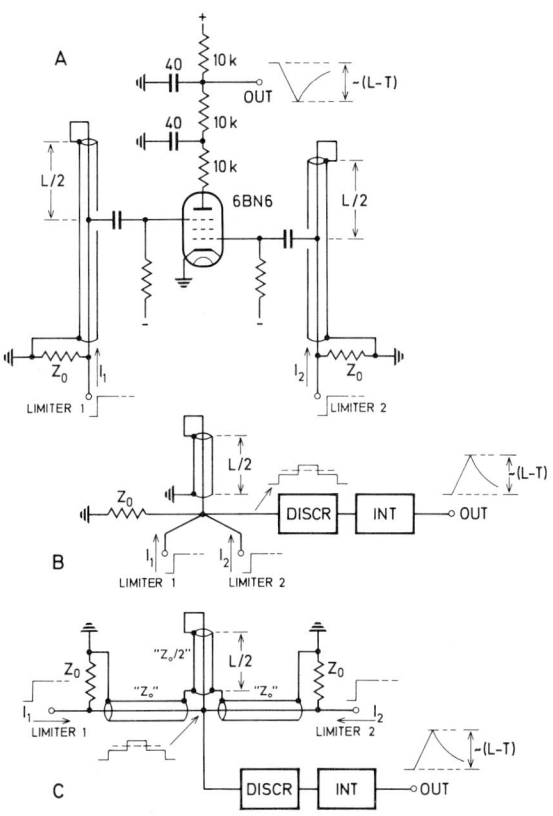

Fig. 5.47 A–C. Overlap converter according to GREEN and BELL [5.082] (A), and two versions of linear mixer type converter (B) and (C)

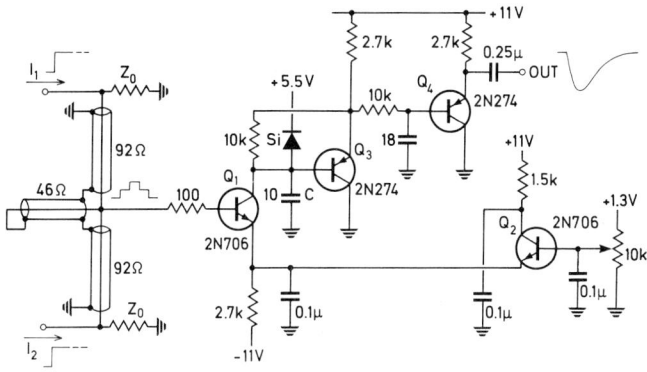

Fig. 5.48. Overlap converter by RODDA et al. [5.063]

BELL [5.082] a double grid tube 6BN6 is used as mixer and discriminator. The 6BN6 is cut off in the quiescent state, it conducts a constant current only during the overlap of the two positive input pulses. The current is integrated by the anode network. Formerly this simple converter was often used [5.083, 5.084].

The addition of two current steps from two limiters on a shorted cable of length $L/2$ (Fig. 5.47(B)) is used in the converter circuits described, e. g. by GORODETZKY et al. [5.085], ASPELUND [5.006] and JUNG [5.086]. The particular circuits differ in the conceptions of the discriminator and the integrator. JUNG loc. cit., for example, used a biased pentode E810F as discriminator. The discriminator ouput pulse cuts off another E810F during the time interval $(L-T)$, the constant anode current is integrated by a capacitor. The system resolution (FWHM) is 4.8 psec.

The last circuit of Fig. 5.47 (C) is derived from the classical coincidence stage according to BELL, GRAHAM and PETCH [5.005] (cf. Fig. 5.24). Again the particular circuits differ in the discriminator and integrator. SUNYAR [5.087] uses a biased semiconductor diode and the parasitic capacity, JONES [5.088] uses a current switch made of two semiconductor diodes and integrates the current in an operational amplifier. RODDA, GRIFFIN and STEWART [5.063] employed a biased fast transistor for discrimination purposes, and integrated the collector current pulse. Their circuit is shown in more detail in Fig. 5.48.

The base potential of Q_3 is held at about 5.5 V by means of a fast silicon diode Si. At rest, transistor Q_1 is cut off and biased by the emitter-follower Q_2 to such an extent, that only input pulses of double amplitude are accepted. The collector current of Q_1 is integrated by the condenser $C \approx 10$ pF and the integration process is linearized with the aid of a bootstrap feedback $Q_3/R = 10$ kΩ. The output pulse is again shaped by another integrator 10 kΩ/18 pF and amplified in the emitter-follower Q_4. Because the output signal exhibits a temperature drift of -55 psec/°C, the whole circuitry is placed in a thermostat 37.0 ± 0.2 °C. The converter accuracy is sufficient for positron lifetime measurements with errors of a few picoseconds.

Fig. 5.49 shows some converter circuits operating with non-linear mixers according to the ROSSI principle. VERGEZAC and KAHANE [5.008, 5.089] investigated the properties of the mixer shown in Fig. 5.49 (A) employing a GARWIN diode D_1. The current of the two transistors is divided between the resistor R (4 mA) and the diode D_1 (6 mA). Only when both transistors are cut off at the same time D_1 also cuts off, and the constant current of 4 mA flows through D_2 into C, where it is integrated.

GRIN and JOSEPH [5.090] proposed a very simple circuit (Fig. 5.49 (B)), based upon a diode AND-gate. Instead of the biased diode D_3, a vacuum tube [5.091] or a transistor [5.092] can be used as an integrating discriminator. Another variant of this converter has been described by SEN and PATRO [5.141].

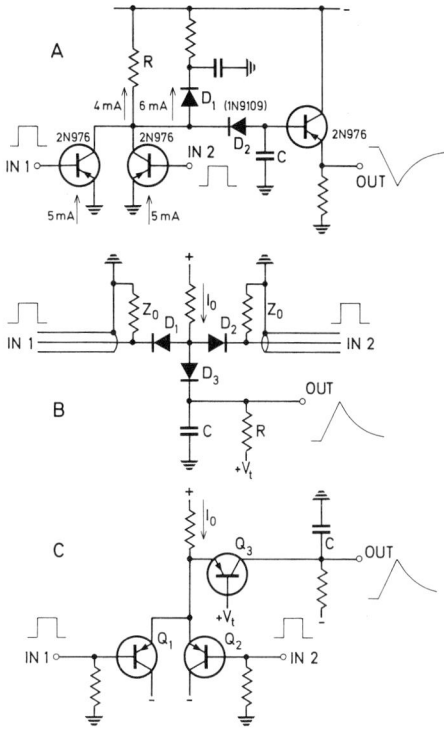

Fig. 5.49 A–C. Overlap converters using non-linear mixers of Rossi type

Finally a circuit investigated by SIMMS [5.093] and JONES and FALK [5.094] should be mentioned, in which the gate diodes are replaced by the base-emitter junctions of transistors (Fig. 5.49 (C)). The transistor Q_3 is slightly cut off ($+V_t$) as long as at least one of the transistors Q_1, Q_2 draws the constant current I_0. If both transistors Q_1, Q_2 are cut off, Q_3 opens and I_0 flows into the integrating capacitor C.

One great advantage of the overlap converters is the fact that a sole start (or stop) pulse does not start any conversion action. However, due to the symmetry of the two inputs, an overlap converter cannot distinguish a positive delay from a negative one (i.e. stop before the start). Hence events with negative delay must be eliminated, e.g. by an additional coincidence stage according to the diagram shown in Fig. 5.50. For the sake of simplicity both start and stop pulses are assumed to

Evaluation of the Time Information

have the duration L. The two pulses are mixed in the overlap converter TAC, and in addition the start pulse is delayed by L and mixed with the stop pulse in a coincidence stage C. The pulse lengths being L, the resolving time of the stage is $2L$. Obviously the coincidence stage will respond to events only when start precedes stop. The coincidence output signal controls the linear GATE of the multichannel analyzer, so that only the analysis of events with positive delay will be performed.

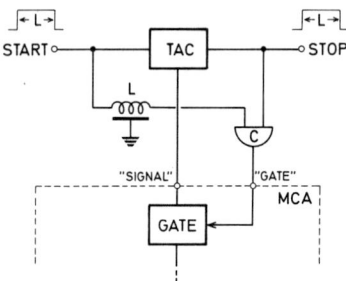

Fig. 5.50. Suppression of the "negative time interval" events in an overlap converter

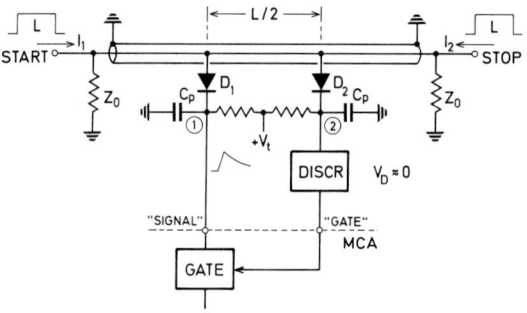

Fig. 5.51. Overlap converter with suppression of the "negative time" events according to BONITZ and BERLOVICH [5.095]

Of course, the principle illustrated in Fig. 5.50 can be realized in many ways. An interesting circuit variant using two identical mixers for TAC and the coincidence stage C has been described by BONITZ and BERLOVICH [5.095] (Fig. 5.51). The diodes D_1 and D_2 are biased $(+V_t)$ so as to suppress sole pulses. The amplitudes of the resulting pulses in points (1) and (2) correspond to the particular overlaps. Whereas the pulse (1) is fed into the amplitude analyzer, pulse (2) is discriminated in a discriminator with a very low threshold ($V_D \approx 0$). This integral discriminator together with the cable mixer forms a coincidence stage with the resolution $\approx 2L$.

5.4.5. The Vernier Principle

The vernier principle can easily be understood by referring to Fig. 5.52. The start pulse triggers a periodic pulse sequence having the repetition frequency f_1, and the stop pulse triggers another pulse sequence having a slightly higher repetition frequency f_2. A fast coincidence stage indicates the instant of first coincidence between the pulses of the start and stop sequences. Obviously the time interval T' between this coincidence and the stop pulse is proportional to the original start-stop time interval T

$$T' = T \cdot \frac{\tau_2}{\varDelta \tau}, \qquad (5.15)$$

as long as $0 < T < \tau_1$. Here $\tau_1 = 1/f_1$ and $\tau_2 = 1/f_2$ denote the periods of the two pulse sequences, and $\varDelta \tau = \tau_1 - \tau_2$ their difference. Since $\varDelta \tau \ll \tau_2$, the time interval T is enlarged considerably.

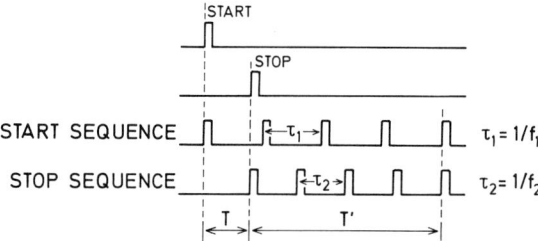

Fig. 5.52. The vernier principle

The vernier techniques can be evaluated in two ways. Either the interval $T' \gg T$ serves as the input magnitude of a conventional time-to-pulse-height converter and the final analysis is performed in a pulse-height analyzer, or the extension factor T'/T is made high enough to digitize T' directly using the techniques described in Chapter 5.4.1. In the second case pulses of either sequence can be used as clock pulses. Hence it is sufficient to count the number of pulses of the stop sequence, for example, in order to get the digital equivalent of T. Counting stop pulses is preferred to the counting of start pulses for eliminating events

Evaluation of the Time Information

with sole start pulses. (The first pulse of any sequence (dotted line) must not be counted!)

Fig. 5.53 shows the block diagram of the original arrangement by COTTINI and GATTI [5.096], which used harmonic oscillations instead of pulse sequences. The start and the stop pulses trigger two resonance circuits OSC 1 and OSC 2 having frequencies $f_1 = 20$ Mcps and $f = 20.2$ Mcps respectively. In the mixer M the superposition of both waves is performed. The zero crossover of the superposition wave which has a period of 5 μsec indicates the exact instant when the phase difference of the two waves disappears ("coincidence" of the corresponding half-waves). The zero crossover instant is indicated by means of a zero crossing discriminator ZCD. The magnified time interval T' is analyzed with the aid of a slow time-to-pulse-height converter TAC. The extension factor amounts to 100. The small delay t_D allows for internal delays in the particular circuits.

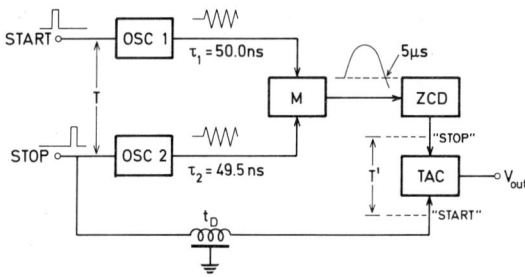

Fig. 5.53. Vernier encoder by COTTINI and GATTI [5.096]

The vernier techniques have been investigated by various authors, and papers have recently been reviewed by BONITZ [5.048] and AMRAM [5.097]. The vernier principle based upon measurement of the phase difference according to Fig. 5.53 can also be used for interpolation purposes in direct digital encoding of time intervals (cf. Chapter 5.4.1) (DE LOTTO, GATTI and VAGHI [5.098, 5.099]). This technique yields a non-limited number of channels and channel widths as small as 50 psec [5.047].

Vernier interpolation techniques can also be used for fast digital encoding of the gating pulse (D) of a WILKINSON amplitude-to-digital encoder, thus yielding an equivalent oscillator (E) frequency of e.g. $1/50$ psec = 20 Gcps!

LEFEVRE and RUSSELL [5.100] realized the vernier technique in an assembly consisting of two ring oscillators, known under the name "vernier chronotron" (Fig. 5.54). The arrangement operates as described in Fig. 5.52. The ring oscillator consists of a delay cable and a gate which also regenerates the circulating pulse. The start and stop pulses are read into the respective delay loops, where they circulate with respective periods of $\tau_1 = 300$ nsec and $\tau_2 = 299$ nsec until the fast coincidence stage C indicates their coincidence and clears the cable loops by cutting off the gates. The number of cycles in the stop loop is counted in a digital scaler and yields the digital equivalent of T (ADDRESS). The channel width is determined by $\Delta\tau = \tau_1 - \tau_2 = 1$ nsec, the channel number is 300. However, due to the attached 256-digit scaler, only 256 channels are employed. The accuracy of the assembly depends upon the constancy of the circulation periods. Further developments of vernier chronotron circuits are reviewed in the papers [5.048] and [5.097].

The measuring range T_{max} of a vernier circuit operating according to Fig. 5.52 is limited by the period τ_1, and the channel number is hence limited to $\tau_1/\Delta\tau$. The measuring range can be increased beyond τ_1

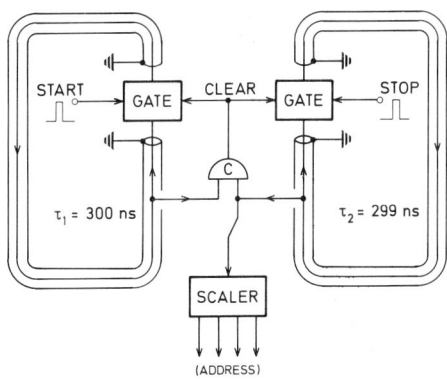

Fig. 5.54. Vernier chronotron according to LEFEVRE and RUSSELL [5.100]

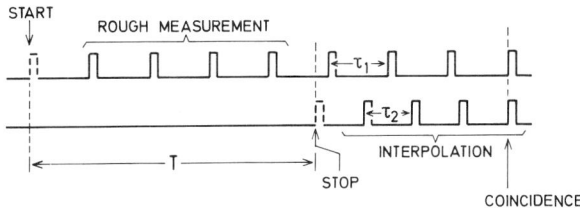

Fig. 5.55. Interpolation of the clock pulse interval by means of the vernier technique

if the start pulse sequence is used as a clock for the preliminary rough time-interval estimation, followed by interpolation using the vernier techniques (Fig. 5.55). The start sequence pulses are counted in a scaler until the arrival of the stop pulse (ROUGH MEASUREMENT). Hereafter the pulses of the stop sequence are counted in another scaler until the coincidence is indicated (INTERPOLATION). CRESSWELL and WILDE [5.101] described such an assembly with conversion frequencies of about 200 Mcps and with a channel width of 0.5 nsec (interpolation 1:10).

5.5. Auxiliary Circuits

Suitable devices must be on hand for testing the time-to-digital-converters and for measuring the time resolution and the integral and differential linearity.

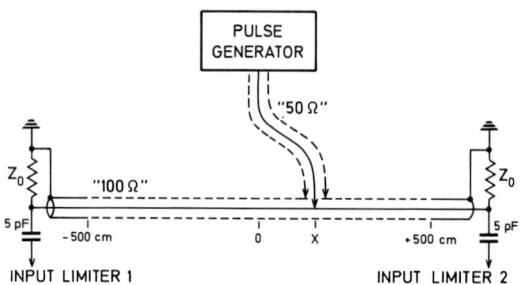

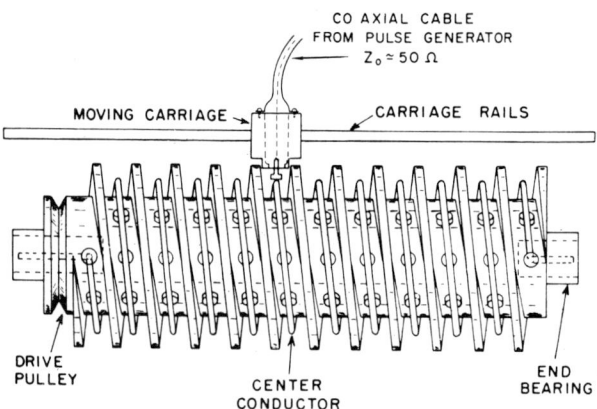

Fig. 5.56. Delay line of continuously variable length, as described by GRAHAM et al. [5.102]

The resolution of coincidence stages and the integral linearity of the converters is commonly measured by applying test pulses from a single test pulse generator to both circuit inputs through delay lines of different lengths. For convenience, delay lines of continuously variable lengths are used. Such a delay line, as described by GRAHAM et al. [5.102], is shown in Fig. 5.56 together with the block diagram of the whole test assembly.

The test pulse from a fast generator is branched from one 50 Ω coaxial cable into two open delay lines with $Z_0 = 100\,\Omega$, which themselves supply the inputs of the two limiters of a coincidence stage or a converter. By moving the cable connection over the whole length (10 m) of the open delay line, the relative lenghts of the left and right parts of the delay line can be varied by 50 nsec. The helical delay line is milled in a brass or aluminium cylinder, the groove profile measuring 11 × 11 mm. The center conductor (∅ 2 mm copper) is supported by lucite nipples at 60° intervals, and its distance from the groove ground is 4.8 mm. Only the centre sliding contact of the moving carriage is shown in the drawing, there being in addition two sliding contacts to the groove edges. The mechanical reproducibility of the relative position of the cable contact is better than 1 mm, i.e. about 1 psec. The delay is a very linear function of the position of the moving carriage, the deviations from absolute linearity being a few tens of psec.

Anyway, it is very important to inject the branched and delayed test pulses into the limiter *inputs* (or into the inputs of other pulse shaping circuits) and not to delay the already shaped limiter pulses by delay lines of variable length, since the length-dependent damping of the delay lines affects the pulse height and shape, and therefore may introduce additional errors.

WEBER [5.103] developed a simple method for testing coincidence stages, which is especially suitable for testing a large number of identical circuits (Fig. 5.57). The frequencies of the two ring oscillators (which are fixed by the cable lengths and which therefore can be regarded as very stable) differ by a few ppm, hence $\tau_1 \approx \tau_2$. The output pulses of the two oscillators are applied to the inputs of the coincidence stage (C). The output pulses of the stage or of the stage mixer are displayed on the screen of a cathode ray oscilloscope CRO, the time base of which is synchronized with the superposition frequency of the two pulse sequencies. Due to the non-synchronism of the two oscillators ($\tau_1 \approx \tau_2$ but $\tau_1 \neq \tau_2$), the pulses shift relatively by $\Delta \tau = \tau_1 - \tau_2$ per coincidence (i.e. per cycle). Hence the device under consideration automatically varies the relative delay of test pulses, which normally requires a manual operation. The envelope of the pulses displayed on the CRO screen therefore yields directly the resolution curve $V_M(t_D)$ of the mixer (cf.

Fig. 5.18). The time scale is defined by the interval $\Delta\tau$ between two adjacent pulses. In Fig. 5.57 $\Delta\tau = 50$ psec. The "channel width" $\Delta\tau$ must be determined by precise measurement of the two oscillator frequencies $1/\tau_1$ and $1/\tau_2$ with the aid of a frequency standard.

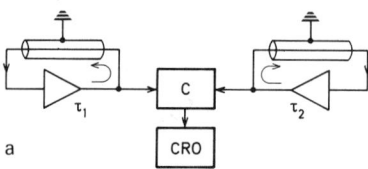

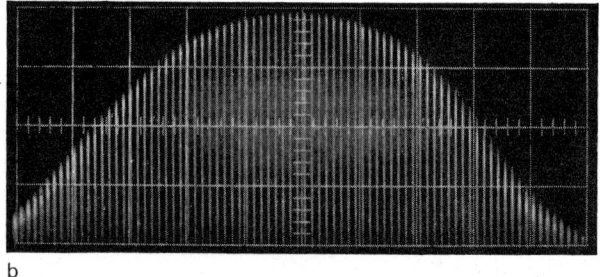

Fig. 5.57 a and b. Operating principle of the test assembly by WEBER [5.103] and a typical CRO display (channel distance 50 psec)

The differential linearity of time-to-digital converters is measured conveniently by applying two non-correlated statistical pulse sequences (e. g. pulses from two separate source-detector systems) to the start and stop inputs: The time interval distribution of the chance coincidences between 0 and T_{max} must be a constant. One of the statistical pulse sequences can be replaced by a periodic signal [5.071, 5.104, 5.142].

Finally, two papers dealing with the adjustment of zero crossing pulse shapers will be discussed shortly. SEYFRIED and DEIKE [5.105] apply periodic pulses with amplitudes modulated between V_{min} and V_{max} to the input of the amplifier followed by a zero crossing discriminator. The discriminator output pulses are displayed on a CRO screen, and the time base is synchronized with the test pulse frequency. When the zero crossover threshold is properly adjusted the discriminator output pulses do not fluctuate despite the amplitude modulation, and the CRO picture is sharp. When it is maladjusted, the leading edge of the pulses becomes blurred.

LANGKAU [5.106] describes a similar arrangement, which can easily be improvised using any cathode ray oscilloscope, and which delivers a *dc* voltage indication directly proportional to the triggering delay.

KIESLER and RIGHINI [5.143] discussed the reliability of measurements of the delays of coaxial cable delay lines. FRANKE and FRITZ [5.144] developed a simple delay circuit, in which the incoming pulse triggers a ramp generator and the output pulse is generated when the ramp reaches a preselected value. By simply changing the slope of the ramp, the delay can be varied between 10 nsec and 100 µsec. BAKER et al. [5.145] discussed the calibration of time-to-pulse-height converters, NADAU et al. [5.146] that of multiple coincidence systems.

6. Digital Circuits

In the foregoing chapters we have shown how to shape, amplify or otherwise process the signals from nuclear radiation detectors in order to conserve an optimum of information during the process of digital encoding. This and the following chapter deal with the processing of digital pulses, each of which carries only the minimum possible information amount of 1 bit. Exactly speaking, we now leave that part of signal processing which is characteristic of nuclear physics and enter the field of industrial electronics and, predominantly, computer techniques.

There are various comprehensive reviews and textbooks on digital electronics, such as [6.001] to [6.007] and [6.068]. A detailed discussion of the basic digital circuits (which is outside the scope of this booklet) with references to the original papers would merely increase the number of review books to be quoted. However, a short treatment of digital electronics is unavoidable for two reasons. The first reason is didactic: any discussion of digital data processing devices without a basic knowledge of digital techniques is impossible. The absence of this chapter would considerably affect the intelligibility of the whole text. The second reason is historical: In the field of fast flip-flops with high time resolution, nuclear electronics could often follow a more direct course than computer techniques. This is due to the fact that in nuclear electronics the flip-flop has a solely counting function (toggle flip-flop), whereas in computer techniques the function of a flip-flop is of a more general nature. Hence we must describe these specific nuclear counting circuits.

In any case, with a few exceptions, only review papers and textbooks are referred to in this short introduction to digital circuits.

> The meaning of the "digital" concept can best be explained by comparing it with the "analog" concept. Analog means the representation of a given magnitude on one scale alone (i.e. by only one voltage value, etc.). The maximum number n of different magnitude values which can be represented in a limited scale range is given by the accuracy of reading the position of the instrument needle (or measuring the particular voltage value etc.). For example, $n=100$ when the scale resolution amounts to 1% of the measuring range. Any particular position of the indicator needle therefore shows an

information amount I, which is (6.01) when measured in bit [6.008]

$$I = \log_2 n. \tag{6.01}$$

The same information I, however, can be represented by using more than one scale, which in turn can exhibit a less fine subdivision. As can easily be seen from (6.01), the subdivision (reciprocal accuracy) must be $n^{1/m}$ when using m scales. This situation is illustrated in Fig. 6.01. A representation is called digital if more than one scale is used.

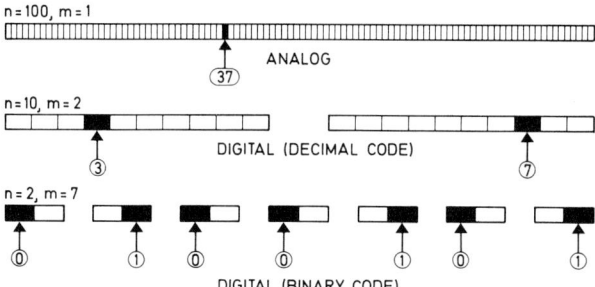

Fig. 6.01. Analog and digital representations of the number 37

The economics of the representation must be assessed. According to WIENER [6.009], the economic effort required to make a scale can be assumed to be proportional to its subdivision n, or better to $(n-1)$, since a scale with $n=1$ (no subdivision) is pointless. Accordingly, the cost A_1 of the analog representation on a single scale is

$$A_1 = c(n-1), \tag{6.02}$$

and the cost A_m of the digital representation of the same information on m scales is

$$A_m = m \cdot c(n^{1/m} - 1). \tag{6.03}$$

A_m exhibits a minimum for $m \to \infty$, i.e. representation on an infinite number of scales without any subdivision ($n^{1/m} = 1$), which is impossible. The nearest approach to this theoretical optimum is the binary representation on scales with only two value ranges (yes — no). In practice the binary optimum is still favoured by the binary character of the physical switches (open, cut off). Therefore the word "digital" is often used in the narrower sense of "binary digital", i.e. the representation of information by signals with only two possible values.

6.1. Basic Digital Circuits

6.1.1. Fundamentals of Boolean Algebra, Gates

Digital electronics deals with bivalent signals. The two possible signal values are denoted by the symbols 0 and 1. In any system the special current or voltage ranges corresponding to the logical 0 and 1 must first be defined. Fig. 6.02 shows an example of such a definition. 1 corresponds to voltages $V \geqslant 2$ Volt, 0 corresponds to voltages $V < 2$ Volt. In order to achieve a clear distinction between 0 and 1 even in the presence of noise, spurious signals etc., a security zone is left on both sides of the 0/1 boundary. This forbidden zone must be wider than the maximum expected noise signals; of course, it must not lie symmetrically in regard to the 0/1 boundary. In Fig. 6.02, for example, 1 corresponds to $V \geqslant 3$ Volt and 0 to $V \leqslant 1$ Volt. Moreover, the voltage ranges corresponding to 0 and 1 are limited downwards (V_{min}) and upwards (V_{max}) by the maximum ratings of the circuit components used.

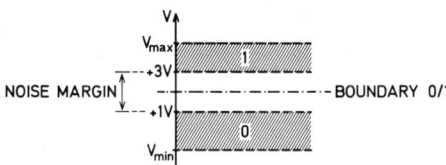

Fig. 6.02. Definition of the logical 0 and 1 (example). The values V_{min} and V_{max} are due to component ratings

Besides the voltage, 0 and 1 can too be represented by current signals. A hybrid representation is also possible, e.g. 1 = input current > threshold I_0; 0 = input voltage < threshold V_0. If the voltage range 1 is more positive than the voltage range 0, the representation is called positive logic (e.g. Fig. 6.02). If 0 is more positive than 1, the logic is called negative.

Fig. 6.03 shows the most general digital circuit. It consists of a finite number m of inputs and one output, the output signal W being a general function of all input signals: $W = f(A, B, ..., M)$. Due to the bivalence of the input signals there are only 2^m different input signal combinations.

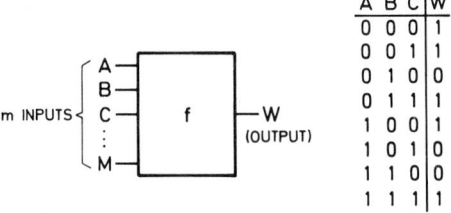

Fig. 6.03. A general digital circuit and the example of a function table for three-input circuit

Fundamental of Boolean Algebra, Gates

Hence the function relationship f can be represented using a discrete table with 2^m rows. This function table shows the desired output signal W for each combination of the input signals. Fig. 6.03 shows an example of such a function table for $m = 3$. However, for $m > 3$ the function table soon becomes obscure.

The rules of calculus valid for systems with bivalent variables have been investigated by GEORGE BOOLE (1815—1864). His algebra laid

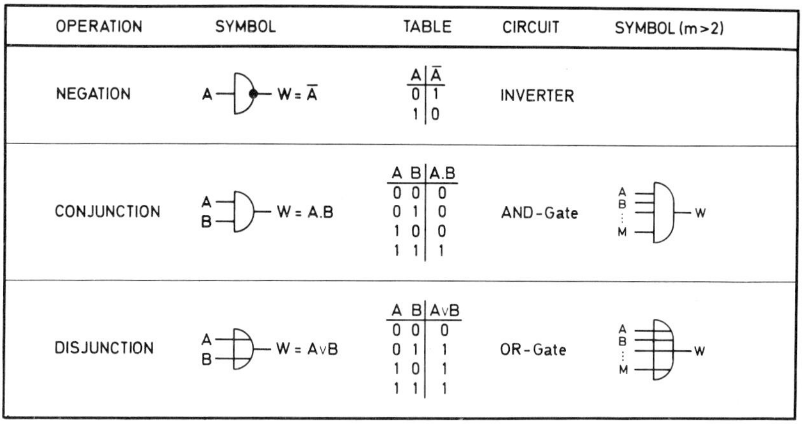

OPERATION	SYMBOL	TABLE	CIRCUIT	SYMBOL (m>2)
NEGATION	A ─▷○─ W = $\bar{A}$	$\begin{array}{c\|c} A & \bar{A} \\ \hline 0 & 1 \\ 1 & 0 \end{array}$	INVERTER	
CONJUNCTION	A,B ─▷─ W = A.B	$\begin{array}{cc\|c} A & B & A.B \\ 0 & 0 & 0 \\ 0 & 1 & 0 \\ 1 & 0 & 0 \\ 1 & 1 & 1 \end{array}$	AND-Gate	
DISJUNCTION	A,B ─▷─ W = A∨B	$\begin{array}{cc\|c} A & B & A\lor B \\ 0 & 0 & 0 \\ 0 & 1 & 1 \\ 1 & 0 & 1 \\ 1 & 1 & 1 \end{array}$	OR-Gate	

Fig. 6.04. Definition of the basic algebraic operations

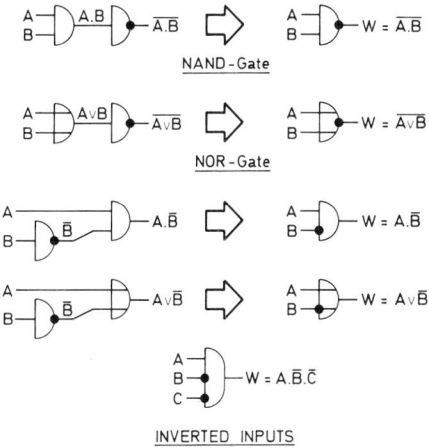

Fig. 6.05. NAND and NOR gates, inverted inputs

271

Digital Circuits

the foundation for modern logical calculus, which was applied later by SHANNON [6.010] to digital networks. A simple introduction to Boolean algebra is given by WEYH [6.011]. The rudiments of Boolean algebra can be reviewed as follows:

Three basic operations can be defined — negation, conjunction and disjunction. Their function tables are summarized in Fig. 6.04. The corresponding circuits for the realization of these operations are denoted as INVERTER, AND-gate and OR-gate, respectively. The definition of conjunction can be extended to more than two inputs: $W = A \cdot B \cdot C \cdots M$ (read W equal to A and B and C and $\cdots M$) is equal to 1 if and only if *all* input variables are equal to 1. Disjunction too can be generalized: $W = A \vee B \vee C \vee \ldots \vee M$ (read W equal to A or B or C or $\ldots M$) is equal to 1, if *at least one* of the input variables is equal to 1. From these three elementary operations more complicated functions can be built up.

Fig. 6.05 shows some frequently used combinations of AND and OR gates with inverters. An AND gate with an inverted output is called a NAND gate (= NOT + AND), an OR gate with an inverted output a NOR gate (= NOT + OR). For simplicity, the inversion of one or more inputs or of the output is denoted by a black dot.

The last row of Fig. 6.05 shows a gate for the realization of the function $W = A \cdot \bar{B} \cdot \bar{C}$. W is equal to 1 only if the combination (A, B, C) of the input variables is equal to $(1, 0, 0)$. Obviously, for any given combination of the input variable values, an AND gate with correspondingly inverted inputs can be found, the output of which just indicates the existence of the particular combination. It is only necessary to invert those inputs whose variables are equal to 0. Using this principle any function table can be expressed using the three elementary operations, and the corresponding logical circuit can be built up using AND and OR gates and inverters. Here each combination (i.e. each table row) leading to $W = 1$ is realized by an AND gate, and the outputs of the AND gates are mixed in an OR gate. For example, the function table of Fig. 6.03 gives

$$W = \bar{A} \cdot \bar{B} \cdot \bar{C} \vee \bar{A} \cdot \bar{B} \cdot C \vee \bar{A} \cdot B \cdot C \vee A \cdot \bar{B} \cdot \bar{C} \vee A \cdot B \cdot C, \quad (6.04)$$

which is realized by the logical circuit shown in Fig. 6.06. The form of expressions such as (6.04) is called disjunctive normal form. In principle, using the disjunctive normal form for every problem given in the form of a function table, it is possible to synthesize the corresponding logical circuit.

However, the disjunctive normal form is in most cases so unwieldy, that a reduction using basic logical rules becomes necessary. The basic rules of Boolean algebra are very simple:

$$X \cdot 1 = X, \quad X \vee 1 = 1,$$
$$X \cdot 0 = 0, \quad X \vee 0 = X,$$
$$X \cdot X = X, \quad X \vee X = X,$$
$$X \cdot \bar{X} = 0, \quad X \vee \bar{X} = 1, \quad \overline{(\bar{X})} = X,$$

$$\left.\begin{array}{l} X \cdot Y \cdot Z = Y \cdot X \cdot Z = Z \cdot Y \cdot X = \text{etc.} \\ X \vee Y \vee Z = Y \vee X \vee Z = \text{etc.} \end{array}\right\} \text{(the commutative law),}$$

$$\left.\begin{array}{l} (X \cdot Y) \cdot Z = X \cdot (Y \cdot Z) = X \cdot Y \cdot Z \\ (X \vee Y) \vee Z = X \vee (Y \vee Z) = X \vee Y \vee Z \end{array}\right\} \text{(the associative law),}$$

$$\left.\begin{array}{l} X \cdot (Y \vee Z) = (X \cdot Y) \vee (X \cdot Z) \\ X \vee (Y \cdot Z) = (X \vee Y) \cdot (X \vee Z) \end{array}\right\} \text{(the distributive law).}$$

Brackets denote the operation to be carried out first, e.g. $A \cdot (B \vee C)$; $(A \cdot B) \vee C$. Conjunction has priority, hence the brackets enclosing the AND operation can be omitted: $(A \cdot B) \vee C = A \cdot B \vee C$. With regard to the algebraic rules, the operation AND recalls the usual multiplication, the operation OR recalls the usual addition, but with one important exception: in Boolean algebra the distributive law holds for both operations.

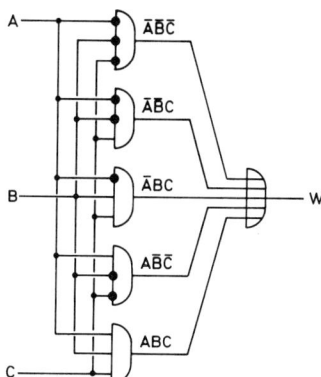

Fig. 6.06. Representation of the logical function W (6.04) by means of elementary gates

As an example of the application of the rules of logic, the expression of the distributive law for disjunction is derived using the basic rules:

$$(X \vee Y) \cdot (X \vee Z) = X \cdot (X \vee Z) \vee Y \cdot (X \vee Z) = X \cdot X \vee X \cdot Z \vee Y \cdot X \vee Y \cdot Z$$
$$= X \vee X \cdot Z \vee X \cdot Y \vee Y \cdot Z = X \cdot 1 \vee X \cdot Z \vee X \cdot Y \vee Y \cdot Z$$
$$= X \cdot (1 \vee Z \vee Y) \vee Y \cdot Z = X \cdot 1 \vee Y \cdot Z = X \vee Y \cdot Z.$$

Digital Circuits

Hence $X \vee (Y \cdot Z) = (X \vee Y) \cdot (X \vee Z)$.

Another very important rule is

$$\overline{X \cdot Y} = \overline{X} \vee \overline{Y}; \quad \overline{X \vee Y} = \overline{X} \cdot \overline{Y} \quad \text{(DE MORGAN theorem)}$$

as well as the generalization by SHANNON:

$$W = f(A, B, C, \ldots, M) \Rightarrow \overline{W} = f^*(\overline{A}, \overline{B}, \overline{C}, \ldots, \overline{M}).$$

Here f^* denotes a function arising from f by replacing each conjunction by a disjunction and vice versa, as in the following simple example:

$$W = (A \cdot B \vee \overline{C}) \cdot D \Rightarrow \overline{W} = (\overline{A} \vee \overline{B}) \cdot C \vee \overline{D}.$$

The SHANNON theorem is very useful for the following situation. Very often (cf. Fig. 6.06) the outputs of AND gates must be combined in an OR gate

$$W = A \cdot B \vee C \cdot D \vee E \cdot F. \tag{6.05}$$

The same operation can be performed easily if only NAND gates are available, as can be seen by a small transformation of (6.05)

$$W = \overline{\overline{(A \cdot B)} \cdot \overline{(C \cdot D)} \cdot \overline{(E \cdot E)}}. \tag{6.06}$$

The corresponding circuit is shown in Fig. 6.07. The same NAND gate serves first to perform the operation of conjunction and subsequently the operation of disjunction.

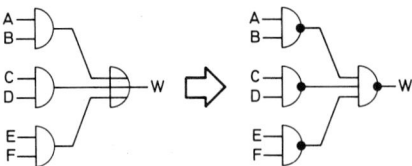

Fig. 6.07. Replacement of AND and OR gates by NAND gates

As can also be easily seen, any circuit performing the AND operation in a given positive logic will perform the OR operation in the corresponding negative logic arising from changing the denotations $0 \leftrightarrow 1$, and vice versa.

All three elementary operations (negation, conjunction and disjunction) can be performed using the NAND gate exclusively:

$$\overline{X} = \overline{(X \cdot X)}; \quad X \cdot Y = \overline{\overline{(X \cdot Y)} \cdot \overline{(X \cdot Y)}}; \quad X \vee Y = \overline{\overline{(X \cdot X)} \cdot \overline{(Y \cdot Y)}};$$

or the NOR gate:

$$\overline{X} = \overline{(X \vee X)}; \quad X \cdot Y = \overline{(X \vee X) \vee (Y \vee Y)}; \quad X \vee Y = \overline{\overline{(X \vee Y)} \vee \overline{(X \vee Y)}}.$$

Fundamental of Boolean Algebra, Gates

Hence, at least theoretically, any digital device can be built up using the NAND or the NOR gates exclusively. To simplify the practical design, however, several different gate modules are preferably used.

Cipher	Dual Code $2^3\ 2^2\ 2^1\ 2^0$	1-2-4-2 Code 2 4 2 1	AIKEN Code 2 4 2 1	Biquinary Code 5 0 4 3 2 1 0
0	0 0 0 0	0 0 0 0	0 0 0 0	0 1 0 0 0 0 1
1	0 0 0 1	0 0 0 1	0 0 0 1	0 1 0 0 0 1 0
2	0 0 1 0	0 0 1 0	0 0 1 0	0 1 0 0 1 0 0
3	0 0 1 1	0 0 1 1	0 0 1 1	0 1 0 1 0 0 0
4	0 1 0 0	0 1 0 0	0 1 0 0	0 1 1 0 0 0 0
5	0 1 0 1	0 1 0 1	1 0 1 1	1 0 0 0 0 0 1
6	0 1 1 0	0 1 1 0	1 1 0 0	1 0 0 0 0 1 0
7	0 1 1 1	0 1 1 1	1 1 0 1	1 0 0 0 1 0 0
8	1 0 0 0	1 1 1 0	1 1 1 0	1 0 0 1 0 0 0
9	1 0 0 1	1 1 1 1	1 1 1 1	1 0 1 0 0 0 0

Fig. 6.08. Various binary codes

When multivalent magnitudes are to be represented using digital means, multi-digit codes must be used. Of course, the most appropriate is the representation of numbers in pure dual code. However, a concession is often made to our conventional calculus, and the numbers are decimal coded, with binary representation of the particular ciphers 0...9 within each decimal digit. The binary code used can either be a pure dual one (i. e. in series of powers of two), or a general binary one[13]. The choice of an actual code depends on the problem to be solved; in counting circuits different codes are used than those in circuits performing algebraic operations. Some frequently used codes are summarized in Fig. 6.08. Since $2^3 < 10 < 2^4$, a binary code for the numbers 0...9 must have at least 4 digits. 6 out of the $2^4 = 16$ possible tetrads of a four-digit code must remain unused (so-called pseudo-tetrads). The 1-2-4-2 code is originated by pulse counting in decade scalers with feedback (Chapter 6.2.1), the corresponding pseudo-tetrads being 1000 to 1101 (8 to 13 of the dual code). The AIKEN code arises when tetrads 0101 to 1010 (i. e. 5 to 10) are omitted. In the AIKEN code the so-called

---

[13] The expression "dual" is used only with regard to a numerical system with the base 2, in analogy to the "decimal" system having the base 10. On the other hand, the expression "binary" merely denotes any representation using bivalent symbols, independent of the code used.

nines-complement $(9-n)$ of any cipher $n=0\ldots 9$ is performed by simple inversion; this effect can be useful in computers using decimal internal organization. The biquinary code is a seven-digit code, and allows a simple error check to be made by performing the cross sum, which must be 2. The whole problem of code choice is discussed in more detail in the reviews [6.007] and [6.012].

The particular digits of an encoded magnitude can either appear simultaneously at the outputs of different lines or sequentially, one digit after another, at a single line (Fig. 6.09). The first case is called parallel representation, the second one series representation. In the series representation a clock-pulse sequence must define the time intervals belonging to the particular digits of the code. A hybrid series-parallel encoding can also be used, e. g. by representing the tetrads of a series decimal code by a parallel binary code using four lines.

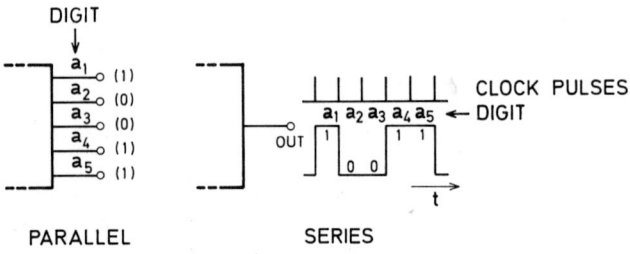

Fig. 6.09. Parallel and series representation of (say) 11001

Regeneration, delay and differentation of digital signals. Besides the logical operations discussed above, a few practical operations are necessary, which do not have any logical character.

When the gate circuits are built up of passive components only, the signal is progressively degraded, and its amplitude becomes smaller. Hence after a few gates the signal may be outside the voltage range corresponding to 0 or 1, and the correct function of the subsequent gates is not assured. Therefore the digital signal must be regenerated by means of active elements (non-linear amplifier) at regular intervals, preferably after each gate. Moreover, any gate output can be loaded only by a limited number of gate inputs. In a higher fan-out is needed, additional regenerative amplifiers (impedance converters, buffer amplifiers) must be introduced. In what follows we shall assume that all gate symbols in the circuit diagrams already include regenerative operations. Wherever the operation of regeneration must be pointed out explicitly, amplifier symbols are used.

Fundamental of Boolean Algebra, Gates

In addition to amplitude regeneration, restoration of the signal shape or of its time position is necessary, especially in series operation. Also, signals belonging to different digit time intervals must often be combined into one logical circuit. Both operations can be performed by using delay circuits which delay the signal synchronously with the clock pulses by one, two or more digit times. In these circuits the original signal is reproduced in the standard shape in the first, second, third etc., following digit time interval, independent of its original degenerate shape or of a possible small non-controllable delay. Some practical delay circuit diagrams will be discussed in the following chapters. In block diagrams the delay circuits are denoted by a rectangle containing a number corresponding to the delay expressed in digit times (Fig. 6.10).

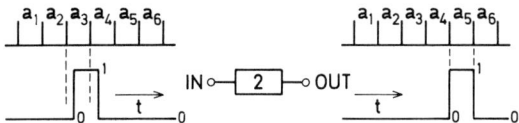

Fig. 6.10. Action of a synchronous delay circuit (delay by 2 digit times)

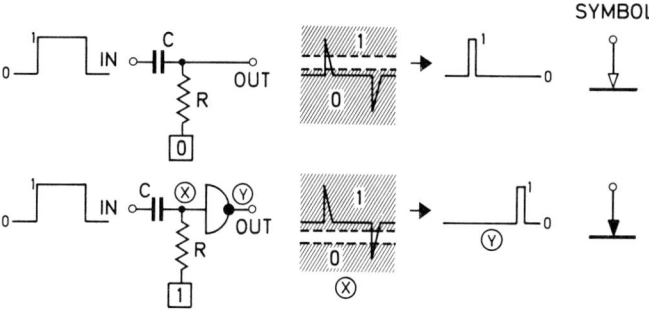

Fig. 6.11. Differentiation of a digital signal by a short time constant

The differentiation of a digital signal by a short time constant RC (i. e. $RC \ll$ digit time interval etc.) is shown in Fig. 6.11. Two cases are possible depending on whether R is connected to a fixed voltage in the voltage range 0 or 1. The upper circuit produces a short 1-pulse (the length of which is given by the value of RC) upon transition of the input signal from $0 \rightarrow 1$. The lower circuit, completed by an inverter, produces a short 1-pulse upon transition from $1 \rightarrow 0$. The differentiation

Digital Circuits

is indicated by an arrowhead at the particular gate input preceded by the differentiator. A white arrowhead indicates action upon transition $0 \to 1$, a black one upon transition $1 \to 0$. Such inputs are called pulse inputs or dynamic inputs.

6.1.2. Circuitry of Different Logics

The most convenient technique for standardizing signals consists in using the transistor states "saturated"—"cut off". In this chapter some of these so-called saturating logics are discussed briefly. All circuits operate with positive logic signals. Of course, after the inversion of the supply voltage polarity and the replacement of the transistors by the

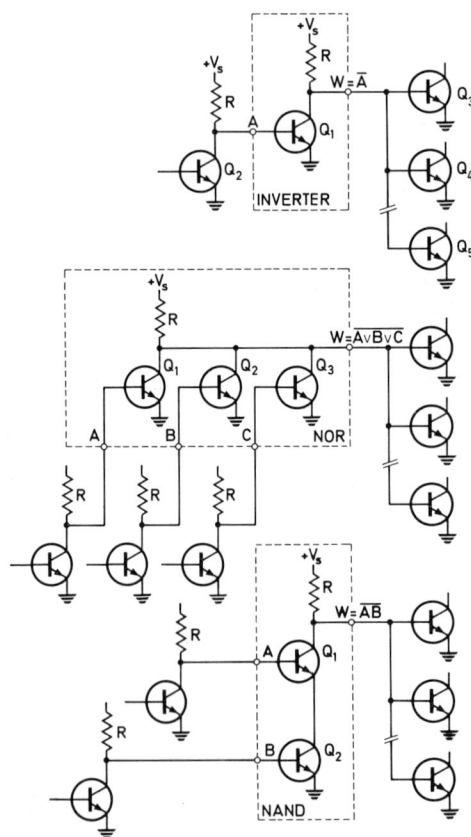

Fig. 6.12. DCTL inverter, NOR and NAND gates

complementary ones, the circuits also accept negative logic signals. With few exceptions the circuits of different logics are not compatible, i. e. gates of different origins would not work satisfactorily if combined in one system. When discussing the operation of particular circuits, we always assume that input signals are delivered by pre-connected gates or inverters of the same type, and that the output is also loaded by gate inputs of the same logic. Different logics are reviewed e. g. by KHAMBATA [6.013] or in the literature [6.002, 6.004, 6.005, 6.068], quoted at the beginning of this chapter.

DCTL (direct-coupled transistor logic). Direct coupling of transistors is possible if the collector-emitter saturation voltage $V_{CE,\,sat}$ is sufficiently lower than the base-emitter voltage $V_{BE,\,on}$, below which the transistor remains cut off. The circuit diagrams of an inverter, a NOR and a NAND gate are shown in Fig. 6.12. For the sake of clarity the preceding and following gate circuits are also shown in this first figure in more detail. Logical 0 denotes voltage $< V_{BE,\,on}$, logical 1 denotes current $> I_{B,\,sat}$ flowing in the particular input. $I_{B,\,sat}$ denotes the minimum base current necessary to saturate the transistor. A logical 0 (i. e. Q_2 saturated) at the input of the inverter cuts off Q_1, hence a current of about V_s/R flows in the bases of the transistors Q_3, Q_4, Q_5 etc. and saturates them (i. e. $W=1$). On the other hand, $A=1$ means Q_2 cut off, a current V_s/R flowing into the base of Q_1 and saturating it, yielding at W an output voltage $V_{CE,\,sat} < V_{BE,\,on}$ (i. e. $W=0$). In the NOR gate obviously the saturation of one out of the three transistors Q_1, Q_2, Q_3 is sufficient for $W=0$. Only if all transistors are cut off, will the current V_s/R flow out of W in the subsequent gates (i. e. $W=1$). This behaviour corresponds to $W = \overline{A \vee B \vee C}$. By connecting two transistors in series, a NAND gate can be realized; only if both Q_1 and Q_2 are saturated is the voltage at W low enough to cut off the following gate transistors ($W=0$). However, instead of the condition $V_{CE,\,sat} < V_{BE,\,on}$, the stronger one $2 \cdot V_{CE,\,sat} < V_{BE,\,on}$ must hold. For three and more inputs the corresponding conditions are difficult to fulfill, therefore DCTL NAND gates with fan-in >2 are seldom used.

The greatest advantage of the DCTL technique is its simplificity. In principle only currents are switched between the gate transistor collector and the bases of the subsequent gate transistors, so that the amplitudes of the voltage pulses remain small and the circuit speed is limited only by the transistors used. However, the transistors are drawn heavily into saturation and the stored charge tends to retard their cutoff. Another advantage is the need for only one supply voltage $+V_s$, which can be relatively low, due to the small amplitude of the voltage signal (e. g. 3...4 V). A great disadvantage is the small security zone

Digital Circuits

between 0 and 1. Since the transistor is not cut off by a negative voltage, it remains sensitive to noise signals of about 0.1 V amplitude and the corresponding noise margin is often insufficient. Another disadvantage is the effect known as "current hogging": Due to the different input resistances of the following transistors, the output current V_s/R is not divided equally under all inputs, some of them drawing more current than others. Since, however, even the transistors with the lowest input current must be saturated, the current is wasted. Hence the fan-out factor (the number of gate which may load a gate output) is limited and much smaller than that which would correspond to the optimum equal current division.

Current hogging can be avoided by introducing resistors $R_1 > R$ in the base connections of the gate transistors (Fig. 6.13) which can be shunted by speed-up capacitors C_1. The collector output now acts as voltage generator, and the base currents are defined by R_1. Of course, the collector voltage amplitude is much higher than in the pure DCTL, consequently the supply voltage must be higher. This modified DCTL technique is sometimes called RTL or RCTL (resistor or resistor capacitor transistor logic). However, the abbreviation RTL also denotes another circuit technique which is described in the following chapter. Although Fig. 6.13 shows only a RCTL inverter, it can be seen immediately that NOR and NAND gates can also be realized in analogy to Fig. 6.12.

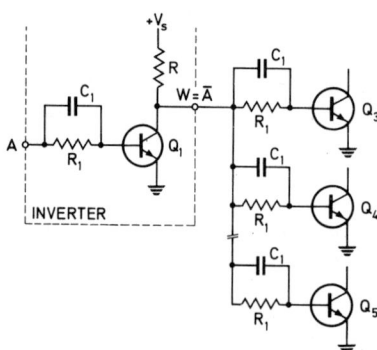

Fig. 6.13. RCTL inverter

RTL (resistor transistor logic). The basis idea of RTL is to bias the bases of the cut-off transistors to an additional negative voltage, in order to enlarge the noise margin (security distance between 0 and 1). The basic inverter circuit is shown in Fig. 6.14. The input signal is

Circuitry of Different Logics

applied to the transistor base through a voltage divider R_1/R_2, if necessary with a speed-up capacitor C_1. $A=0$ hence yields $V_{BE}<0$ and the transistor Q_1 is cut off. On the other hand, the component values of the voltage divider are chosen to yield a positive V_{BE} for $A=1$, which suffices for base current high enough to saturate Q_1. In principle all parallel and series gate configurations of Fig. 6.12 can be rebuilt using this basic circuit. However, the logical operation is mostly performed by resistor networks and the transistor is used only as a non-linear regenerative amplifier. If, for example, the resistor values $3 \cdot R_1$ and R_2 in the lower circuit of Fig. 6.14 are adjusted to always cut off Q_1, except when $A=B=C=1$, then the circuit performs the NAND operation (of course, any speed-up capacitors must be omitted). By changing component values also the NOR characteristics can be realized. Obviously the voltage level corresponding to the logical 1 must be very well defined, and stable, if gates with higher fan-in are desired. Therefore the collector resistors R are completed preferably by limiter diodes D, holding the 1 output voltage equal to V_0 ($V_0<V_{s1}$), independent of the load.

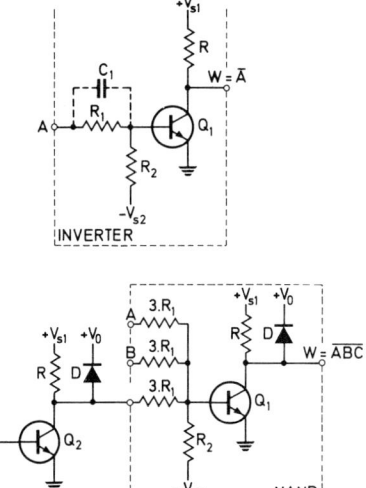

Fig. 6.14. RTL inverter and NAND gate

Relatively cheap, medium-fast logical circuits can be built in RTL techniques. A great disadvantage is the need for two or even three different supply voltages, at least two of which (namely V_0 and $-V_{s2}$) must be extremely stable.

281

DTL (diode transistor logic). The linear mixer of the RTL technique can be replaced by a non-linear one, in connection with which different gate properties are improved. In principle AND and OR gates can be realized using diodes and resistors only (Fig. 6.15). All three inputs of the AND gate must exhibit a logical 1 when the output W must be positive, i.e. when a current V_s/R_D must flow into a load connected to the output ($W=1$). In the OR gate obviously one diode is sufficient to carry over a logical 1 signal to the subsequent circuit. Due to the voltage drop over the diodes, the signal is severely degenerated and cascading a larger number (e.g. of AND gates) is impossible. Moreover the output impedance is different for the two different current directions — therefore it is hardly possible to connect an AND gate to the output of an OR gate. For that reason diode gates necessitate periodic signal regeneration, preferably after each gate.

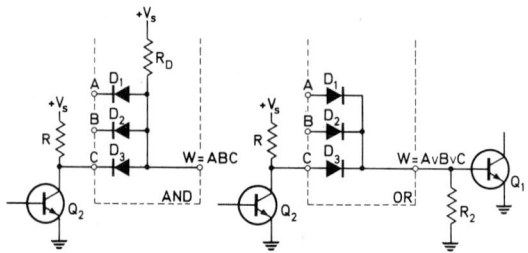

Fig. 6.15. AND and OR gates in the pure diode logic

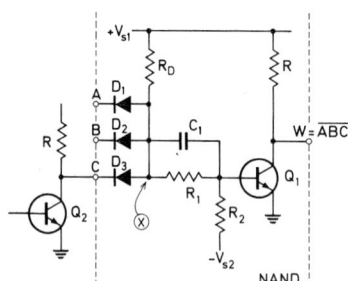

Fig. 6.16. DTL NAND gate

A typical DTL circuit is shown in Fig. 6.16. Exactly speaking, it is a combination of a diode logic AND gate and a RTL inverter for signal regeneration. The voltage divider R_1/R_2 ensures that the transistor Q_1

Circuitry of Different Logics

is well cut off, but a high positive voltage amplitude is needed at (X) in order to open and saturate Q_1. Hence the circuit is slow, the supply voltage V_{s1} must be high, and the ratio speed/power consumption is low.

If the resistor R_1 is replaced by a low impedance voltage source the whole amplitude of the voltage signal at (X) is available for the control of Q_1. Hence V_{s1} can be reduced. Mostly one or two forward biased diodes are used for the required voltage source. Fig. 6.17 (A) shows a

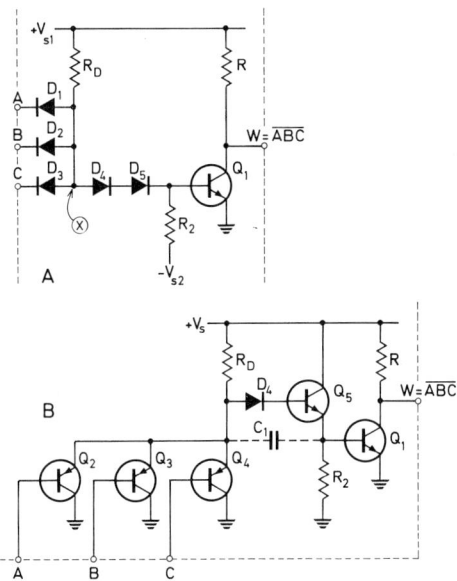

Fig. 6.17 A and B. Two versions of the DTL NAND gate

circuit with two coupling diodes D_4, D_5. The voltage drop over D_4 and D_5 often overcompensates the positive voltage corresponding to logical 0 at (X) ($=V_{CE,\,sat}+$ voltage drop over D_1, D_2 or D_3), hence a slightly negative V_{BE} of Q_1 results. Mostly $V_{BE}\approx 0$ is aimed for, since this is a good compromise between low switching delay of Q_1 and high noise margin. Often a separate negative voltage supply $-V_{s2}$ can be omitted and R_2 is simply connected to ground. Because of the small dynamic forward resistance of D_4+D_5, a small R_2 can be chosen, thus eliminating possible adverse effects due to I_{CBO} of Q_1. Since short switching times and low power consumption are achieved when using the modified DTL technique according to Fig. 6.17 (A), this technique is sometimes denoted by LLL (low level logic).

The gate diodes of circuits Fig. 6.16 and Fig. 6.17 (A) can be replaced by emitter-base junctions of transistors. Due to the current amplification of the resulting emitter-follower configuration, smaller input control currents are needed. The diodes D_4, D_5 can also be partly or entirely replaced by an emitter-follower (Fig. 6.15 (B)). Although built up almost entirely of transistors, the resulting circuit technique is still denoted by DTL, due to its origin. The switching speed of the circuit is somewhat improved by the capacitor C_1, which draws additional current out of the base of Q_1 into the low impedance of Q_2, Q_3 or Q_4 when cutting off Q_1. The fan-in and fan-out factors of the circuit Fig. 6.17 (B) are very high.

TTL (transistor-transistor logic). The diodes of the DTL gate shown in Fig. 6.17 (A) can also be replaced by transistors in another manner, as demonstrated in circuit (B). If the base of a npn-type transistor is used as the common anode of D_4 and D_3 (or D_2 or D_1), the circuit shown in Fig. 6.18 arises. The gate transistors Q_2, Q_3 and Q_4 operate in the extreme saturation, both emitter-base and collector-base junctions are forward biased. Where $A=B=C=1$ the current I_B flows through the collector-base junctions of Q_2, Q_3 and Q_4 into the base of Q_1 and saturates this transistor, yielding $W=0$. If a resistor R is used in the collector circuits of the inverter transistors Q_1 and Q_5, I_B is magnified by the amplifying action of the inversely operated transistors Q_2 to Q_4.

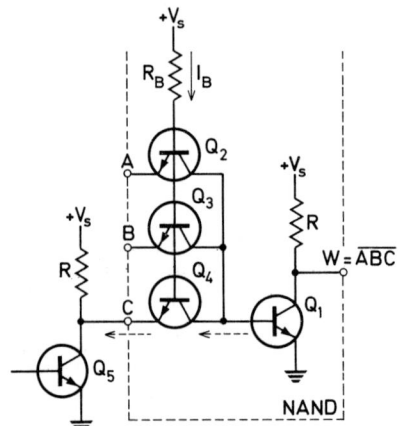

Fig. 6.18. TTL NAND gate

However, the resistor R is not decisive for the circuit function and can be omitted. If now at least one of the inputs is shorted, e. g. $C=0$, the current I_B is directed through Q_5 to ground and Q_1 is cut off (i. e. $W=1$).

Circuitry of Different Logics

Unlike a diode combination, the transistor Q_4 can draw even a high current in the backward direction of the base-collector junction. This current enables the stored base charge of Q_1 to be reduced rapidly. Hence the cut-off delay is very low. Moreover, the gate transistors remain saturated independent of the input state, always conducting in either the conventional or the inverse direction. Since, there is no stored charge to reduce during current inversion, the gate transistors are extremely fast. FOGLESONG [6.014] reports, for example, that the contribution of 2N709 gate transistors at $I_B \approx 3$ mA to the total propagation time of the whole NAND gate is only 0.3 nsec. The TTL technique is used mainly for fast applications, where a large number of active elements can be tolerated. The inverter transistor of the TTL circuits can be combined with DCTL inverter circuits to yield more complex gate systems. Fig. 6.19 shows such a combination of a TTL NAND gate and a DCTL NOR gate.

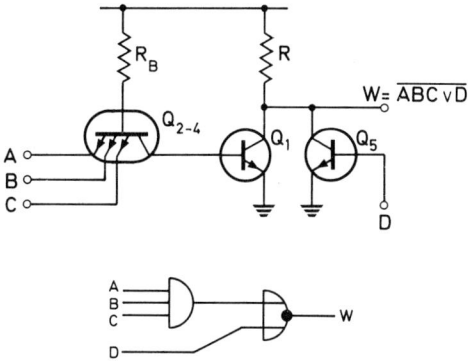

Fig. 6.19. Combination of a TTL NAND gate and a DCTL NOR gate

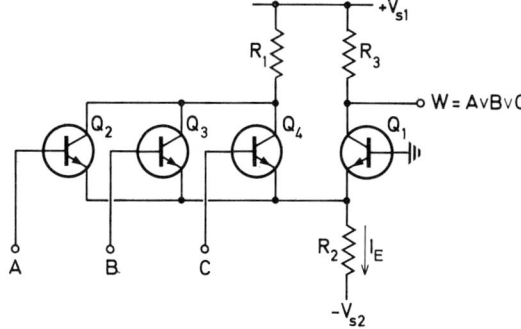

Fig. 6.20. ECTL OR gate (positive logic)

285

Digital Circuits

In saturating logics, the switching speed is limited mainly by the stored base charge of the saturated transistors. Stored charge effects can be avoided if the transistors are not saturated. There are various circuit logics using non-saturated transistors, known as CML (current mode logic) or ECTL (emitter-coupled logic). As an example, Fig. 6.20 shows an ECTL OR gate. If all inputs are $A=B=C=0$ (slightly negative) the transistors Q_2 to Q_4 are cut off, Q_1 conducts and the output voltage is slightly positive (i. e. $W=0$). If, for example, $A=1$ (positive) holds, Q_2 takes over the current I_E, Q_1 is cut off and the output voltage is highly positive (i. e. $W=1$). The worst disadvantage of these circuits is that the voltage ranges corresponding to logical 0 and 1 at the gate input and output differ by a fixed amount. Hence this difference must be allowed for by using Zener diodes to cascade the gates. Or gates with npn and pnp transistors must be used alternately, the voltage range definition being different for both. For more details refer to the quoted review literature.

The choice of a particular circuit technique depends upon various factors, such as signal propagation delay, fan-out factor, noise margin, power consumption per logic function, and economic considerations. Different criteria apply under different conditions. In conventional circuit techniques, for example, the aim is to reduce the number of transistors required for a logical function and a sophisticated RTL circuit may be realized, whereas in the integrated circuit techniques DTL or TTL is preferred, since it is cheeper to manufacture a transistor in a silicon wafer than a precise resistor.

In any case, the choice of a particular circuit technique is not important for an understanding of the operating principles of digital circuits. Having discussed the fundamentals of circuit techniques in this chapter, we shall use only the gate block symbols (Fig. 6.04, 6.05) in what follows. Practical circuits will be discussed only exceptionally, for instance with fast counting type flip-flops.

6.1.3. The Flip-Flop

The bistable character of a *dc* amplifier with positive feedback has already been pointed out in Chapter 4.1. Here the flip-flop — bistable multivibrator — will be investigated from the point of view of logic functions.

A flip-flop arises when two inverters are connected in series (Fig. 6.21a). Due to the limiting conditions $A=\bar{B}$, $B=\bar{A}$, only two (01 or 10) out of four (00, 01, 10 or 11) possible combinations of A, B can exist. Hence the circuit exhibits two stable states. The two outputs are complementary. Therefore the definition of the state of one output is sufficient to indicate the state of the whole flip-flop. For simplicity, one arbitrary output is denoted as the main output Q, and the indication of the flip-flop state relates to this output (viz. $Q=0$ or $Q=1$).

The flip-flop state can be changed by external signals if two NOR gates are used instead of inverters (Fig. 6.21b). The two inputs E, F carry logical 0 in the state of rest. A short 1-pulse at E causes $Q=0$, independent of the original flip-flop state; a short 1-pulse at F sets $Q=1$. Incidentally, the block symbol of Fig. 6.21b indicates a flip-flop of the type described.

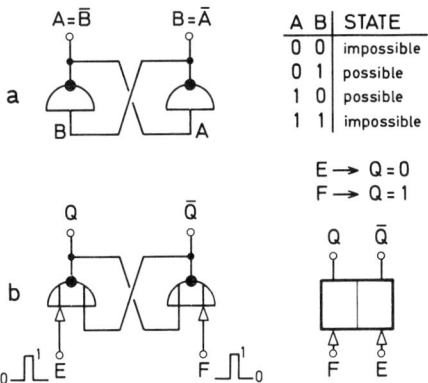

Fig. 6.21 a and b. Flip-flop made of two series-connected inverters or NOR gates

However, the flip-flop state is normally changed by routing one single clock pulse by choice to the inputs E or F using AND gates. A gate-controlled flip-flop is shown in Fig. 6.22a. The clock pulse CP is applied to two AND gates through differentiators acting upon the $0 \to 1$ transition, for example. Depending on whether $S=1$ or $R=1$, the flip-flop is set to $Q=1$ or reset to $Q=0$. When $R=S=0$ the flip-flop state is not influenced (i. e. $Q_{ante}=Q_{post}$). When $R=S=1$ the clock pulse is applied to both NOR gates, leading to an indefinite state (i. e. $Q_{post}=?$). The auxiliary inputs C (clear) and P (preset) allow the flip-flop to be cleared ($Q=0$) or preset ($Q=1$) by external 1-pulses, respectively. This circuit is known as a *RS flip-flop*. It is one of the most popular digital circuits; the related block circuit symbol is shown on the left of Fig. 6.22c. Note that S is situated below Q (and not below $\bar{Q}$ as in the detailed circuit diagram).

The RS flip-flop becomes a scale-of-two (T *flip-flop*), if the inputs R, S are connected to the outputs $Q, \bar{Q}$, respectively (Fig. 6.22b). With $Q=1$ the T pulse ($T=$ toggle or trigger) is routed into the left NOR gate and causes $Q=0$; with $Q=0$ into the right one causing $Q=1$. Hence every T pulse inverts the original flip-flop state ($Q_{post}=\bar{Q}_{ante}$). A differentiator is commonly preconnected to the T-input. For example, Fig. 6.22b shows a T flip-flop acting upon $1 \to 0$ transition.

Digital Circuits

When discussing flip-flop operation, we have neglected the signal propagation delay in the gates. This can be done, since with the exception of a corresponding signal delay, the switching speed of the actual gates does not affect the logical function of a correctly laid out flip-flop.

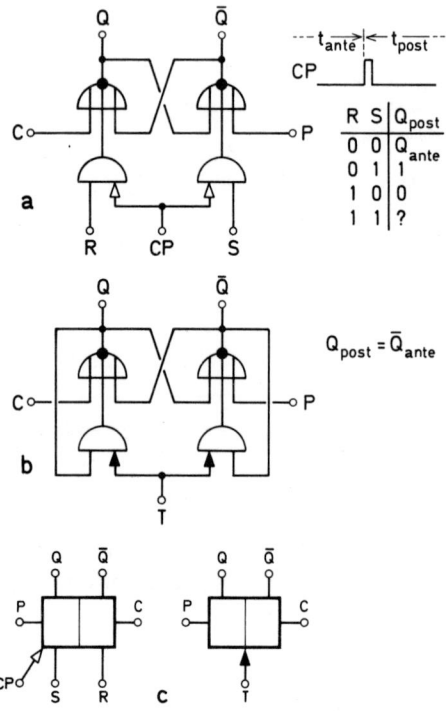

Fig. 6.22a–c. RS flip-flop (a), T flip-flop (scale-of-two) (b), and circuit symbols for both flip-flops

However, in one case the propagation delays can be used to perform logical operations. Fig. 6.23a shows a flip-flop with propagation delays t_R and t_S indicated by separate symbols in the R and S inputs. Of course, the actual delay still depends on the flip-flop state, e. g. $t_R < t_S$ when $Q = 0$ and $t_R > t_S$ when $Q = 1$. The delay dependence on Q or $\bar{Q}$ is symbolized by a dotted line. By a corresponding differentiation of the T signal, the control pulses are made shorter than the minimum delays. Hence, assuming e. g. $Q = 0$, the control pulse at first acts at the left NOR gate, and its state remains unaffected due to $Q = 1$. Thereafter $(t_S > t_R)$ the control pulse reaches the right NOR gate, and toggles the flip-flop. An analogous process is performed if the initial state was $Q=1$. Hence, despite the absence of control gates, the circuit of Fig. 6.23 behaves like a T flip-flop or as a scale-of-two. In practical circuits the actual signal delays are of course much more complex and cannot readily be localized. Anyway, the version in Fig. 6.23 under discussion is representative of the function of almost all practical counting flip-flop circuits.

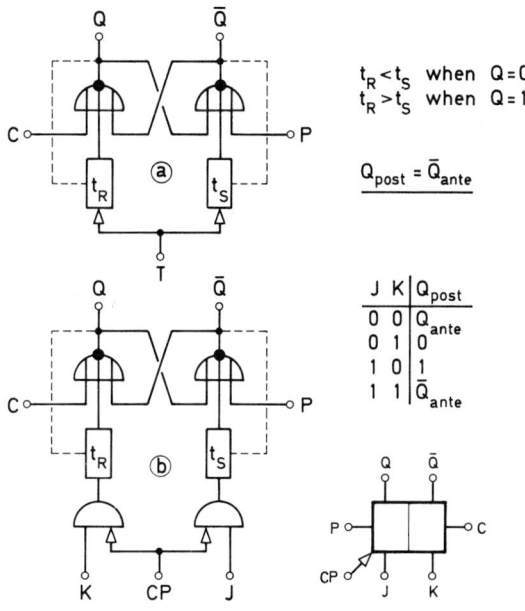

Fig. 6.23a and b. T flip-flop without gate control (a), and JK flip-flop (b)

By adding control gates to a T flip-flop the *JK flip-flop* arises. Its logical organization, the function table and the block diagram circuit are shown in Fig. 6.23b. Instead of the delay-type T flip-flop a gate-controlled one can also be used. The difference between RS and JK flip-flop is apparent only in the last row of the function table. In the last one, the combination $J=K=1$ also yields a significant switching operation, namely the state inversion $Q_{post}=\bar{Q}_{ante}$. Connecting the inputs J and K to a constant voltage in the 1 range, the CP input becomes a T input, and the flip-flop becomes a scale-of-two. The auxiliary inputs C and P are self-explanatory.

289

6.1.4. Practical Flip-Flop Circuits

This chapter will be devoted above all to fast T flip-flops, as required for fast pulse scalers. Most of the T flip-flops are executed in the RTL techniques using two inverters, the triggering action being performed by means of condensers, diodes or emitter-followers connected to the bases or collectors of both inverter transistors. Two conventional circuit diagrams are shown in Fig. 6.24. The one on the left uses relatively

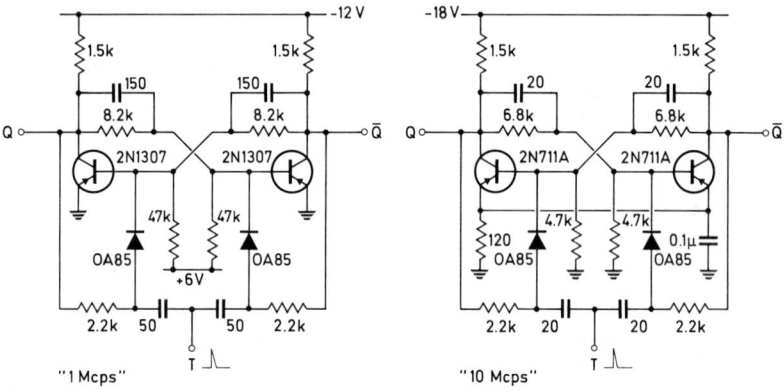

Fig. 6.24. Two practical RTL flip-flop circuits

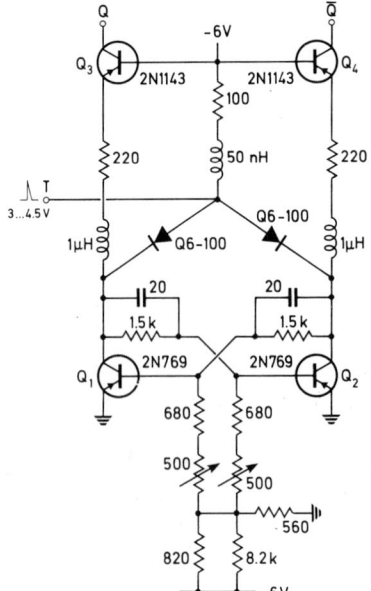

Fig. 6.25. Practical 100 Mcps T flip-flop according to VERWEIJ [6.015]

slow transistors and operates satisfactorily down to 1 μsec time resolution (statistical pulse distribution). The positive trigger pulse is directed by two diode gates in the base of the respective conducting transistor; it cuts off the transistor and hence toggles the flip-flop. Instead of two supply voltages ($+6\,\text{V}$, $-12\,\text{V}$), a single one ($-18\,\text{V}$) can be used when the emitter potential is raised by means of a $120\,\Omega$ resistor as shown in the right-hand circuit. Obviously the counting resolution is given solely by the transistors used: despite having the same load resistors as in the left-hand circuit ($1.5\,\text{k}\Omega$), the right-hand circuit operates reliably over 10 Mcps using the transistors 2N711A. Fig. 6.25 shows an example of a faster circuit according to VERWEIJ [6.015] performing well for pulse repetition frequencies up to 100 Mcps. The circuit technique corresponds to the one used in fast pulse amplifiers (cf. Chapter 3.6). The load of the fast transistors 2N769 consists of a resistor $220\,\Omega$ and a choke $1\,\mu\text{H}$, the trigger pulse is coupled by means of two fast diodes into the collectors of Q_1 and Q_2 and then through two capacitors 20 pF into the corresponding transistor bases. Obviously the diode combination serves as a routing gate, applying the greater part of the cutting off pulse T to the base of the respective conducting transistor. Due to the signal delays the flip-flop changes its state if the pulse T is sufficiently short. The load resistors are connected to the supply voltage $-6\,\text{V}$ by means of the common base stages Q_3, Q_4. The decoupled output signals Q and $\overline{Q}$ are picked up from the collectors of Q_3 and Q_4, respectively. With the aid of $500\,\Omega$ potentiometers the triggering thresholds for the set and reset actions can be adjusted to the same value, thus enabling the flip-flop to be used as a fast discriminator.

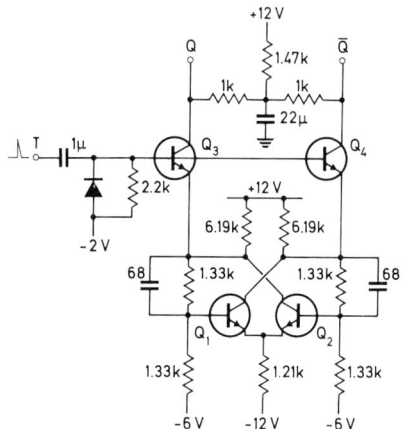

Fig. 6.26. A CML-similar flip-flop according to JACKSON et al. [6.016]

Digital Circuits

The last example shows a circuit using non-saturating transistors according to JACKSON et al. [6.016] (Fig. 6.26). The positive trigger pulse is directed by the two emitter-followers Q_3, Q_4 particularly into the base of the respective cut-off transistor and toggles the flip-flop. Again the collectors of Q_3 and Q_4 can be used as decoupled signal outputs. Using the relatively inexpensive transistors 2N706, counting frequencies of 10 Mcps are achieved.

In all circuits additional direct inputs of the clear or preset type can be realized by means of biased diodes connected to the bases of the inverter transistots.

6.1.5. Tunnel Diode Circuits

The use of tunnel diodes in multivibrators has been discussed in detail in Chapter 4.1.6. Two properties of TD multivibrator are of importance for digital circuits:

1. A TD multivibrator stage exhibits a well-defined triggering threshold and can thus be used for discriminating between the logical 0 and 1 (this corresponds to the regeneration of the digital signal). Hence logical circuits similar to the RTL technique can be realized, using a TD one-shot multivibrator, for example, instead of the output inverter transistor. We have already utilized this principle in the lay-out of various TD coincidence circuits (Chapter 5.3.2), because a coincidence stage is none other than an AND gate. The use of tunnel diodes in digital circuits has been reviewed, e. g. by SPEISER [6.012], so original literature references will be limited to two papers [6.017, 6.018].

2. A (bistable) TD multivibrator exhibits two stable states and can be used directly as a flip-flop. The principle is shown in Fig. 6.27. As is well known, (A) and (B) indicate the two stable states of the circuit. A positive input pulse sets the circuit into the state (B), while a negative one resets it into the state (A). However, a TD flip-flop differs from a RS one, since both set and reset inputs are at the same terminal, set and reset pulses differing only in their polarity.

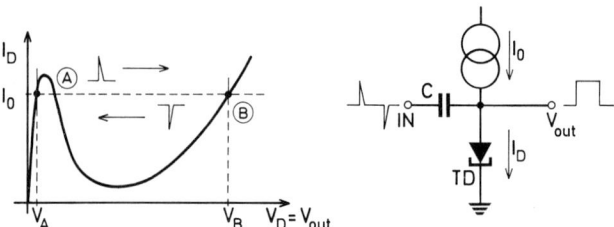

Fig. 6.27. Operating principle of a tunnel diode flip-flop

Separate "R" and "S" inputs can be realized by connecting the tunnel diode in series with two load resistors R ($2R \gg |R_{neg}|$), as shown in Fig. 6.28 a. The diode voltage V_D can again exhibit two different values corresponding to (A) and (B), however, any change in V_D is distributed equally among both resistors R. Hence two complementary outputs Q and $\bar{Q}$ exist (of course, the definition of the voltage ranges corresponding to 0 and 1 is different for Q and $\bar{Q}$). Input pulses at S cause $Q \to 1$, input pulses of the same polarity at R cause $Q \to 0$. HAZONI [6.019] completed the circuit by two diode gates D_1, D_2 (Fig. 6.28 b), which route the input pulse respectively to the circuit terminal at logical 0. Hence the circuit operates as a T flip-flop. The bias voltages V_1 and V_2 of the gate diodes must lie between the 0 and 1 voltages of the respective output. Similar circuits are described by other authors [6.020, 6.021, 6.069]. The counting speed is limited by the coupling diodes D_1, D_2, rather than by the tunnel diode. The output signal can be picked up at the anode or cathode of the tunnel diode. However, in order to avoid disturbances of the flip-flop by the subsequent circuits, a decoupling by a suitable transistor stage is to be recommended. One decoupling possibility is shown in Fig. 6.28 b. Here the supply voltage is applied via the emitter-base junction of the transistor Q_2. At the collector of Q_2 the whole current signal $(V_B - V_A)/2R$ is available without any feedback to the flip-flop.

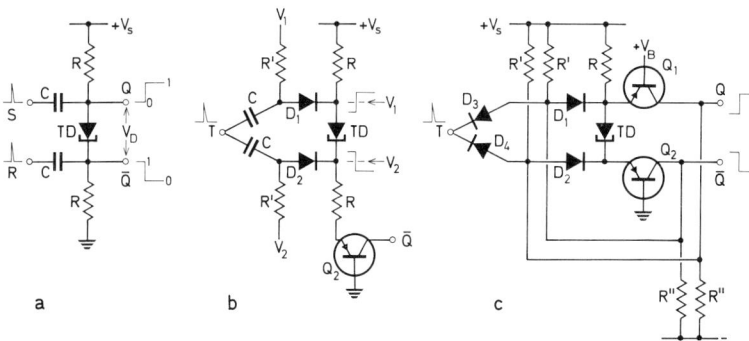

Fig. 6.28 a–c. Tunnel diode flip-flop circuits

RADEKA [6.022] and TAN [6.069] use two transistors in common base configuration for decoupling both outputs Q and $\bar{Q}$. The routing diode gates are controlled by the decoupled output signals. The RADEKA circuit (Fig. 6.28 c) operates satisfactorily up to 250 Mcps.

Another circuit which is often used is based on the proposal made by GOTO [6.023] to replace the load resistor of TD 1 by another tunnel

Digital Circuits

diode (TD 2, Fig. 6.29a). Two out of the three points of intersection of the two diode characteristics are stable. In the point (A) almost the whole voltage drop V_s lies over TD 2 ($Q=0$), in (B) over TD 1, hence making $V_D \approx V_s$ ($Q=1$). If the circuit is completed by a storage element such as a small choke L (Fig. 6.29b), it is toggled by input pulses between the two stable states (the choke conserves the initial current direction during the triggering process). With this circuit counting frequencies up to 400 Mcps have been achieved [6.024]. Sugarman et al. [6.025] made the injection of input pulses symmetrical using a pulse transformer with two separate secondary windings and one separate coupling diode per tunnel diode. In Sugarman's circuit the counting speed seems to be limited mainly by the coupling diodes used. The sensitivity of the Goto circuit of Fig. 6.29b can be improved by introducing another small inductivity at point (X). Baldinger [6.026] uses shorted coaxial cables instead of chokes (Fig. 6.29c) for high count rates (~ 300 Mcps).

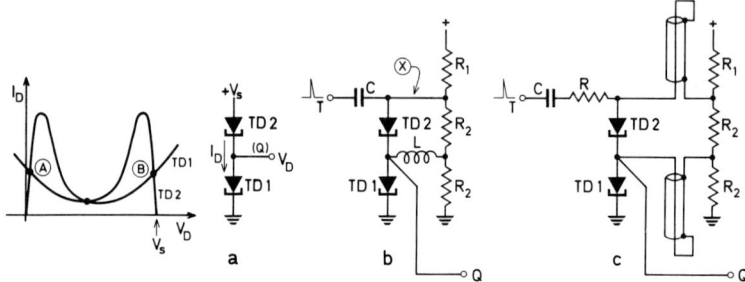

Fig. 6.29a–c. Goto twin-TD flip-flop according to Goto [6.023]

Murata [6.070] presented a detailed analysis of the facing-coupled TD pair circuit.

Weber [6.027] uses the simple flip-flop configuration of Fig. 6.27 for counting purposes, which requires an inverter producing a pulse of opposite polarity for the resetting action. The inverter itself consists of another TD multivibrator and an inverting transformer. Because of the high switching speed, a simple piece of coaxial cable acts as the transformer. The self-explanatory circuit — laid out for negative logic — is shown in Fig. 6.30. A single flip-flop operates reliably up to about 400 Mcps. However, due to the necessary decoupling, the counting speed drops to about 250 Mcps when several flip-flops are connected in series.

Somewhat outside of the systematics, the possibility of making tunnel diode circuits with more than two stable states shall be pointed out. Circuits with ten stable states are of special interest for decimal scalers. Spiegel [6.028] used 10 tunnel diodes connected

Tunnel-Diode Circuits

in series according to Fig. 6.31. This configuration yields I_D versus V_D characteristics with ten negative resistance voltage ranges. When supplied by constant current I_0, the circuit exhibits eleven stable states 0...10. The last state is made metastable by introducing a discriminator (Q_1) with triggering threshold between states 9 and 10, which resets the circuit to the state 0 with the aid of Q_2. The reset pulse serves at the same time as output pulse to the subsequent decade counter. The maximum counting rate is limited mainly by the auxiliary transistor circuits used. RABINOVICI [6.029] reported characteristics with ten negative resistance ranges to be exhibited by a series connection of merely four tunnel diodes. Of course, the respective peak and valley currents must conform to certain conditions.

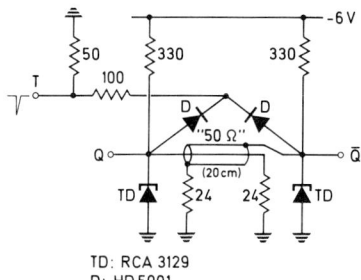

TD: RCA 3129
D: HD 5001

Fig. 6.30. A fast tunnel diode T flip-flop described by WEBER [6.027] (negative logic)

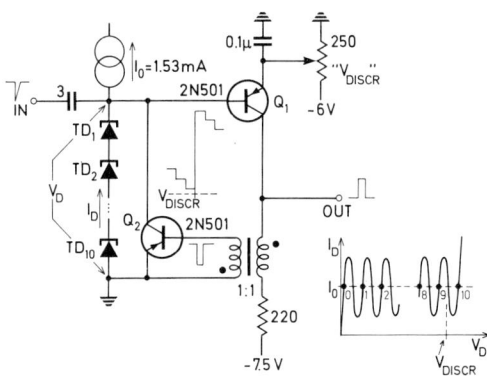

Fig. 6.31. Tunnel diode decimal scaler according to SPIEGEL [6.028]

STANCHI [6.030] pointed out that one simple tunnel diode may yield three stable states if use is made of its dynamic characteristic, which exhibits two and more negative resistance ranges in the frequency range of 1 to 100 Gcps (so-called bitripole).

Despite the possibility of direct realization, decade counters with tunnel diodes are mainly built up in the biquinary configuration (cf. Chapter 6.21), using one T flip-flop and a scale-of-five circuit. For the latter, circuits with five tunnel diodes according to Fig. 6.31 [6.031], or TD ring scalers with five flip-flop stages [6.027] can be employed. BALDINGER and SIMMEN [6.032] described a very reliable ring scaler using tunnel diodes

295

Digital Circuits

decoupled by transistors (Fig. 6.32). If the preceding transistor is cut off, the tunnel diode exhibits two possible stable states (A) and (B). The transistor following after a TD in (A) is cut off, the one after a TD in (B) is saturated. Therefore the TD following after a TD in (B) is in state (C). Hence (B) is always followed by (C), and (C) is followed by (A) or (B).

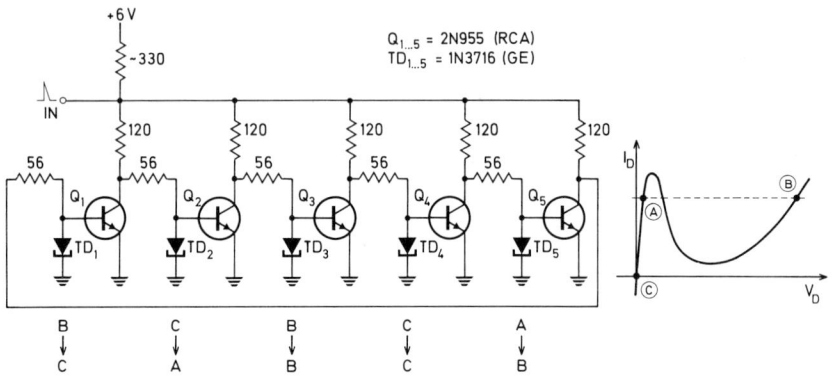

Fig. 6.32. Ring scaler using tunnel diodes and transistors according to BALDINGER and SIMMEN [6.032]

The total current is adjusted to yield e. g. the state BCBCA. A trigger pulse effects the transition (A)→(B), the following stage comes into (C), and the next one performs the transition (C)→(A). Hence (A) is shifted by two positions per triggering pulse (e. g. BCBCA→CABCB→CBCAB→ABCBC etc.). With the component values indicated in the circuit diagram, counting speeds of 70 Mcps have been achieved, with GaAs tunnel diodes and faster silicon transistors up to 140 Mcps. Counting speed of up to 300 Mcps can be reached with modern components [6.071]. A decade counter — TD flip-flop followed by the ring counter — performed well up to 500 Mcps. Another 500 Mcps TD ring counter has been built by TAN and MAXWELL [6.072, 6.073].

6.2. Scalers and Registers

6.2.1. Shift Registers

Shift registers are flip-flop chains capable of storing encoded digital information, in which the information is shifted by one digit per clock pulse. The principle of operation is shown in Fig. 6.33. The *RS* flip-flops used are assumed to switch without any delay. Initially some information is stored in the flip-flops II, for example the word 1011. Clock pulse I transfers this information into flip-flops I, and the following clock pulse II transfers this information into the respective following flip-flops II. Hence two clock pulses, I and II, in the correct sequence, are necessary per shift step. The intermediate storing of the information in the flip-flops I is necessary, since the flip-flops operate without delay, and when they are connected without intermediary register elements, indefinite states may arise.

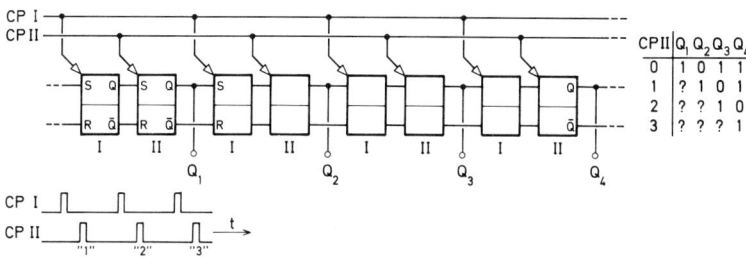

Fig. 6.33. Shift register consisting of two flip-flops per stage

However, the expense of two flip-flops per 1 bit stored information is too high and can be accepted only in very exceptional situations. In practical circuits the intermediary storage is performed almost always by means of some delay elements acting as dynamic storage elements. If the clock pulse is made sufficiently short, e.g. by suitable differentiation, the intrinsic propagation delay of the flip-flops suffices. Fig. 6.34 shows two examples of practical shift register circuits where use is made of the flip-flop propagation delay. In the shift register made of EF flip-flops

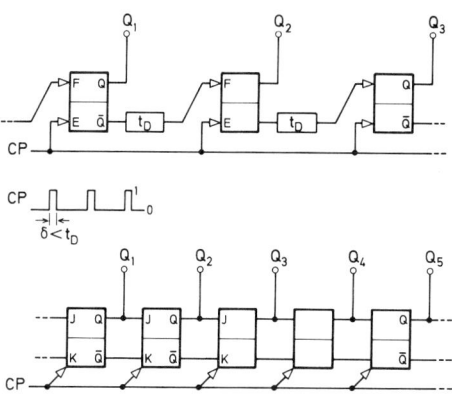

Fig. 6.34. A shift register made of non-gated EF flip-flops (top) and a shift register using JK flip-flops (bottom)

297

Digital Circuits

(without control gates, Fig. 6.21) the delay t_D is indicated by a special diagram symbol, in the JK flip-flop shift register the flip-flop symbol allows for the delay. In both cases the length of the differentiated clock pulse must be smaller than the delay t_D. In the EF shift register the clock pulse at first clears all flip-flops (i.e. $\bar{Q} \to 1$). All stages initially being at 1 (i.e. $Q=0$), produce short 1-pulses upon the transition $0 \to 1$ at $\bar{Q}$, which are delayed by t_D and set the respective following stages to 1. The JK shift register is self-explanatory.

The shifting direction can easily be revolved, if the particular flip-flops are connected together by means of a gate system allowing the sequence of the flip-flops to be changed. Fig. 6.35 shows such an example for the EF shift register of Fig. 6.34. For the sake of clarity, the explicit delay symbols t_D are omitted. Depending on whether the "left" or the "right" AND gates are open, the clock pulses shift the register content to the left or to the right.

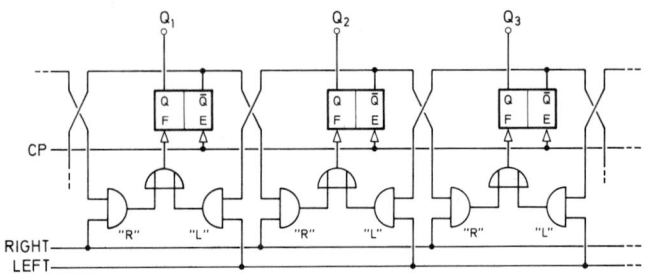

Fig. 6.35. Bidirectional shift register

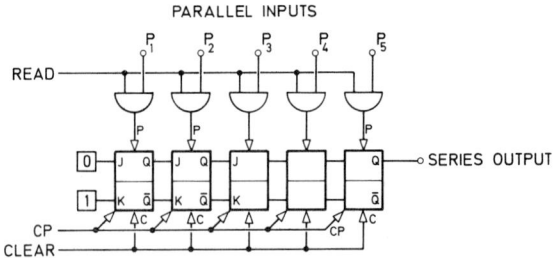

Fig. 6.36. Five-digit parallel-to-series converter

Shift registers are employed, e.g. as scalers or as circulation registers in arithmetical units of computers (this will be discussed briefly later), however, very often a shift register serves for the parallel-to-series conversion of a digital signal. Such a parallel-to-series converter is shown in Fig. 6.36. First all JK flip-flops are cleared by an auxiliary signal at C, then a short pulse READ opens the gates connecting the parallel inputs $P_1, P_2, \ldots, P_5$ with the corresponding preset inputs P of the flip-flops. Hence the flip-flops are set according to the information at P_1 to P_5. A sequence of five clock pulses finally causes this information to appear in series representation at the output of the last flip-flop. If the inputs of the first flip-flop are connected to constant voltages $J = 0$, $K = 1$, during the shifting process, logical 0 is read in all flip-flops, thus making the clearing of the converter unnecessary.

6.2.2. Pulse Scalers

In computer techniques a pulse scaler is mainly required to perform sequential switching operations, but in nuclear metrology pulse counting serves to determine pulse count rates correlated with fundamental physical magnitudes, such as source strengths, transfer probabilities, reaction cross-sections etc. Hence, pulse counting is the most fundamental operation of nuclear electronics.

Ring scalers. Any shift register, or to be precise, any circulation register, can be used as a scale-of-n circuit, with an arbitrary counting base n. Here, prior to counting, all flip-flops except one $(n-1)$ are cleared and one is set to 1. A decade scaler of this type is shown in Fig. 6.37. $Q_0 = 1$ and $Q_1 = Q_2 = \cdots = Q_9 = 0$ is set by the pulse CLEAR. The pulses to be counted are applied to the clock pulse inputs, each input pulse shifts the stored 1 by one position to the right. The tenth input pulse restores the initial state, at the same time producing a transition $0 \to 1$ at Q_9 which is differentiated and used as the output pulse for a subsequent decade. The state of the scaler appears at the ten outputs Q_0 to Q_9 in the simplest digital code, the "1-out-of-10" code.

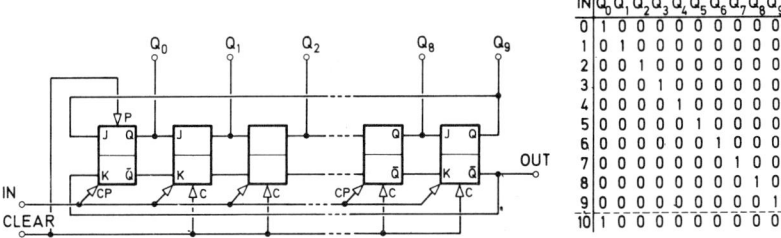

Fig. 6.37. Circulation register used as decimal ring scaler

Digital Circuits

If, for example, some external disturbance causes more than one flip-flop to be set to 1, this is not automatically corrected. This disadvantage can be avoided if the flip-flops are additionally *dc* coupled in order to enable one, and only one, out of *n* flip-flops to be in the state 1, or if an additional gate system is used to suppress unwanted states (Fig. 6.38). The first measure — the introduction of resistors with different values into the common emitter circuits of the respective left and right transistors of all flip-flops — can be applied to flip-flops of the type shown in Fig. 6.24, right-hand side. The higher resistance emitter circuit, in which only one transistor conducts, can be used as the counting input. The input pulse polarity must be chosen so as to cut off the respective conducting transistor. When flip-flops without access to the emitters have to be used (e.g. integrated circuits), the state of the first $(n-1)$ flip-flops is supervised by an AND gate which enables a 1 to be read into the first flip-flop only if $Q_0 = Q_1 = \cdots = Q_8 = 0$.

If bidirectional shift registers such as in Fig. 6.35 are used, bidirectional ring scalers for performing pulse addition and subtraction result.

The number of flip-flops per scale-of-n can be reduced to $n/2$ by using the so-called twisted-ring (Möbiusstrip) configuration. A corresponding decade scaler with five flip-flops is shown in Fig. 6.39. Due to the crossed connections between FF 5 and FF 1, the register at first fills up with 1 and thereafter with 0, as can be seen from the function table. Again the propagation of an ambiguous, false initial state can be avoided by means of an AND gate, which allows a logical 1 to be read into the FF 1 first after all five flip-flops carry a logical 0. A fast twisted-ring saler has been described e.g. by TAN and MAXWELL [6.074].

Binary scalers. Connecting *n* T flip-flops in series produces a dual scaler with 2^n positions (i.e. positions $0\ldots2^{n-1}-1$). An example with $n=5$ is shown in Fig. 6.40. Before counting the pulses the scaler is cleared to $Q_0 = Q_1 = \cdots = Q_4 = 0$ by a CLEAR pulse. Input pulses toggle the flip-flop FF 0. Upon every transition $\bar{Q} = 0 \to 1$ an input pulse is directed to FF 1. In the following stages the counting process repeats accordingly. The state of the scaler is indicated at the parallel outputs $Q_0, Q_1, \ldots, Q_4$ in dual code.

A decade scaler using four flip-flops can be constructed if precautions are taken to suppress 6 out of the $2^4 = 16$ possible states. The simplest way to do this is to generate 6 additional pulses by some artificial means in the course of counting the first 10 pulses. Considering that setting the second flip-flop corresponds to two input pulses, and setting the third to four input pulses, we need only set the FF 1 and the FF 2 one additional time in the course of counting 10 pulses. In the often used circuit of Fig. 6.41 this is effected by a feedback from Q_3 to the *P* inputs of FF 1 and FF 2.

Pulse Scalers

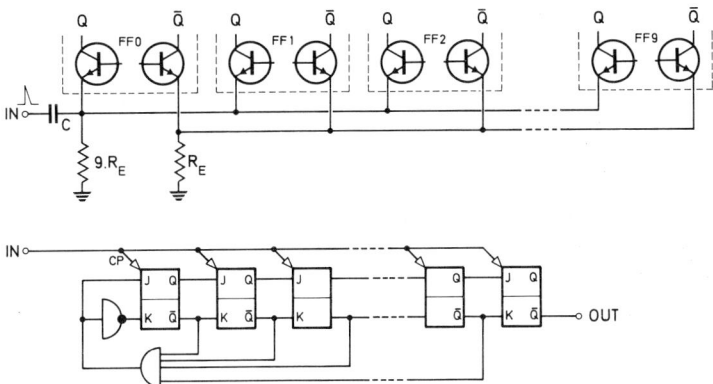

Fig. 6.38. Two ways of suppressing the propagation of false, ambiguous initial register states

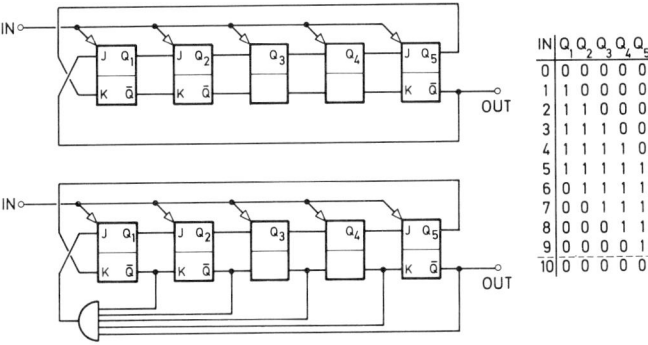

Fig. 6.39. Twisted-ring scaler

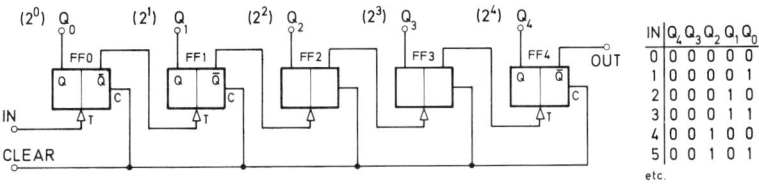

Fig. 6.40. Dual scaler consisting of nT flip-flops

301

Up to the seventh input pulse the circuit operates in pure dual mode. The eighth input pulse causes $Q_3 \to 1$, the differentiated (and delayed) transition $0 \to 1$ sets also $Q_1 \to 1$ and $Q_2 \to 1$ via the feedback connection. Hence after ten input pulses the initial state is reached and an output pulse for the next decade is generated.

Obviously the feedback principle is not limited to the counting base of 10. By feeding back the eighth pulse to none, one, two or all of the first three flip-flops, additional $0, 1, 2, \ldots, 7$ input pulses are simulated. Hence, using four flip-flops, scalers with counting base of $9, 10, 11, \ldots, 16$ can be realized. The generalization of this technique to n flip-flops yields scalers with counting bases between $2^{n-1} + 1$ and 2^n. Moreover, several different feedback loops can be used. The general case of different feedback loops has been discussed by MAXWELL and MARAZZI [6.033].

The feedback pulse can also be derived from any other input pulse besides the eighth. For example, the AIKEN code arises from the pure dual code when adding 6 to all dual expressions, starting with 5 ($=0101$). Hence in the corresponding decimal scaler Fig. 6.42, the state 0101 is controlled by means of an AND gate $X = Q_0 \cdot Q_2 \cdot \bar{Q}_3$. With the fifth input pulse, X becomes 1, the transition $0 \to 1$ sets $Q_1 \to 1$ and $Q_2 \to 0$ by suitable auxiliary inputs. By the last transition $Q_3 \to 1$ is also set. In general a four input AND gate is necessary for the indication of a particular scaler state (e.g. $X = Q_0 \cdot \bar{Q}_1 \cdot Q_2 \cdot \bar{Q}_3$). Since $X = Q_0 \cdot Q_2 \cdot \bar{Q}_3$, except after the fifth input pulse, always yields $X = 0$, the input $\bar{Q}_1$ can be omitted in this particular case.

Besides feedback, any given counting code can be realized by employing gates which control the path of the input pulses to the particular flip-flops. This idea is illustrated in Fig. 6.43 showing a dual coded decimal scaler. Up to the eighth input pulse gate X is open and the scaler operates in pure dual code; after the eighth pulse, X is cut off and Y is open. The tenth pulse causing the transition $0 \to 1$ at $\bar{Q}_0$ resets the FF 3 to $Q_3 = 0$ through Y, hence restoring the initial state 0000. At the same time an output pulse appears at $\bar{Q}_3 \to 1$.

Difference scalers. Occasionally, e.g. when subtracting the background from the overall count rate, a scaler is required which can change its state by $+1$ as well as by -1. Fig. 6.44 illustrates the principle of such a scaler; for the sake of simplicity a pure dual scaler has been chosen. According to whether the right or the left AND gates are open, the binary chain counts in the forward or in the backward direction. The short input pulses (ADD and SUBTR) are lengthened in two monostable multivibrators MMV. They trigger the first flip-flop FF0 and at the same time open the gates for forward or backward counting respectively. The duration of the multivibrator output pulses must be longer than the maximum possible signal propagation delay through the binary

Pulse Scalers

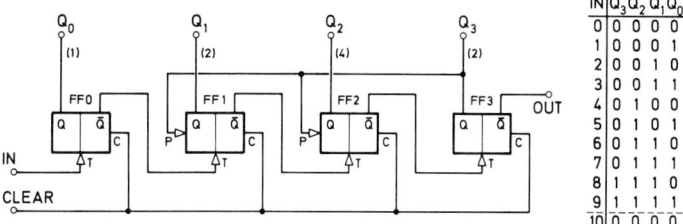

Fig. 6.41. 1-2-4-2 decimal scaler with feedback

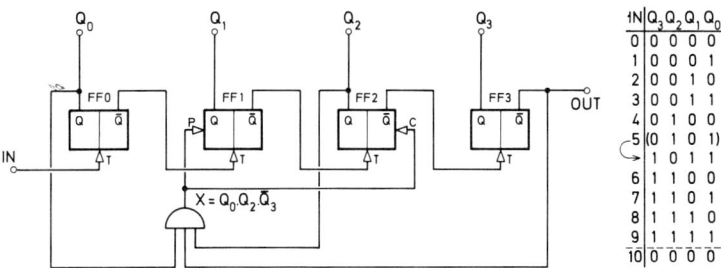

Fig. 6.42. AIKEN code decimal scaler

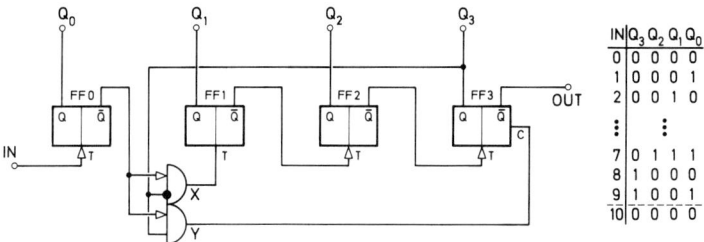

Fig. 6.43. Dual-coded decimal scaler with gate control

Digital Circuits

counting chain. Mutual inhibition of the addition and subtraction inputs protects the scaler from coincident ADD and SUBTR pulses.

If the scaler is to be organized in one of the decimal codes, the dual scaler of Fig. 6.44 must be completed by suitable feedback or control gates which, of course, will be different for forward and backward counting.

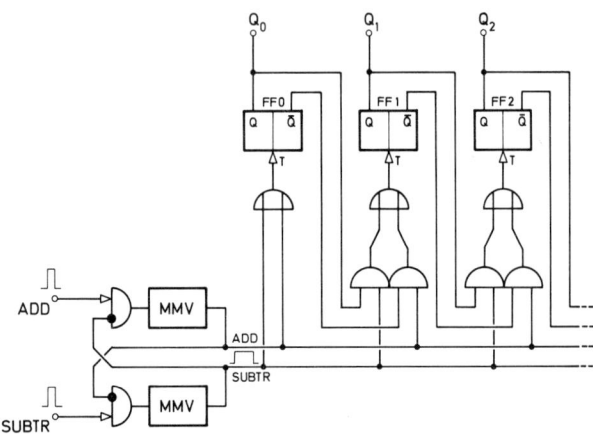

Fig. 6.44. Dual-coded bidirectional scaler

Obviously the counting speed of a scaler according to Fig. 6.44 is not determined by the first-stage resolution but by the much greater signal propagation delay from the first to the last counting stages (the dead time introduced by the monostable multivibrator is even slightly higher. Therefore synchronous scalers are used preferably as difference scalers.

Synchronous scalers. In synchronous scalers all counting stages are triggered together, synchronously with the input pulse. Besides difference scalers, the problem of instant formation of all counting digits arises especially when the state of a free-running scaler must be read out in parallel into a register, e. g. as shown in Fig. 6.36, without interrupting the counting process. This is required in connection with direct digital encoding of time intervals etc. In the read-out process the read pulse must be applied between two counting pulses, but long enough after the preceding counting pulse to ensure a well-established equilibrium, even in the highest counting stages.

The principle of a synchronous scaler is very simple (Fig. 6.45): the input is applied to all flip-flops through gates which are controlled by the states of all preceding flip-flops. Hence all flip-flops toggle synchronously with the input pulse. By suitable additional conditions the dual scaler of Fig. 6.45 can easily be converted into a decimal coded scaler. With the dual scaler, the number of inputs of the particular control gates is increased by one for each counting stage and very soon reaches an unacceptable value. For bigger scalers, preferably only isolated subunits of 3 to 4 flip-flops (e. g. decades) are built into the synchronous mode, the overall organization of the scaler being asynchronous. Ring scalers, too, operate synchronously.

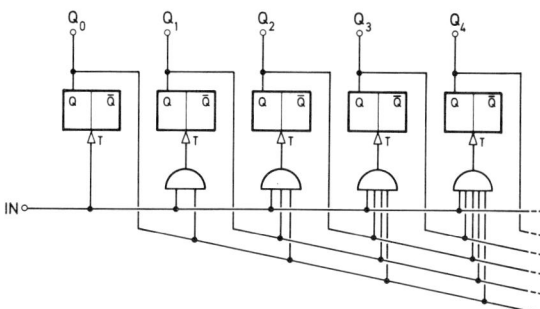

Fig. 6.45. A synchronous dual scaler

Scalers consisting of a register and an adder. Fig. 6.46 shows a so-called dynamic scaler consisting of a circulating register (accumulator) and an adder. The content of the accumulator circulates either continuously or once for each input pulse through the adder. According to the state of the other adder input, the content either remains unchanged or is increased or decreased by 1. The time demand for the simple addition (or subtraction) of 1 is disproportionately high. Nevertheless, such a system may offer advantages if more registers having merely a storage function are combined with a single adder into a multiscaler assembly.

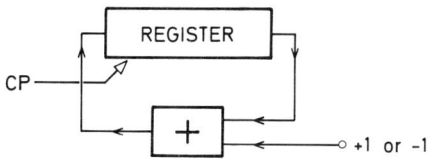

Fig. 6.46. Dynamic scaler consisting of circulating register and an adder

Digital Circuits

Indication of the scaler state. For the optical indication of the actual scaler state, it is sufficient to connect small pilot lamps — if necessary with suitable amplifying components — to the outputs Q or $\bar{Q}$ of all flip-flops. The indication is hence performed in the binary code used. Often the indicator triodes DM 160 (or equivalent) with anode coated by fluorescent compounds, or small gas discharge lamps, are used for indication purposes (Fig. 6.47). Due to their high starting voltage ($\sim$100 V) the discharge lamps must be connected to the logical outputs by means of an amplifier with a high-voltage transistor Q_1 (driver). The bias V_t of the cathode of DM 160 or of the emitter of Q_1 is chosen somewhere between logical 0 and 1. Hence DM 160 or Q_1 conducts only at 1. For the supply voltages V_1 and V_2 the condition $V_2 > V_1$ holds. The difference $V_2 - V_t$ must be higher than the starting voltage, and $V_2 - V_1$ must be smaller than the quenching voltage of the discharge lamp.

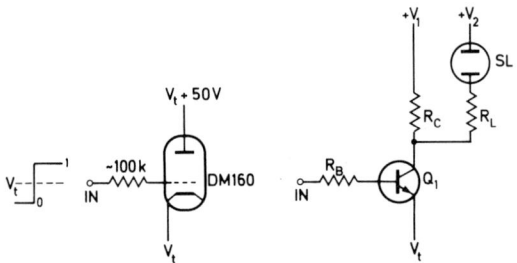

Fig. 6.47. Indication of the flip-flop state by indicator triodes (e. g. DM 160) or by a transistor-driven discharge lamp

A state indication in the binary tetrad code is sufficient if it serves merely as a check, and if the counting results are otherwise automatically decimal-encoded and printed, tape recorded or further processed. However, normally a decimal indication in normal Arabic numerals is required. The simplest technique is to use ten pilot lamps denoted by $0, 1, 2, \ldots, 9$ per decade. Discharge indicators with ten electrodes having the shape of numerals 0 to 9 (so called NIXIE indicators) have also found widespread use. The inputs of the ten indicator lamps or the ten electrodes of a NIXIE tube are connected to the four flip-flop outputs of a decade by means of an encoding matrix. The example in Fig. 6.48 is connected in accordance with the 1-2-4-2 code. In principle the encoding consists of forming the ten disjunctive normal forms (6.04), all of which have only one conjunction term. In the general case, ten AND gates with four inputs each are necessary for the encoding matrix.

However, since not all of the 16 tetrads are used, some conjunctions can be reduced without introducing ambiguities, as can readily be seen. The encoding principle can easily be applied to all codes.

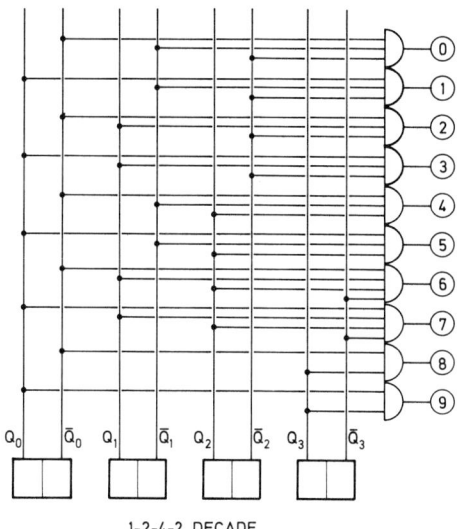

Fig. 6.48. Encoding matrix for the circuit of Fig. 6.41

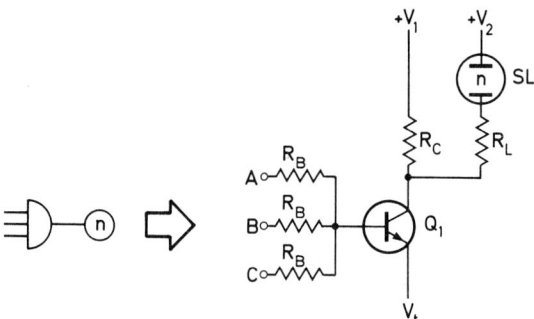

Fig. 6.49. A RTL AND gate with state indication

The AND gates can be performed in any circuit logics, for example, using diodes. However, since the time resolution is not critical, the inexpensive RTL technique is preferred. As shown in Fig. 6.49, for example, the non-linear transistorized amplifier of Fig. 6.47 can be completed by two, three or four preconnected resistors to and AND gate.

Digital Circuits

The bias V_t is adjusted to light the pilot lamps only when $A=B=C=1$. When the voltage states belonging to the logical 0 and 1 are not well defined, such RTL gates with four inputs may cause difficulties.

The general review literature quoted at the beginning of this chapter describes many examples of practical decade scaler circuits. Here only few references will be made to papers describing inventive solutions. ALEXANDER and HEYWOOD [6.034] investigated a decade with non-saturated transistors (CML techniques) for 20 Mcps. KUCHELA [6.035] described a simple feedback circuit for the 1-2-2-4 code. BONDAR [6.036] constructed a bidirectional decimal scaler. COOKE-YARBOROUGH et al. [6.037] used only two flip-flops and a tristable circuit for a decimal scaler, hence somewhat reducing the number of circuit components. Very often decimal scalers are made up of a combination of a flip-flop with a five-member ring scaler. These so-called biquinary scalers enable high counting speeds to be achieved [6.038, 6.039].

There exist also different special counting tubes, such as various cold cathode tubes (decatrons) or the heated vacuum tube E1T (cf. e. g. [6.040] to [6.042]). However, in modern circuits these tubes are seldom used.

6.3. Logical and Arithmetical Digital Circuits

In this chapter some examples of circuits used in data processing systems will be discussed.

Comparison of digital-encoded numbers. Very often a circuit is needed to decide which of the two given numbers is the greater (or whether both are equal). Fig. 6.50 summarizes the situation in the case of two single-digit dual numbers A and B. According to which of the three relations $A>B$, $A=B$ or $A<B$ holds, an output signal appears at W^+, $W^=$ or W^-, respectively. The relation $W^= = A \cdot B \vee \bar{A} \cdot \bar{B}$ can be further simplified using Boolean algebra. The calculation will be quoted in detail, serving as another example of the use of the basic logic rules:

$$W^= = A \cdot B \vee \bar{A} \cdot \bar{B} = \overline{(\bar{A} \vee \bar{B}) \cdot (A \vee B)} = \overline{A \cdot \bar{B} \vee \bar{A} \cdot B} = \overline{(A \cdot \bar{B}) \cdot (\bar{A} \cdot B)}.$$

Hence

$$W^= = \overline{W^+} \cdot \overline{W^-}. \qquad (6.07)$$

A B		W^+ $W^=$ W^-	
0 0	A = B	0 1 0	$W^+_= A\bar{B}$
0 1	A < B	0 0 1	$W^=_= AB \vee \bar{A}\bar{B}$
1 0	A > B	1 0 0	
1 1	A = B	0 1 0	$W^-_= \bar{A}B$

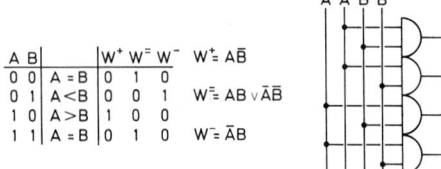

Fig. 6.50. Comparison of single-digit dual numbers A and B

The number of gates in Fig. 6.50 can be reduced by using the relation (6.07). With multidigit dual coded numbers a separate comparison for each digit must be performed, starting with the most significant (highest) one. Once established, non-equality is not affected by the less significant digits (viz. e. g. 10000>01111). Fig. 6.51 shows a comparator circuit for two $(n+1)$-digit dual coded numbers. The circuit makes use of the relation (6.07). Only if the highest, highest but one, etc. digits are equal, is the comparison of the following digit permitted. Equality of the two numbers is indicated only if all corresponding digit pairs are equal. The highest unequal digit pair yields an asymmetry signal blocking the following gates and causing $W^+ = 1$ or $W^- = 1$ through the corresponding NAND gate (here performing an OR function). This circuit

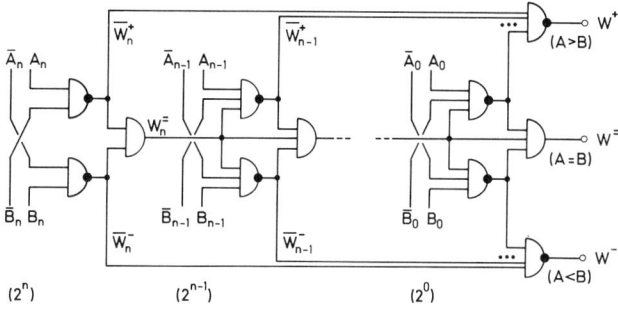

Fig. 6.51. A $(n+1)$-digit number comparator according to the relation (6.07)

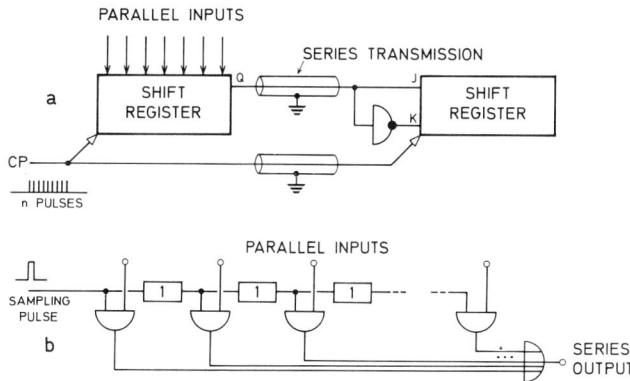

Fig. 6.52a and b. Data transmission between two shift registers (a) and another possibility of parallel-to-series conversion by means of sequential sampling of the parallel outputs (b)

Digital Circuits

has been investigated in detail by VINCENT [6.043], and another one by AMRAM et al. [6.044]. A dual comparator can also be used for binary-coded decimal numbers, if in the tetrad code used a higher dual equivalent corresponds to a higher numeral. Obviously all three tetrad codes shown in Fig. 6.08 fulfil this condition.

Data transmission. As the result of pulse counting, the interesting data are normally present in the parallel representation at n outputs Q or $\bar{Q}$ of the n scaler flip-flops. Though parallel transmission over n lines is the fastest, it necessitates the use of numerous gates an lines. Somewhat more time-consuming but considerably more economical, is the parallel-to-series conversion of the information with subsequent transmission in series representation over a single line. Shift registers (Fig. 6.52a) are especially well suited as input or output devices for series transmission. Both shift registers must be driven by the same clock pulses CP. (When the transmission goes over longer lines causing significant signal delay, the clock pulse sequence of the receiver shift register must be delayed correspondingly.) The series representation can by achieved without using the bulky parallel-to-series converter shown in Fig. 6.36, if the parallel outputs of the scaler are sampled successively (Fig. 6.52b). The sampling pulse is delayed by just one digit time between the inputs of the adjacent AND gates. This delay can be performed by means of a flip-flop chain driven by the clock pulses. In this case a shift register results, acting as a delay line in which a logical 1 runs from left to right ($\rightarrow$ring scaler). However, passive delay lines can be used, too [6.045]. Another shift register, or a reciprocal series-to-parallel converting system using delay lines, acts as the "receiver".

MCNAUGHT and PEARSON [6.046] and HORSTMANN [6.047] report the use of normal telephone links for digital data transmission over longer distances.

Special problems arise, when high-voltage interfaces must be crossed, for example, in course of the transmission of signals from counting equipment located in the high-voltage head of an accelerator to grounded processing equipment [6.075, 6.076]. It is preferable to use light links (light source→receiver) in this case (BUDGE [6.077]).

Addition. The mathematical operation of the addition of two numbers can in principle be performed with the aid of a scaler counting successively the first and the second number of pulses. Using bidirectional scalers, the operation of subtraction can also be performed. When the numbers to be added are not present as pulse sequences, but are already digital-encoded in parallel or series representation, the addition, can be performed by means of a logical network. Using dual code, the addition is very simple, as the following example of adding 12 (1100) and 14 (1110) shows

Logical and Arithmetical Digital Circuits

$$
\begin{array}{r}
12 \\
+14 \\
\hline
26
\end{array}
\quad
\begin{array}{cccccc}
(1)\leftarrow & (1)\leftarrow & (0)\leftarrow & (0)\leftarrow & & C\ (\text{carry}) \\
\downarrow 1 & 1 & 1 & 0 & 0 & A \\
\downarrow 1 & 1 & 1 & 1 & 0 & B \\
\hline
1\ (1) & 1\ (1) & 0\ (0) & 1\ (0) & 0 & S\ (\text{sum})
\end{array}
\quad (6.08)
$$

In (6.08) the carries to the next higher digits are indicated explicitly. In the last significant digit there can be no carry, hence $C_0 = 0$, and the sum has only two terms A_0, B_0. The sum S_0 and the carry C_1 to the next higher digit are formed in a so-called half-adder, whose function table, circuit symbol and two possible circuits are shown in Fig. 6.53. Using the rules of Boolean algebra, we can easily check that $S_0 = \bar{A}_0 \cdot B_0 \vee A_0 \cdot \bar{B}_0 = (A_0 \vee B_0) \cdot \overline{(A_0 \cdot B_0)} = \overline{(\bar{A}_0 \cdot \bar{B}_0)} \vee \overline{(A_0 \cdot B_0)}$. Which of the given circuits will be used depends on whether both A_0 and $\bar{A}_0$ and B_0 and $\bar{B}_0$ are available, etc. Of course, other gate combinations can also be used.

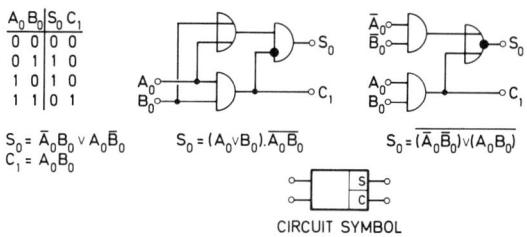

Fig. 6.53. Half-adder

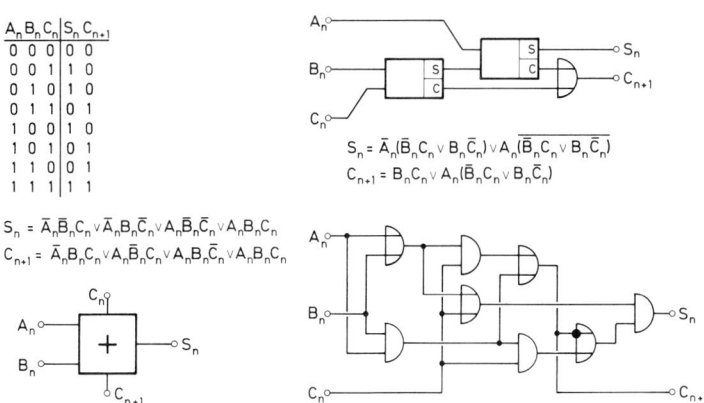

Fig. 6.54. Full-adder

Digital Circuits

The so-called full-adder for the higher digits must have three inputs (A_n, B_n, C_n) and two outputs (S_n, C_{n+1}). The function table, circuit symbol and some practical circuit alternatives are shown in Fig. 6.54. If half-adders are available as building modules, the full-adder can easily be realized using two half-adders (therefore the name "half-adder"). If the adder must be built up using particular AND, OR, NAND or NOR gates, the number of gates can be somewhat reduced by suitable transformation of the S_n and C_{n+1} expressions. One alternative is shown having only 4 OR, 4 AND gates and 1 inverter, and the reader may try to derive for himself the corresponding logical expressions for S_n and C_{n+1}.

For the addition of two dual numbers in *series* representation, one full-adder is necessary (Fig. 6.55), the carry C_{n+1} is delayed in a flip-flop (= one-stage shift register) by one digit time and fed into the input C_n. Instead of employing a separate shift register SR 3 for the sum, the output S_n is commonly reconnected to the input of SR 1. Hence, after the addition is performed, the result $S = A + B$ appears in the shift register SR 1, called the accumulator.

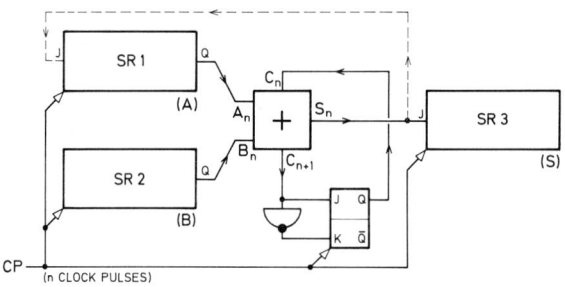

Fig. 6.55. Addition of dual numbers in series representation

The subtraction is performed by adding the so-called ones-complement of the number to be subtracted. The ones-complement brings the original number to ...1111. For example: original number 10110111 yields a ones-complement of 01001000. Obviously, forming the ones-complement consists in the simple inversion of the original number and can be performed by a single inverter in series representation. The result of the subtraction still has to be corrected by 1 in the last significant digit if the addition of the ones-complement yields an "overflow":

Logical and Arithmetical Digital Circuits

$$\begin{array}{r} 23 \\ -9 \\ \hline 14 \end{array} \Rightarrow \begin{array}{r} 10111 \\ -01001 \\ \hline ? \end{array} \Rightarrow \begin{array}{r} 10111 \\ +10110 \\ \hline (1)01101 \\ \llcorner\!\!\rightarrow 1 \\ \hline 01110 \quad (=14) \end{array} \qquad (6.09)$$

For a detailed discussion of the numerical mathematical relationships, or of the measures which must be taken when the accumulator overflows during the pure addition, etc., refer to [6.012] or any other monograph on digital computers

In *parallel* representation the sum is formed instantaneously, the carry output of each adder being connected to the carry input of the next higher one. If the numbers are decimal-coded using a binary tetrad code, the addition must first be performed within each decade. Any decimal carry runs to the next higher decade. With most of the tetrad codes, even the addition within decades is preferably performed in the dual mode, and any false, so-called pseudo-tetrads, are first corrected in a correction network. With the AIKEN code, for example, the dual addition of two numbers always yields the correct result (including the decimal carry), except when one of the following pseudo-tetrads appears: 0101, 0110, 0111, 1000, 1001, 1010. The interim result must be corrected by dual addition of 0110 when no decimal carry is present (= addition

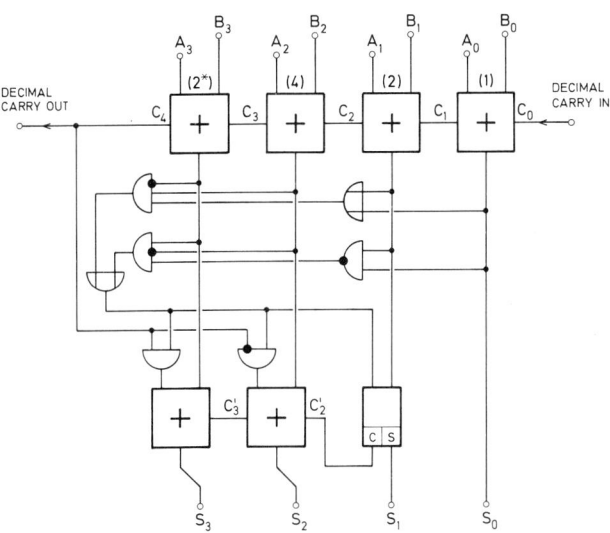

Fig. 6.56. Decimal adder for AIKEN code in parallel tetrad representation

of 6), or by dual addition of 1010 when decimal carry is present (= subtraction of 6). Any carry resulting from the correction addition must be suppressed:

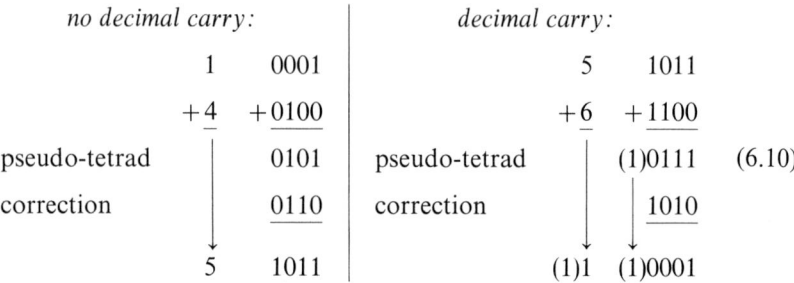

$$\begin{array}{ll}
\textit{no decimal carry:} & \textit{decimal carry:} \\
1 \quad 0001 & 5 \quad 1011 \\
+4 \quad +0100 & +6 \quad +1100 \\
\text{pseudo-tetrad} \quad 0101 & \text{pseudo-tetrad} \quad (1)0111 \\
\text{correction} \quad 0110 & \text{correction} \quad 1010 \\
5 \quad 1011 & (1)1 \quad (1)0001
\end{array} \qquad (6.10)$$

Fig. 6.56 shows a decimal adder for parallel-represented AIKEN-coded digital numbers. The upper row of the four full-adders forms the dual sum, the gate network checks the presence of any pseudo-tetrad and initiates a corresponding correction addition of either 0000, or 0110, or 1010. Such an adder can also be used for series-parallel representation, i. e. when the corresponding tetrad bits are transmitted in parallel, but the particular decimal digits are transmitted in series.

For decimal subtraction, the addition of the so-called nines-complement (completing the original number to …9999) is used [6.012], e. g.

$$\begin{array}{r}
6382 \qquad\qquad 6382 \\
-\ 2931 \ \Rightarrow\ +\ 7068 \\
\hline
3451 \qquad\quad (1)3450 \\
\hookrightarrow 1 \\
\hline
3451
\end{array} \qquad (6.11)$$

Multiplication and division are performed by sequential addition and subtraction respectively within each digit. Multiplication and division by powers of the base (i. e. 2^n in the dual and 10^n in the decimal representation) are performed by shifting to the left or to right by a corresponding number of digits. For the right shift by one digit, one pulse is applied to the clock pulse input of a shift register. The left shift is performed by applying $n-1$ clock pulses to a n-stage circulation register. For mathematical and circuit technique details we shall once more refer to SPEISER [6.012] and other computer monographs.

Dual-to-decimal conversion. Very often a dual-coded number must be translated into the decimal numerical system, and vice versa. Various techniques can be used for solving this problem. In what follows, two examples will be discussed.

Fig. 6.57 shows a circuit converting dual-coded numbers into the dual-coded decimal representation. Use is made of the fact that setting of the first, second, third and fourth flip-flop of a decade scaler corresponds to 1, 2, 4 and 8 input pulses, respectively. The carries are fed into the first flip-flop of the respective following decade. This circuit has still to be completed by a system of delay lines in the dual inputs, in order to ensure that at no time a logical 1 exists at more than one input of any decade. The signals from the particular powers of 2 are fed to the corresponding inputs of the converter scaler, which adds the particular contributions (e. g. $2^7 = 128 = 8 \cdot 10^0 + 2 \cdot 10^1 + 1 \cdot 10^2$). BRINI et al. [6.048] described a converter for nuclear applications using the principle of Fig. 6.57. OXLEY [6.049] investigated another converter type for nuclear applications with selector switches and vacuum tubes.

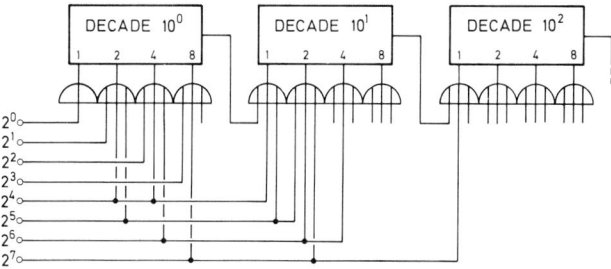

Fig. 6.57. A dual-to-decimal converter

ROWLES et al. [6.050] employ the technique of doubling the scaler state and adding 1 or 0 according to the respective digit of the dual-coded number. This method is illustrated by the conversion example $110101 \rightarrow 53$:

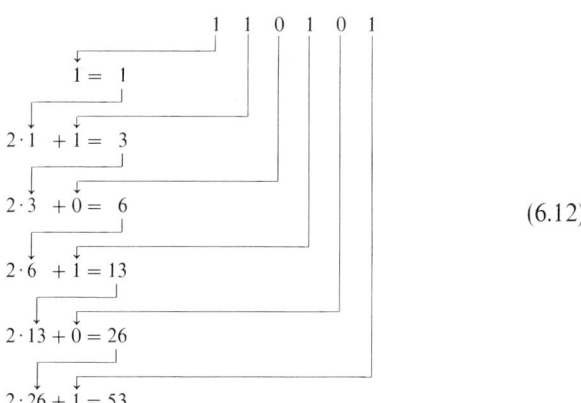

(6.12)

Digital Circuits

This known mathematical method is based upon the following: by sequential doubling, the first digit of the $(n+1)$ digit dual number is multiplied by 2^n, the second one by 2^{n-1}, and so on, down to the last one, which is multiplied by $1 = 2^0$. For the circuit we need a decimal scaler with counting decades equipped with an additional gate system allowing the decade state to be doubled by a common pulse. With decade states >5 a carry to the next decade arises. However, since every doubling forms even numbers, which are terminated by 0 in the dual-coded decimal representation, the initial state of the first flip-flop of every decade just after doubling is 0. Hence only the first flip-flop of the next decade is set to 1 by the carry pulse, and there is no propagation of higher-order carries through the counting chain. The dual number is applied to the input of the first flip-flop of the first decade, starting which the most significant digit, synchronously with the doubling pulses. This technique operates very fast, since only one logical operation has to be performed per bit of the dual-coded information.

The reciprocal conversion, i.e. decimal-to-dual, is performed by similar means (e.g. [6.051]).

6.4. Memories

In principle any element with two stable states can be used for storing digital-coded information. Hence the flip-flop is actually a storage element, scalers are memories with the additional capability of performing the logical operation of adding 1, shift registers are also memories, etc.

However, the construction of high-capacity memories from flip-flops or other active elements is too expensive. Moreover, in the case of failure of the supply voltage, the stored information would be lost. When no additional capabilities, such as the above-mentioned addition etc., are needed, or when no extreme write and read speeds are asked for, memories using cheap passive components are employed. Memories utilizing magnetic storage elements are used mainly in data processing systems.

Magnetizable continua (tapes, drums, plates etc.) are preferred for storage capacities over $10^5 \ldots 10^6$ bits, lower information amounts being stored preferably in assemblies consisting of lumped elements (ferrite cores). In general, a higher memory capacity must be paid for by a longer access time. A distinction must be made between two types of memories: random access memories, where each memory cell can be contacted directly, e.g. by energizing only the selected group of ferrite cores; and memories with sequential read-out where access to a specified memory cell may lead over a great number of other cells (e.g. magnetic tape). Since, in memories with sequential read-out, the selection of the specified cell is performed mainly by mechanical means (tape position), their access time is relatively long. Nevertheless, sequential memories can successfully be used for storing information which enters at high rates, if adjacent cells are used for storing subsequent incoming information, and if the random fluctuation of the information rate is reduced by means of a buffer register (so-called derandomizer).

According to the object of the particular application, different types of memories will be preferred. Ferrite core memories are used almost without exception in multichannel analyzers, in multiscaler assemblies or as buffer registers. Magnetic tape memories or other sequential systems are used in analyzers with a very large number of channels, in multi-parameter analyzers, or for storing of results prior to their "off-line" processing in a digital computer. In digital computers or in other more complex systems there is a whole hierarchy of memories of different capacity and access time.

A detailed discussion of memory systems is outside the scope of this booklet and can be omitted, since this field is covered in a vast number of computer textbooks (e. g. [6.012, 6.068]). In what follows only the operating principle of the ferrite core memory will be briefly described, especially in order to explain some important concepts such as address, address register, read, write, etc.

Ferrite is a magnetizable material with an almost rectangular hysteresis loop and with small electric losses up to very high frequencies. For simplicity, we shall assume that the idealized rectangular characteristics of Fig. 6.58 apply. The two values $+B_r$ and $-B_r$ of the remanent magnetic flux density correspond to the states 1 and 0, respectively (this co-ordination is an arbitrary one). The core can be brought into state 1 by applying to it a magnetic field strength H higher than the critical value of the so-called coercive field H_c; the same field of opposite sign resets the core into the state 0.

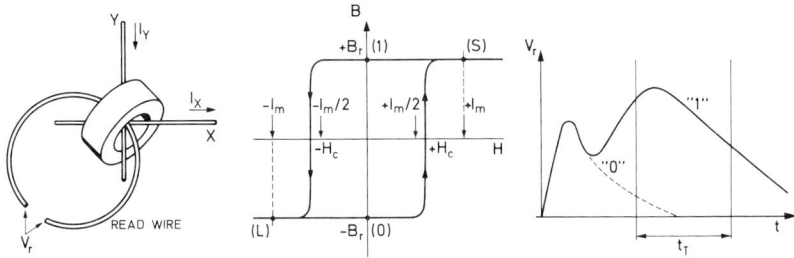

Fig. 6.58. Operating principle of ferrite cores

Due to the well-defined square hysteresis loop, the core is able to discriminate the magnetic field H, or the field inducting current I. Field strengths lower than H_c do not change the state of the core. This can be utilized for the selection of a specified core in the core matrix. The field strength H is induced by the partial currents $I_m/2$ flowing in two perpendicular wires X and Y. The condition $I_m/2 < I_c < I_m$ holds. Here I_c denotes the current corresponding to the coercive field H_c. Since one partial current $I_m/2$ is not high enough to activate the core, only that core lying at the intersection of two energized wires X and Y is always activated.

An additional read wire is necessary to read out the core state. For reading-out, the core is brought into the point L by a negative current pulse $-I_m$. A core initially in state 1 changes its flux density by $-2 \cdot B_r$ and induces a voltage pulse in the read wire, the time integral of which is $2 \cdot F \cdot B_r$ (F = cross-section of the core). $2FB_r$ lies at about 10^{-7} Vs. The actual voltage pulse amplitude still depends on the transition time. A core initially in state 0, on the other hand, does not induce any voltage pulse.

The transition time is approximately proportional to the control current I_m. The proportionality constant is almost the same for all known ferrites. For high read pulses, high control currents I_m are needed, hence ferrites with high coercive field H_c must be used due to the condition $I_m/2 < I_c(H_c) < I_m$. With H_c of a few A/cm, the transition time comes to a few 0.1 μsec and the read pulse voltage amplitude is some 0.1 V.

The characteristic shape of the read pulse $V_r(t)$ for a current step starting at $t=0$ is shown in Fig. 6.58. Due to various non-ideal core properties and due to the induction of disturbance voltages in the read wire by the selection currents, cores in state 0 also produce a pulse, which, however, is smaller and above all shorter than a 1 pulse. The discrimination between 0 and 1 is made either by amplitude discriminators, or better by sampling the read pulse during the time interval t_T delayed in regard to the selection current steps, in so-called read amplifiers.

The read process clears the core. Hence, if a continuation of the information storage is desired, the information must be rewritten in the cores immediately after being read out. In all memories therefore an invariable read-write cycle must be observed for read or write processes. For example, Fig. 6.59 shows the internal organization of a 16 "number" memory, for "numbers" having 1 bit each.

Due to the wiring techniques, the positive direction of the selection currents I_x, I_y changes from wire to wire. The direction of the positive partial currents $+I_m/2$ is thus indicated for each particular selection wire by an arrow. The two current generators I_x and I_y are connected to the selection wires via gates, which are controlled from the address register by means of suitably encoded outputs (for instance, the access to the core "33" is shown in the figure). The read wire passes the memory core in both directions, thus read pulses of both polarities are induced. Therefore the read signal must first be rectified in the read amplifier.

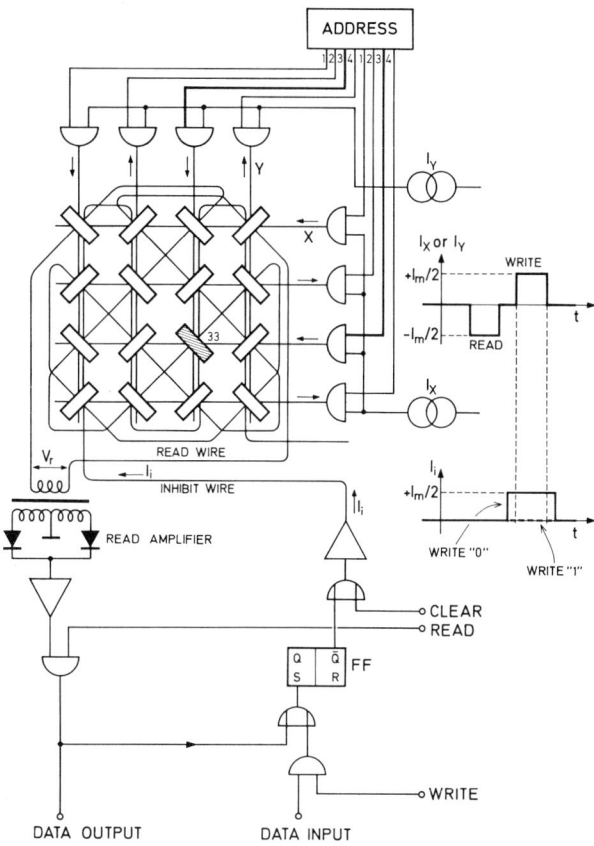

Fig. 6.59. Example of the internal organization of 16 1-bit-number ferrite core memory

The selected core is reset to 0 by means of the current pulse READ $(-I_m/2)$ at X and Y. Its content appears at the read amplifier output, from where it is fed to the external circuits of the assembly (OUTPUT) and to an auxiliary storage register FF (consisting of a single flip-flop

Digital Circuits

in this special case of a 1-bit-number memory). From FF the content is re-written into the core by the subsequent WRITE pulse.

For the write process an additional so-called inhibit-wire is provided, which passes the cores in the opposite direction to the respective Y wires. The current pulse WRITE $(+I_m/2)$ sets the selected core to 1, except when the inhibit-wire carries a current $I_i = I_m/2$. The current I_i just neutralizes the currents I_y, hence if I_i is present, the selected core remains at 0 despite the write pulse.

For data input into the memory, the external data are first transferred to the auxiliary register FF, then the complete read-write cycle is applied. The selected core is first cleared and then either set to 1 or left at 0. The selected core can be cleared during a read-write cycle by activating the input CLEAR.

For storing numbers (words) having more than 1 bit content, many matrix planes with separate read and inhibitor wires, but with common X and Y wires, are correspondingly stacked together. The data input and output then occur in parallel representation. Of course, many other assemblies can be used, having another internal organization, though most of them are based upon the principle of current coincidence selection. The cycle times of the ferrite core memoires in general are between one and a few microseconds (for more details cf. [6.012], [6.052] to [6.054], [6.068]).

Shorter cycle times are achieved in memories based on thin magnetizable films [6.068]. Besides lowering the losses even for the highest frequencies under consideration, thin ferromagnetic films exhibit some interesting physical effects which favour their use for storage purposes. ALEXANDRE, ANTIER and GRUNBERG [6.055] investigated the properties of thin-film memories and their possible applications in nuclear electronics. The cycle time of a practical memory came to 200 nsec. EMMER [6.056] described the use of a fast thin-film memory in an extremely fast multichannel analyzer.

6.5. Data Output

Besides the optical indication of the content of various registers, memories or scalers, very often a data output device is required to record the measuring results in the form of a measuring protocol for later use or for later processing. Various electro-mechanical printers are employed for this purpose. For intermediate data storing prior to later off-line processing in digital computers, punched-tape records have proved to be advantageous. Different printer and punch systems are reviewed in the literature [6.007, 6.012, 6.054, 6.068].

With few exceptions data output is performed in the decimal code. The corresponding output circuit must therefore translate the data from the machine code used into a "1-out-of-10" code, as in Fig. 6.48, though now instead of the indicator lamp the coil of the particular type-bar electromagnet is activated. A distinction must be made between parallel printers, printing at once a complete row of the protocol (e. g. all digits of the scaler, etc.), and series printers, in which only one numeral (or in general one alphanumerical symbol) is printed at once. Most of the tape printers operate in parallel; electric typewriters are series printers. Electromechanical desk computers, which can be used to perform simple calculations or arithmetic operations on the output data, such as totalizing, background subtraction, etc., often need a serial input of the data, whilst the subsequent printing process occurs in parallel (row by row).

A widespread variant of the parallel printer uses type-wheels with the numerals arranged around the wheel circumference. There is one type wheel for each decimal digit, and it can be switched over to the next higher numeral by means of an electromechanically controlled bolt. If the initial wheel position corresponds to 0, it is switched by one input pulse to 1, by two to 2, etc. When all wheels are in the correct position the printing is effected. After or before each printing process all wheels are returned to the initial position.

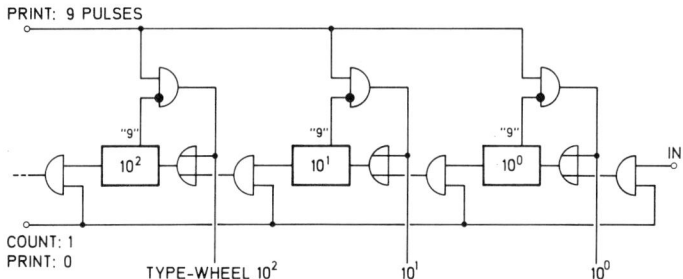

Fig. 6.60. Control of printer type-wheels by means of unitary pulse trains

Putting the wheels in the printing position can be performed very easily, if 9 is chosen as the initial position and each pulse switches the wheels to the next *lower* numeral. Hence, $9-n$ pulses are necessary to put the wheel in the position n. Fig. 6.60 shows a circuit for coupling this printer to a decimal scaler. When the counting process is finished, the decade interconnections are interrupted by an 0-signal (PRINT). Hereafter 9 pulses are fed into the inputs of the particular decades and into the coils of the corresponding type wheels through AND gates, which are cut off as the decade states reach 9. When the content of scalers or registers is to be printed without destroying the flip-flop

Digital Circuits

states, it must be transferred to a buffer register prior to printing. A non-destructive print-out method for BCD encoded numbers by means of unitary pulse trains has been described by POLYCHRONAKIS and PHILOKYPROU [6.078].

When typewriters are used for data output, the printing process must be preceded by (automatic) setting of the tabulator, by which the carriage or the type head is brought to the correct column. Then the data are applied in series representation at a rate of about 10 cps, determined by the mechanical repetition frequency of the type-rods. For series encoding, shift registers or a sequential sampling of the decimal digits as in Fig. 6.52b can be used. However, unlike the internal information flow, the printing process must start with the *most significant* digit. The clock pulse frequency can be delivered by an auxiliary internal oscillator, but it is possible to maximize printing speed by using the termination pulse of one type stroke as the releasing pulse for the next one. When data from several scalers are to be read out, a small programming unit by means of which the number of columns can be programmed offers advantages, since it enables the return transport of the carriage to be activated at the beginning of each row.

With the aid of the simple circuits in Fig. 6.61 non-significant zeros at the beginning of the number can be suppressed. At the beginning of each printing process the RS flip-flop is reset to $Q=0$. Hence the pulses, which otherwise would activate the type 0, are fed to the blank space key. As soon as a numeral which is not zero appears (for which at least one of the four tetrad outputs carries a logical 1), the flip-flop is set to $Q=1$. The following zeros, lying in the middle of the number and thus significant, are printed normally.

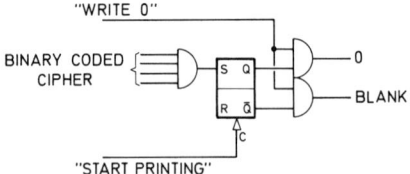

Fig. 6.61. Suppression of non-significant zeros at the beginning of the number

Digital-to-analog conversion. For monitoring digitally encoded magnitudes without high requirements on accuracy, the digital information can be converted into analog and displayed on a measuring instrument or a CRO screen. In principle, all digital-to-analog converters are based upon the addition of currents or voltages corresponding to the weights of the particular digital outputs. The principle is illustrated in

Fig. 6.62. The digital 0 is assumed to correspond to exactly 0 Volt, and the digital 1 to some well-defined standard voltage V_N. The accuracy of the conversion depends upon the validity of this assumption and on the accuracy of the resistors R to $400 R$. As can easily be seen, even with the most stable resistors a conversion of decimal numbers with more than three digits becomes inaccurate.

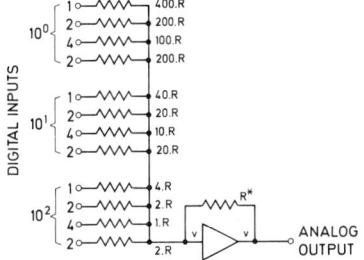

Fig. 6.62. Digital-to-analog conversion of a three-digit long BCD coded decimal number

The condition of "0"↔0 Volt and "1"↔V_N is commonly realized, e.g. by connecting a single highly stable standard voltage source V_N to all digital inputs of the converter through transistors, which are controlled (saturated or cut off) by the digital signals. Another possibility of standardizing the "0" and "1" voltages is offered by diode networks as in Fig. 4.45, or emitter-follower chains as in Fig. 6.63. In the circuit of Fig. 6.63a the partial currents are defined by V_N and the particular emitter resistor R_n; the particular R_n must be graduated to correspond with the weights of the code digits. The partial currents add at the common collector resistor R. The necessary gradation with codes having

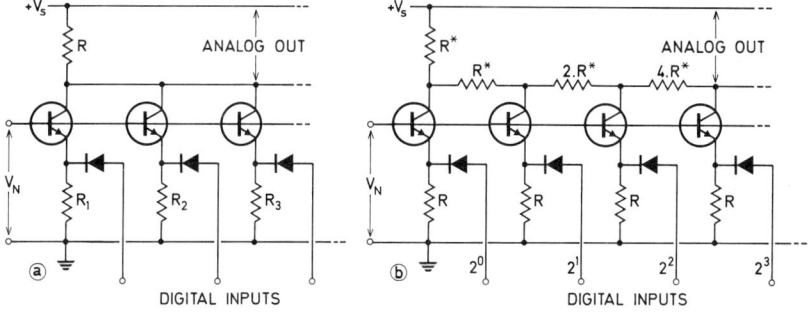

Fig. 6.63a and b. Digital-to-analog converters using emitter-followers

Digital Circuits

monotonously increasing digit weighting can also be performed by a graded collector resistor; all emitter resistors remain equal. Fig. 6.63b instances this principle for the case of pure dual code.

Other circuits, as well as the potential error sources and the accuracies which can be reached, are discussed in detail (e.g.) by BORUCKI and DITTMAN [6.007] and DAKIN and COOKE [6.057].

6.6. Count Rate Meters

Besides pulse counting in digital scalers, the pulse repetition frequency of periodic pulse sequences or the mean count rate of pulses with random distribution can be measured in integrating count rate meters, indicating the count rate by analog representation (e.g. as an analog voltage).

The operating principle of the integrating count rate meter is as follows: a standard value (e.g. a standard charge) is applied to a memory element per input pulse. The particular values are added. At the same time the memory is depleted exponentially with the time constant τ. Hence, the stored value $Q(t)$ becomes

$$Q(t) = \sum_{t_i \leq t} Q_0 \cdot \epsilon^{-\frac{t-t_i}{\tau}}, \tag{6.13}$$

where Q_0 denotes the standard value and t_i denote the instants of incoming pulses. As can easily be seen, the mean value $\bar{Q}$ of $Q(t)$ is

$$\bar{Q} = Q_0 \cdot r \cdot \tau, \tag{6.14}$$

proportional to the count rate r. The mean square deviation σ_Q of $Q(t)$ for pulse sequences obeying the Poisson statistics is

$$\frac{\sigma_Q}{\bar{Q}} = \frac{1}{\sqrt{2r\tau}}. \tag{6.15}$$

The memory can be most simply realized by using a condenser C, to which a small standard charge Q_0 is applied per input pulse by means of the so-called diode pump circuit [6.001, 6.058]. The condenser is discharged by a resistor R, the time constant being $\tau = RC$ (Fig. 6.64). The monostable multivibrator MMV delivers negative pulses of constant amplitude V_0. By means of these pulses the condenser C_0 is charged through the diode D_1. At the trailing edge of each pulse D_1 cuts off and the standard charge $Q_0 = V_0 C_0$ flows through D_2 into $C \gg C_0$. The voltage $V_{out} = Q/C$ across the storage capacitor is measured by means of a suitable impedance converter and a measuring instrument.

The indication remains linear only if $V_{out} \ll V_0$. When this condition is not valid, D_2 is negatively biased and only a part of the charge Q_0 is transferred to C, thus making the relationship between V_{out} and r non-linear. In addition to the basic circuit, Fig. 6.64 shows two linearizing techniques. By means of the bootstrap feedback b) a potential difference V_{out} is conserved over C_0, which just neutralizes the D_2 bias. Another possibility is offered by the Miller integrator c), in which the potential of the point (X) is always zero ("virtual ground"). Therefore D_2 is not biased, irrespective of V_{out}. The same effect is accomplished by using a transistor in common base configuration instead of D_2 and integrating its collector current by C [6.067]. For circuit details we shall refer to the quoted review literature.

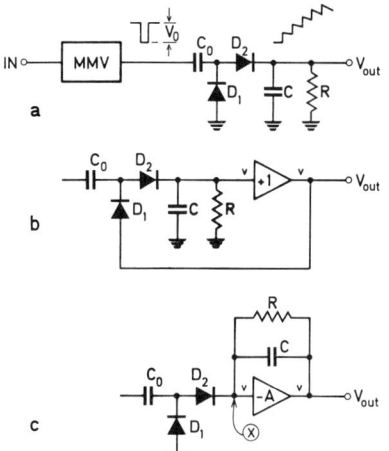

Fig. 6.64 a—c. Operating principle of the diode-pump pulse rate meter (a) and two possibilities for improving its linearity ((b) = bootstrap feedback, (c) = Miller integrator)

In counting practice, besides diode pumps, other circuits are also employed for the additive transfer of the standard charge Q_0 to a condenser C. Complete count rate meter circuits have been published by various authors [6.059] to [6.061]. With the exception of the described linear indication, any integrator of suitably non-linear characteristic may be used. Very often circuits are desired which yield an indication proportional to the logarithm of the count rate [6.062, 6.066, 6.079].

The accuracy of the analog count rate meters is limited by the stability of V_0 and by the voltmeter circuit to about 1%. Difficulties arise when higher time constants of about 1 hour are required (which may be necessary for low-level count rate monitoring).

VINCENT and ROWLES [6.063], WERNER [6.065] and TOLMIE and BRISTOW [6.080] described a digital count rate meter, whose accuracy and time constant is — at least in principle — unlimited. Instead of storing a standard voltage, the pulse number is stored in digital representation in a pulse scaler, raising its state by 1 per input pulse. The exponential depletion is performed by periodic subtraction of a fraction of the scaler state from the same. A power of the scaler base is preferably chosen for the fraction, e.g. 1/1 000 in a decimal sacler, or 1/1 024 in a dual scaler. In this case the division of the scaler state simply consists of shifting digits to the left by a corresponding number. The principle is illustrated in Fig. 6.65.

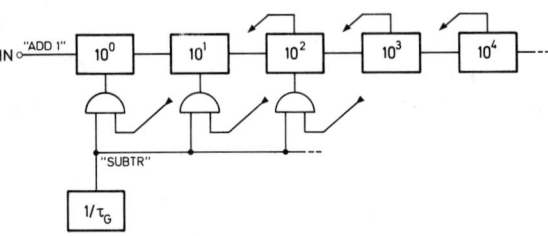

Fig. 6.65. Operating principle of a digital pulse rate meter

A scaler, consisting of five counting decades, for example, is equipped with an auxiliary circuit allowing the content of its three upper decades to be subtracted from the three lower decades upon a command pulse SUBTR. Of course, a bidirectional scaler must be used for this device. The clock pulses SUBTR are delivered from a pulse generator at the count rate of $1/\tau_G$. Hence each τ_G the scaler content is decreased by 1/100 (shifting by two decades!). The corresponding time constant τ amounts to $\tau = -\tau_G/\log(1-1/100) \approx 100 \cdot \tau_G$. Hence the time constant τ can easily be changed by proper choice of $1/\tau_G$.

By the way, the number of clock pulses required to clear entirely a given scaler content is proportional to the logarithm of the same. This possibility of digital logarithm taking has been used by VINCENT et al. [6.064] for the construction of a digital period meter.

7. Data Processing

In this chapter we shall discuss all types of digital devices for the acquisition and processing of nuclear data, from the simplest scaler up to a system including on-line computers. The primary interest commonly consists in the determination of the probability of some specified events, which in turn always leads to the measurement of the count rate of the digitized pulses. Hence, the counting of events for a given time is the key operation in nuclear metrology. Further logical and mathematical operations on the results of pulse rate counting are not specifically a task for nuclear metrology, but can be performed manually, using desk calculators or a general electronic computer, at any time after the results have appeared. Nevertheless, the integration of suitable preliminary data processing, or even of the final processing in an on-line computer, into the actual measuring assembly has great practical importance. It permits a significant interpretation of an experiment still in progress and the physicist is able to make the necessary changes immediately, adjusting certain parameters of the experiment. Hence, optimum results can be obtained even in very complex experiments, the duration of which would otherwise be prohibitive.

The analysis of the visual information in bubble and spark chambers, stored on photographic films, is an important exception to the above generalization. Here the count rate of specified events is less important than the measurement of the energetic and kinetic parameters of a single rare event. Of course, the digitizing techniques are different from those in pulse detector systems. This will be briefly pointed out in chapter 7.5.

7.1. Simple Counting Systems

The simplest counting system consists of a pulse scaler and an input gate, opened for the measuring time t_M. Dividing the count number N by t_M gives the pulse rate r: $r = N/t_M$. The time interval t_M can be measured by any stop-watch, or occasionally by an electromechanical control mechanism. However, even in this rudimentary form of data processing, an electronic timer consisting of a standard frequency

Data Processing

generator and a pulse scaler is generally used. The complete array is shown in Fig. 7.01.

The generator GEN is stabilized by means of a piezoelectric crystal, a tuning fork, or — when no high accuracy is claimed — be means of the mains frequency of 50cps or 60cps. The generator frequency is first divided in a suitable flip-flop chain. Hence a pulse sequence with (e.g.) 1 cps appears at its output. Before counting, both scalers are cleared by the push-button CLEAR. The control flip-flop is also reset to 0. The START signal sets $Q = 1$, both AND gates open and both scalers register the input pulses. The counting process is terminated by the push-button STOP. The state of the timer-scaler yields t_M in units of the reciprocal generator frequency (e.g. in seconds), the state of the pulse scaler yields N.

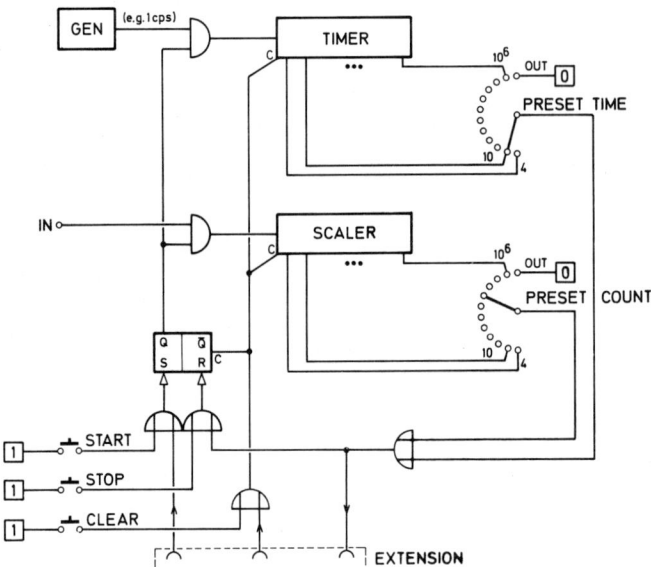

Fig. 7.01. Scaler-timer assembly

The counting process can also be terminated automatically after a preset time interval. Here the outputs of the particular decades of the timer are fed to the reset input of the flip-flop by means of a selector switch PRESET TIME. For finer graduation the outputs of the particular flip-flops within one decade are used. If, for instance, the outputs of all second flip-flops as well as the decade outputs are used in a six-decade timer with 1 cps generator, the following preset times result: $4 - 10 - 40 - \cdots - 4 \cdot 10^5 - 10^6$ sec.

Since the statistical accuracy of the measurement depends upon the total count number N, it is often desirable to preset this value. This can be accomplished by similar means (selector switch PRESET COUNT). If neither of the two preset selectors is set to 0, the measurement is terminated at the condition which applies first. Hence, for example, a set of samples can be counted with constant mean square deviation (preset count), but the single sample measuring time is limited by a preset time, in order to avoid interruption by a possible non-active sample.

Apart from the push-buttons, the scaler-timer system can be controlled also by external signals, for example, from an automatic sample changer. For external control the preset output and the inputs START and CLEAR are fed to a multiterminal connector EXTENSION, to which any auxiliary device must be connected. The preset output pulse then originates (e.g.) the process of sample changing, printing out the results etc. After the external process is terminated, the auxiliary device must activate CLEAR and somewhat later START, thus initiating another pulse counting process.

COLOMBO and STANCHI [7.047] and LÖVBORG [7.048] describe some practical scaler-timer arrays.

7.2. Multiscaler Arrays

The assembly of Fig. 7.01 can be extended to any desired number of scalers controlled by a common timer. Preset count control is also possible, and the stop signal can be delivered by any scaler (usually the scaler with the lowest pulse count rate is chosen for preset count control of the array).

In multiscaler arrays the use of separate scalers may become uneconomic. PECINA and WEIDEMANN [7.001] therefore proposed the construction of a counting system consisting of a central ferrite core memory and an adder. The inputs of the particular channels are connected to buffer scalers of very limited capacity, the contents of which are periodically transfered to corresponding locations of the memory and totalized.

Multiscaler arrays are seldom read out manually by the operator, but almost always automatically with the aid of a printer, tape puncher etc. In this case the optical indication of the decade states in decimal code can be omitted, thus saving the expensive encoding matrix. For test purposes the outputs of the flip-flops are equipped with binary indicators, such as DM 160, displaying the state in the actual binary code.

Data Processing

Often one central encoding matrix with decimal display by means of Nixie tubes is used; this can be connected to any particular scaler by means of pushbutton switches [7.002].

For read-out purposes the scalers are connected by means of multi-wire cables to a central control unit which itself is connected to the printer [7.003]. The control unit samples by means of a gate system the scaler outputs in a preselected order and connects them to the printer input. The read-out process is initiated by a start pulse PRINT. The maximum number of scalers which can be read out is limited by the number of available control unit inputs.

The disadvantage of the system described is the necessity of connecting each scaler to the control unit by a separate, often very long cable. This can be avoided by locating the sampling gates in the particular scalers. In this case all the scalers must be connected together and only the first need be connected to the control unit. Since the scalers are normally situated one over the other in vertical racks, the interconnections between them are simplified. The particular sampling gates are controlled by signals from the central unit in the correct sequence. Alternatively, each scaler can be fitted with an additional control flip-flop, to initiate the read-out process. The flip-flop of the first scaler to be read out is set to 1 by a command pulse ($\rightarrow$PRINT) from the control unit. After the read-out process has been terminated, the flip-flop is reset to 0, at the same time setting to 1 the control flip-flop of the next scaler. This process is repeated until the last scaler has been read out. Its resetting pulse signals the termination of the whole read-out process to the control unit, which itself can start another measurement [7.004]. In principle the number of scalers to be read out in this system is unlimited.

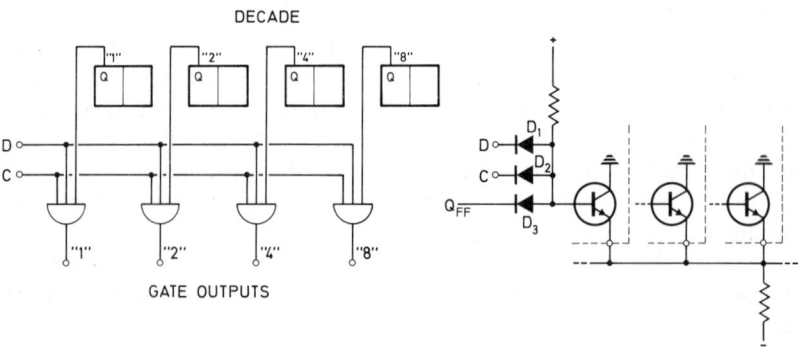

Fig. 7.02. Double-gated decade by McGinnis et al. [7.005], and circuit detail of the AND gate with emitter-follower output

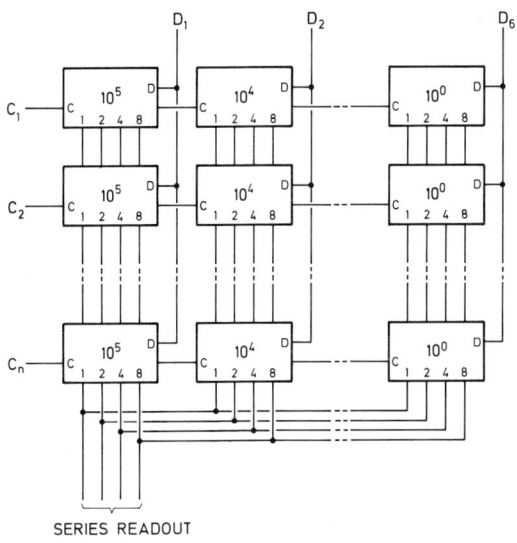

Fig. 7.03. Block diagram of a series read-out system using the decades of Fig. 7.02

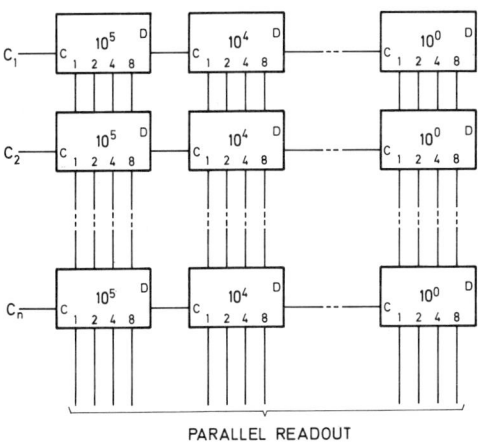

Fig. 7.04. Block diagram of a parallel read-out system using the decades of Fig. 7.02

Data Processing

In order to be able to connect together in parallel the outputs of the sampling AND gates, a diode or an emitter-follower must be used for the output elements of the gates: the interconnection of the outputs then exhibits the properties of an OR mixer (no reciprocal effects of the particular AND gate outputs). The circuit details are described by MCGINNIS et al. [7.005]. Fig. 7.02 shows the principle of their double-gated decade.

When several AND gates are connected together the output emitter-follower yields the desired OR mixing. The control line C (counter) serves the selection of a particular counter (it is common to all decades of the counter), the line D (digit) serves for the selection of a given decimal digit.

Fig. 7.03 shows the block diagram of a series read-out system using the decades of Fig. 7.02. For the sake of clarity, only three six-digit scalers are shown. All gate outputs "1", "2", "4" and "8" of the decade flip-flops are interconnected respectively, and connected through a central encoding matrix to the series printer or tape puncher. A central control unit (commutator) first sets $C_1 = 1$, and then successively $D_1, D_2, ..., D_6 = 1$, whereby the six decimal digits of the first scaler are read out in the correct sequence. The process is then repeated for the second ($C_2 = 2$) and following scalers.

In a parallel read-out system (Fig. 7.04) only the flip-flop outputs of the corresponding decimal digits are interconnected. The 6×4 outputs are applied by means of six encoding matrices to the six output elements of the printer. To all D inputs of the decade gates a logical 1 is applied (in the special case of Fig. 7.02 the diode D_1 is left out). The read-out process of the first scaler is initiated by $C_1 = 1$, the second by $C_2 = 1$, etc. The C-inputs are controlled from a central control unit (commutator), the read-out repetition rate is adapted to the printing speed of the parallel printer used.

A novel routing system for multicounter arrays has been described by SCHIAVUTA and SOSO [7.049].

7.3. Multichannel Analyzers

Very often the problem of amplitude analysis arises, i.e. of classifying the incoming pulses according to their amplitudes. In Chapter 4.2 various converter systems for digitizing the pulse height information have been discussed. In principle, one separate scaler could be used per channel of the converter, and the analyzed events could be counted in the respective scalers. The particular scaler would serve as an arithmetical unit ($+1$) and a storage cell at the same time. However, since any event belongs always to only one channel, the expense of the multiplication of the

Multichannel Analyzers

arithmetical unit is superfluous, and a simple memory with a corresponding number of cells but with only one central arithmetical unit suffices.

The combination of an ADC with a computer memory has been described first by HUTCHINSON and SCARROTT [7.006]. Since the papers by BYINGTON and JOHNSTONE [7.007] and SCHUMANN and MCMAHON [7.008], multichannel analyzers have been built almost without exception with ferrite core memories, though recently magnetic film memories have been used [7.009]. The multichannel analyzers have been reviewed by several authors [7.010] to [7.014].

The recent (1968) contribution to the Hutchinson-Scarrott MCA has been given by TAKABA [7.050].

The block diagram of a multichannel analyzer (MCA) having a typical internal organization is shown in Fig. 7.05. The output pulses of the analog-to-digital converter ADC are counted in the address scaler AS. When the conversion is terminated, ADC activates the CONTROL LOGIC which reads out the content of the specified memory location in the ADD-1-SCALER, magnifies it by $+1$, and rewrites it into the same memory location. Finally the address scaler is reset to zero and the linear gate — which was blocked during the whole analysis process — is opened in order to accept another input pulse.

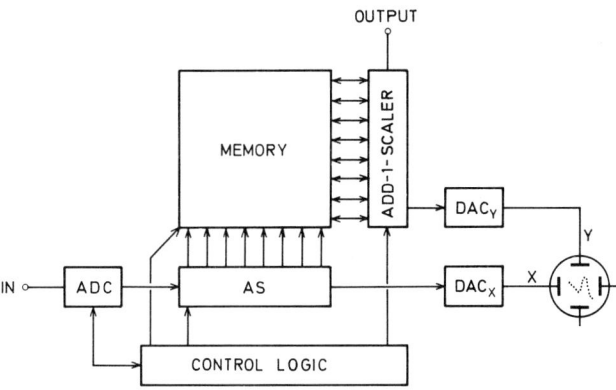

Fig. 7.05. Block diagram of a multichannel analyzer

The working speed can be somewhat increased by means of parallel address transfer from a buffer register in ADC to the address scaler AS in the MCA (cf. Fig. 4.42). The address pulses are counted in the buffer register of the ADC; after the buffer register state has been transferred to AS, a new encoding process in ADC can start, without awaiting the termination of the memory read-write cycle.

For data display two digital-to-analog converters DAC are used, whose outputs control the X and the Y deflection of a cathode ray oscilloscope, respectively. The state of the address scaler corresponding to the channel position is commonly used as the X co-ordinate. By means of beam intensity modulation the oscilloscope spot is made bright just at the instant when both scalers are at rest. The display can run during the analysis, or the stored spectrum can be displayed by driving the address scaler AC at a constant frequency of (e. g.) 10 kcps through all memory locations by means of an auxiliary oscillator. Due to the limited accuracy of the analog display, the Y digital-to-analog converter is connected only to the outputs of 2–3 decades of the "add-1" scaler, the actual display range being selected by a selector switch.

A $X-Y$ plotter can also be connected to the same outputs as for the oscilloscope. Here only the repetition frequency of the address sampling must be reduced to about 10 cps. Alternatively, the plotter pulse indicating the termination of the particular point plotting can be used as the forward pulse for the address increase, hence fully utilizing the plotting speed of the instrument.

Parallel or series printers can be connected to the decade outputs of the "add-1" scaler. As with the $X-Y$ plotter, the read-out speed can be controlled by an internal oscillator of about 10 cps, or by means of the printer back pulse. For series output devices an additional parallel-to-series conversion is necessary using either a shift register or sequential output sampling by means of AND gates.

Most of the analyzers are equipped by an "add-1" scaler, allowing also for subtraction of 1 (hence "add/subtract-1" scaler). With the operation mode selector adjusted for subtraction, the background spectrum can for example, be subtracted for the effect + background pulse height spectrum stored in the MCA.

If the address scaler is allowed to run only up to the state n_2, only that part of the memory is used whose ADDRESS is $n \leq n_2$. Here only a logical circuit is necessary, watching for AS states exceeding n_2 and suppressing the read-write cycle in the positive case. If additionally the AS is not reset to zero, but to n_1 at the beginning of each digitizing process, only the part $n_1 \leq n \leq n_2$ of the memory is activated. Hence by suitably adjusting n_1 and n_2, and by reducing the gain of the analog part of the assembly, the pulse height spectrum can be stored in any part of the memory (of course, with reduced channel number). Very often the memory is divided into two halves or four quadrants. Different spectra therefore can be stored in different quadrants and intercompared.

With the aid of the "add-1" scaler the spectrum stored in one part of the memory can be transferred to another part. Here the original spectrum is read out channel by channel into the scaler and re-written into the original memory location, without resetting the scaler to zero. Thereafter the address is changed as desired, the scaler content is written into the new memory location and the scaler is reset to zero. When transfer between two halves or two quadrants of the memory is considered the address change consists in changing the most significant digit only and can easily be performed (e.g. 157→357, when transferring the second quadrant into the fourth one in a 400 channel MCA).

The ability to perform simple arithmetical operations on spectra from different memory parts is a very useful property of a MCA, which, however, requires relatively expensive circuitry. As an example of such an operation, take again the subtraction of the background spectrum stored in one memory quadrant from the total spectrum stored in another quadrant. Here two registers are necessary, corresponding to the "add-1" scaler, and a parallel adder. The subtraction is accomplished by forming the complement of the number to be subtracted. The result is written into the memory directly from the adder outputs. An assembly with two shift registers and a series adder can also be used, and the result must appear in one of the registers (accumulator) prior to being written into the memory. More complex arithmetic operations such as multiplication and division are seldom required.

> The reason for avoiding these operations in MCA is not the expense of the circuitry, but rather the following: for example, when two numbers are multiplied, the number of digits in the product may often be in excess of the scaler range. Obviously the representation of the data as natural numbers used throughout in MCA systems is not suitable for general mathematical operations. Accordingly, in computers the numbers are represented by the exponent of a given base and by a corresponding constant length mantissa, e.g. $1485 = 10^4 \times 0.14850$; $523 = 10^3 \times 0.52300$ and $1485 \times 523 = 10^6 \times 0.77665$ (cf. e.g. SPEISER [7.015]). However, for the pure cumulative operation mode of a MCA memory this representation offers no advantages.

Multichannel analyzers are variously organized so as to allow data input into the memory from external information carriers (magnetic tape, punched cards). The "add/subtract-1" scaler serves in this case as the input register and, if necessary, as a series-to-parallel converter.

Of course, the MCA memory can also be used for storing data from several independent channels, such as counting pulses from particular detectors of a multi-detector array. If the count rates are low enough to allow the probability of chance coincidences to be ignored, a memory with n locations equipped with a single "add-1" scaler corresponds to an array of n scalers. The "multiscaler" operation mode of the MCA recently achieved importance due to the rising complexity of multi-detector arrays (detector hodoscopes).

Finally, there is another operation mode of the MCA: a slowly changing dc voltage is applied to the input linear gate of the ADC and sampled by periodic or random pulses. This operation mode can be used successfully in Mössbauer experiments. The input voltage is made proportional to the relative velocity v of the source-absorber system, and is sampled by each radiation detector pulse. Hence the MCA stores directly the so-called Mössbauer spectrum, i. e. the detector pulse count rate as a function of the relative velocity v (cf. e. g. [7.016]).

HARMS [7.051] discussed the automatic dead-time corrections in MCA systems. EULER [7.052] described the internal organization of a versatile 16384-channel pulse-height analyzer, FULLWOOD [7.053] the internal organization of an extremely inexpensive simple one.

7.4. Multiparameter Analyzers

In general, more than one parameter characterizes nuclear processes. Even for a simple nuclear transmission with two radiations emitted in coincidence, the emission angles and energies of both radiations must be measured at the same time, i.e. multiparameter analysis of the frequency distribution of the registered events must be performed.

The simplest method of multiparameter analysis is a suitable arrangement of detectors and single-channel discriminators limiting all parameters but one to a narrow value range, only the remaining parameter being analyzed in a MCA. When a statistically significant number of events has been registered, the value range of the fixed parameters is slightly changed and the measurement is repeated, etc.

The time requirement can be substantially reduced if more than one parameter is digitized and registered in the memory. The memory address then consists of the code expressions of all parameters under consideration. Fig. 7.06 shows an example of a two detector array, the event being characterized by the amplitudes of the coincident pulses. The coincidence stage C allows coincident pulse pairs to enter the converters ADC_1 and ADC_2 through the corresponding linear gates LG_1 and LG_2. The converter output pulses are counted in two address scalers AS_1 and AS_2, and the content of these scalers together serve as the memory location address. For example, if 100 channels (00 to 99) are available per parameter, an event causing a pulse height 37 in the detector DET_1 and 6 in DET_2 is registered in the memory location 3706.

Such multiparameter analyzers have been described, e. g. by CHASE [7.017] and ALEXANDRE and ROBINSON [7.018]. By means of small adapters, each multichannel analyzer can in principle be converted into a two-parameter analyzer [7.019, 7.020, 7.054]. However, the number of channels available for each parameter is rather limited. Hence a

512-channel analyzer can be converted into a 4 × 128- or 8 × 64-channel instrument. GUILLON [7.014] discussed some properties of commercially available multiparameter analyzers.

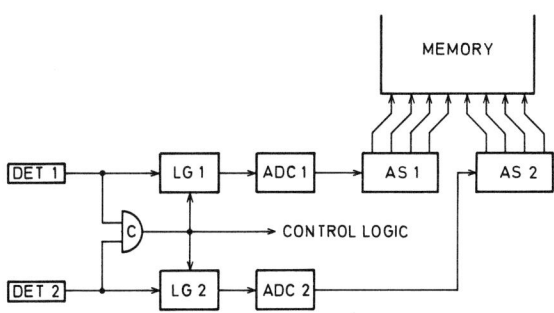

Fig. 7.06. Principle of a two-parameter analyzer

The necessary storage capacity increases exponentially with the number of parameters. A three-parameter system with only 100 channels per parameter already necessitates a memory with $(10^2)^3 = 10^6$ locations. However, core memories with more than about 20 000 locations cannot be made because of economic and technical difficulties. On the other hand, different parts of the memory are differentially utilized. Due to the coincidence condition, the background is considerably reduced and the events relevant from the standpoint of nuclear physics tend to accumulate at lines or peaks, and most of the memory locations remain empty. The following numerical example has been given by NORBECK [7.021]. Assuming that events must be analyzed with regard to 4 parameters with an accuracy of $1^0/_{00}$, $(10^3)^4 = 10^{12}$ memory locations are necessary. If 10^7 events are registered, even in the most unlikely case of one event per location, the efficiency of memory exploitation is 10^{-5}. Considering the event accumulation on resonance values etc., the mean utilization of the memory becomes 10^{-6} to 10^{-7}.

The only remedy consists in limiting the value ranges of the particular parameters or parameter combinations to ranges where physically significant events are expected. However, a sharp limitation is not possible until the result of the experiment is known. In practice this dilemma is solved in two ways, either by successive storing of the digital encoded address of all registered events in a cheap slow memory, such as magnetic tape, and by off-line analyzing of the information in a digital computer (so called descriptor storage), or by means of memories with associative internal organization which automatically perform the selection of physically relevant events according to their frequency.

According to a proposal made by WAUGH [7.055], the effective size of a memory can also be increased by means of storage of differences of, rather than the spectra themselves.

Descriptor storage. In a cumulative memory a specified memory cell is associated with one particular value of a parameter or of a parameter combination, the number of such events being stored in this cell. The digital code word of the event serves as the memory cell address. On the other hand, the same information can be conserved by storing every new code word in a new memory cell (hence every cell is associated with one arbitrary event only). For the sake of clarity, such stored code words are denoted as *descriptors* (and not as address, this term being reserved for the description of the memory cell location). The particular descriptors are stored in the order of arrival on an inexpensive bulk memory (magnetic tape, punched tape, etc.). After the experiment is terminated, the memory content is analyzed in a computer.

Since no classification is made during the descriptor recording, the use of a magnetic tape merely causes a delay in the analysis. However, this delay is favourable for various reasons. Having recorded the events, a preliminary fast analysis can be made allowing the ranges of interest to be defined roughly. Then a detailed — and correspondingly long-lasting — analysis is performed, of events within these ranges. Often the rough analysis is performed on-line, by parallel evaluation of the descriptors in a conventional multiparameter analyzer with a very limited channel number. For example, the most significant digits of the descriptor may serve as address. Another advantage of descriptor recording consists in derandomizing the time of arrival of the events. Hence infrequent events are stored and then fed into the computer at a conveniently fast rate, limited only by the computer speed, thus reducing expensive computer time. The greatest disadvantage of descriptor recording is the very late receipt of the results, adversely affecting the operator's survey of the experiment. However, this disadvantage is offset to a high degree by the use of an on-line multiparameter analyzer for the preliminary analysis. Descriptor recording systems have been described by various authors (e.g. [7.022] to [7.027]). SPINRAD [7.028, 7.029] and EGELSTAFF and RAE [7.030] reviewed the field of descriptor storage.

The maximum descriptor recording speed is limited by the properties of the memory used. Moreover, owing to the random distribution of the event arrival instants, many events are not recorded even when the mean time interval between two registered events is higher than the time necessary for digital encoding and subsequent descriptor recording. The event loss can be reduced by preceding the bulk memory by a small fast buffer register built up of flip-flops or ferrite cores, which serves as a derandomizer. The reduction of the event loss as a function of the mean event count rate, bulk memory speed and the number of locations of the buffer register, has been dealt with by ALEXANDER et al. [7.031] (cfl. also [7.029]).

Systems with associative memories. In an associative memory, adjoining memory locations are associated with similar events. Since, however, the location address does not describe the values of the stored event parameters, each location in addition to the event number must comprise also the corresponding descriptor. Memory locations are reserved in the succession of event arrival. Since the physically relevant events are also the most frequent, their descriptors will appear at first with a high probability and will fill the available memory volume. Hence the limitation of parameter values to the relevant ranges is accomplished automatically. If during the analysis process memory locations with an event count rate below a preselected low threshold are cleared and new descriptors are accepted, the search for the most frequent descriptors is actually improved.

The first multiparameter analyzer with associative memory has been described by HOOTON [7.031]. The same author also reviewed the related problems. The search algorithm according to which the memory is searched through to ascertain whether or not it already contains an incoming descriptor, can be realized by a wired program (i. e. by a fixed organization of the memory control logics) or by suitable programming of a general-purpose computer (SOUČEK [7.032, 7.033] and SOUČEK and SPINRAD [7.034]). The second alternative is the preferred one.

The so-called "tree algorithm" [7.031] is most often used for memory organization. The principle can easily be seen from Fig. 7.07. At first all memory locations are empty. The first descriptor (A) is written into the first memory location. The second descriptor (B) is first compared with A, found to be greater, and stored in the next higher free location (2). The address of this location is noted in location 1 as "higher node address". The third descriptor (C) is smaller than A and is stored in the next free location (3), under recording 3 as the "lower node address" in the location 1. The fourth descriptor D is compared with A, and since $D>A$, it is also compared with the content of the "higher node" (i. e. location 2, descriptor B). Since $D<B$ and the lower node address of location 2 is still empty, D is stored in the next free location (4), and 4 is noted in 2 as the lower node address, etc. etc. A higher operating speed can be achieved, if several trees are allowed to "grow" into the

Data Processing

memory from different starting locations. Despite the very limited number of locations in the memory (i.e. 4000), a virtual resolution corresponding to $>10^6$ channels is achieved by the associative organization.

MEMORY LOCATION	DESCRIPTOR	PULSE COUNTS	HIGHER NODE ADDRESS	LOWER NODE ADDRESS
1	A	1	2	3
2	B,E	2	5	4
3	C	1	empty	empty
4	D	1	empty	empty
5	F	1	6	empty
6	G	1	empty	empty
7	empty	empty	empty	empty
etc.				

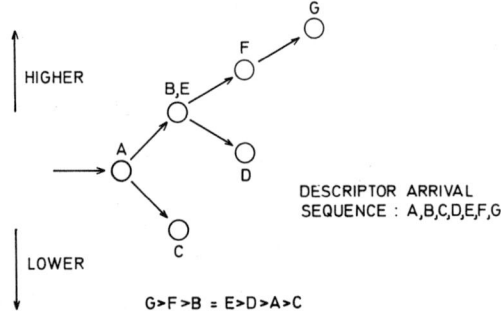

Fig. 7.07. Principle of the "tree algorithm" according to [7.031]

Optical data display. Practical difficulties limit the number of parameters for optical display to two. For the display the corresponding parts of the descriptors and the stored event number N are converted into analog voltages by means of three digital-to-analog converters, which are applied to the deflection system and to the brightness control of a cathode ray tube. Two representation versions are often used.

In the first representation version, one parameter controls the horizontal deflection and the count number N controls the vertical deflection. The voltage signal proportional to the second parameter is applied to both vertical and horizontal deflection systems, thus moving the spot in a diagonal direction. Hence a pseudo-three-dimensional representation arises (Fig. 7.08 a). The spot brightness can in addition be made proportional to N. If only preselected parameter values control the spot brightness, sections through the three-dimensional surface can be displayed (Fig. 7.08 b). In this representation the channel position of a given point cannot be easily estimated.

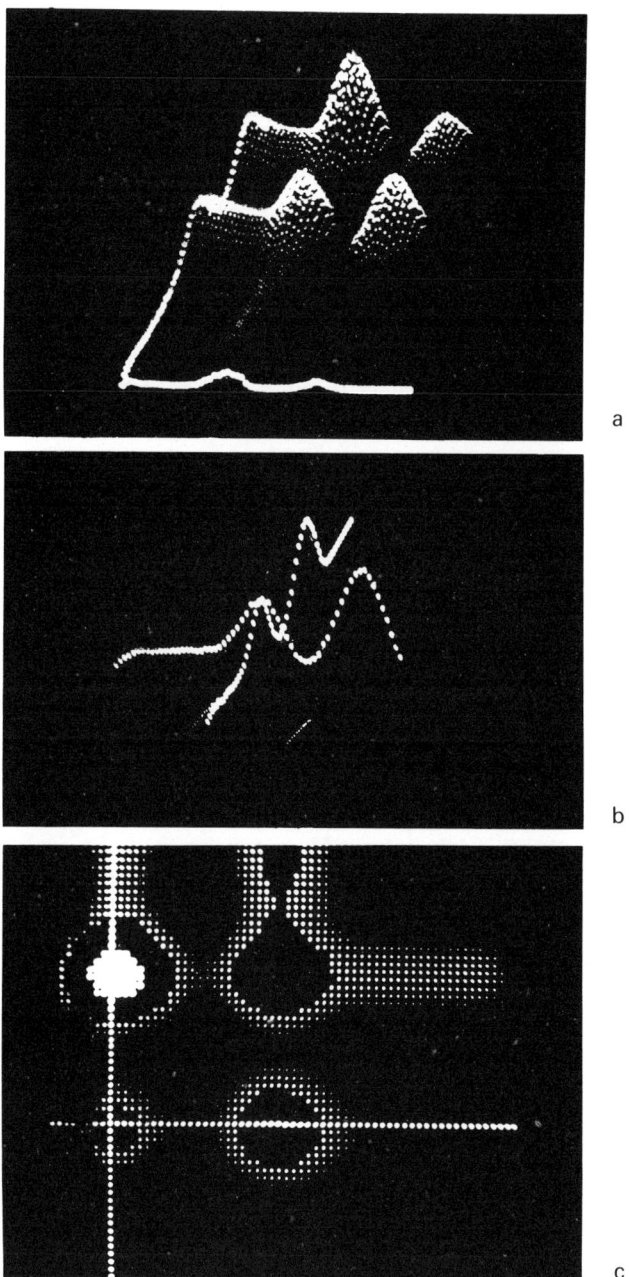

Fig. 7.08 a–c. Various possibilities of optical data display according to BIANCHI et al. [7.035]

In the second representation version, the two parameters control the vertical and the horizontal deflection, respectively, and the count number N controls the spot brightness, thus yielding a two-dimensional brightness-modulated map (Fig. 7.08c). Though the channel position can be determined easily, the information about N is only qualitative. However, the voltage proportional to N can be analyzed by means of a simple single-channel analyzer, reducing the spot brightness when the N-voltage lies within the preselected (narrow) channel. Herewith a dark contour-line is formed, which can be shifted by changing the channel position, thus allowing the map surface to be analyzed in more detail.

Different display systems are discussed by various authors [7.036] to [7.039]. A direct three-dimensional stereo-display has also been attempted [7.040]. The multiparameter displays in isometric as well as in contour map presentation are reviewed in the comprehensive paper by STEGER [7.056].

The identification of the descriptors or of the count number N at a particular point of the optical display is facilitated considerably by the use of the so-called light pen [7.041]. The light pen consists of a photosensitive element and a small light guide shaped to a spike. The spike is brought into contact with the point under consideration. At the instant when the point lights up, the light pen generates a pulse, by means of which the momentary content of the corresponding registers is transferred in parallel to auxiliary registers, the state of which is indicated by some digital means (e. g. NIXIE tubes). Hence by indicating the interesting points on the oscilloscope screen, the light pen enables the corresponding parameters (descriptors) and the number of events to be read out digitally. Moreover, the light pen is also suitable as an input device. For example, the range of physically relevant values for which a more detailed analysis is desired, need only be indicated on the screen by the light pen, etc.

7.5. On-Line Computers

We have already met the concept of a digital computer in the previous chapters, especially in Chapter 7.4 concerning multiparameter analysis, without, however, having explained it in any detail. This omission shall be briefly rectified in this chapter. For a detailed discussion we refer once more to SPEISER [7.015].

The term "computer" today denotes, almost without exception, digital stored-program data handling and processing devices, essentially

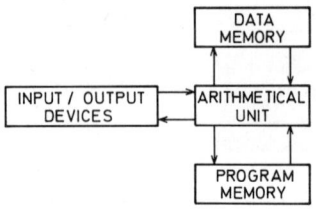

Fig. 7.09. Block diagram of a stored-program computer

organized as shown in Fig. 7.09. The computer consists of an arithmetical unit (of any required complexity), a system of memories, and input/output devices. For the sake of clarity, two separate memories are shown in Fig. 7.09, one for the data (numbers) and one for the program, although no separation is necessary and both data and program orders can be stored in a common memory.

The operating principle is simple: according to a program, consisting of particular orders to be performed in a defined sequence, arithmetical and logical operations are performed on numbers from the data memory or from the input devices; the results are either stored in the data memory or read out by means of the output devices. The program runs step by step, i. e. order by order. The desired program must be transmitted to the computer prior to data processing; a computer with an empty program memory is a "tabula rasa" which must first be fitted to the purpose in hand by programming.

From the viewpoint of circuit techniques, any order consists in activating gates in the arithmetical unit and between the arithmetical unit and the memories, and input/output devices which allow the exact information flow necessary for performing a particular operation. In general the computer is synchronously organized, i. e. the orders are executed one after the other within fixed clock intervals. There is an internal hierarchy of orders: more complex orders, such as multiplication, are resolved in a number of more elementary suborders, like addition, etc. The length of the clock interval is chosen to correspond to the simplest suborders.

In general, any order consists of the specification of the desired mathematical or logical operation (e. g. the four species — addition, subtraction, multiplication and division; data input, data output, transfer memory → arithmetical unit, transfer arithmetical unit → memory, etc.), of the address or addresses of the memory locations in which the operands are stored, and of the specification of the memory location in which the next program order is stored. Especially simple orders, with only one operand address, occur when the result of the operations always remains in a special register of the arithmetical unit, the so-called accumulator (ACC), and serves as one of the two operands for the next order. Hence the addition of the content of memory locations 063 and 178 and subsequent storing of the result in location 624, for example, need the following program (content of registers and memory cells is indicated by rectangular brackets):

$$\text{Transfer } [063] \rightarrow \text{ACC}$$
$$\text{Addition } [\text{ACC}] + [178], \text{ result remains in ACC}$$
$$\text{Transfer } [\text{ACC}] \rightarrow 624.$$

If the orders are stored in the program memory in the sequence of desired execution, the structure of the orders can be further simplified by omitting the address of the next order.

One particular species of orders will be pointed out: the so-called conditional orders. The orders in question have the following general structure: if condition A holds, follow the next order in the sequence, if not, carry out the order stored in memory location XYZ. Either internal (e. g. $[ACC] < 0$?), or external signals can be used for the conditions. Internal conditions allow programs with logical branchings and closed loops to be realized, which are necessary for iterative calculations. External conditions allow (e.g.) for synchronizing the computer to an attached experiment: the computer "waits", until an external signal indicates the presence of new experimental data at the input.

Of course, modern computers need not be programmed step by step using detailed orders in their so-called machine language. There are various transcription systems for translating normal mathematical expressions into a standard, unequivocal form, acceptable by the computer. FORTRAN in particular has found a very widespread application. The computer then translates with the aid of standard transcription programs the FORTRAN orders into its machine language. However, if the full computing speed of the computer is to be utilized, or if the computer is to be coupled to external auxiliary devices, at least a partial programming in the machine language is necessary.

A stored-program computer represents the most general signal "processing" device stipulated in the introduction, Chapter 1, Fig. 1.01. Of course, the original detector signal must already be digitized before it is applied to the computer. In principle, two modes of operation are possible. Firstly, the digitized signals can be stored in a bulk memory (magnetic tape, punched tape) and applied to the computer later, at the end of the experiment. This is known as "off-line" operation. In this case, the computer is completely separate from the remaining electronic set-up and independent of it.

We are more interested in the second operation mode, where the computer is integrated into the experimental set-up and performs the calculations etc. in "real-time" or "on-line" operation. In an on-line computer the information is processed as it is delivered by the system detector-preprocessing. The result of the signal processing is available immediately after the conclusion of the experiment, or — exactly speaking — preliminary results are available at any stage of the experiment. In on-line operation the computer speed limits the rate at which information can be accepted from the detector system. Vice versa, for processing high data rates, fast computers are necessary.

The integration of the computer into the experimental set-up is, of course, not an electronic, but mainly a logical operation of programming the computer.

Since the early sixties, when on-line techniques began to be used for nuclear physics experiments, several application fields have developed. One of them — the programming of a computer as a multiparameter analyzer — has already been described in Chapter 7.9. In this application little use is made of the arithmetical unit of the computer, the computer being used mainly because of its high memory capacity. On the other hand, in high energy physics, in experiments on large particle accelerators, on-line computers are used for immediate calculational evaluation of the scattering processes registered by large counter hodoscopes. This technique is described in the review articles by SPINRAD [7.029] and LINDENBAUM [7.042]. It will be discussed here using the classical example of the Brookhaven experimental set-up [7.043]. Recently, on-line techniques have been reviewed by JONES [7.044], BUTLER and BUTLER [7.045], BROWN and MUELLER [7.046], JONES [7.058] and LIDOFSKY [7.059]. GEMMEL [7.057] gave a detailed description of the Argonne National Laboratory on-line computing systems for low-energy nuclear experiments.

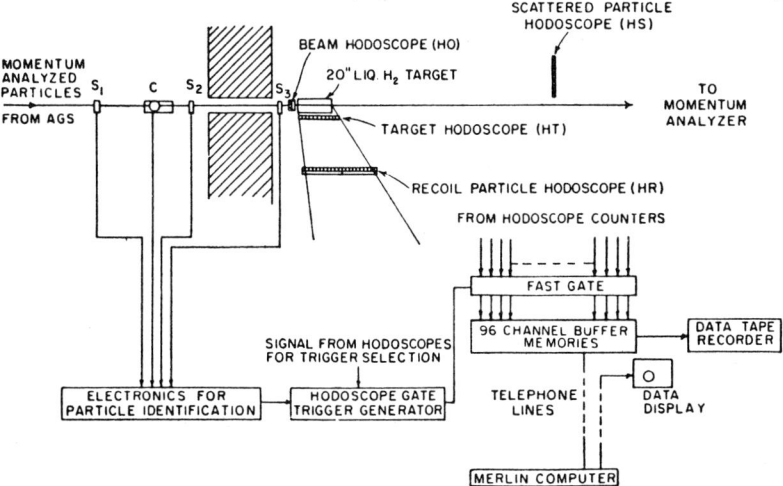

Fig. 7.10. Experimental set-up for the detection of small-angle elastic scattering at the Brookhaven 33 BeV AGS [7.043]

Fig. 7.10 shows the experimental set-up for the detection of small-angle elastic scattering at the Brookhaven 33 BeV alternating gradient synchrotron. The detector system consisted of a counter telescope

(Čerenkov counter C and three scintillation detectors S_1, S_2, S_3) for particle identification, and of several hodoscopes for determining the space location of the incident, scattered and recoil particles. (A hodoscope is an array of adjoining long thin scintillation slab detectors locating one Cartesian coordinate of a particle trajectory. Two perpendicular hodoscopes are necessary for locating both X and Y coordinates.) The digitized information on the selected events was first stored in a 5 µsec buffer memory. The buffer memory content was transmitted periodically, between two accelerator bursts, over telephone lines to the MERLIN computer situated about one mile away. At the same time, this information was recorded on a magnetic tape for later off-line processing upon other criteria. The computer classified all recorded events according to various criteria concerning the number of pulses of each hodoscope and the coplanarity of the particle trajectories and calculated the kinematic space angles. All results were stored in the computer memory. The desired data displays were transmitted almost continuously back to the CRO screens in the control room of the synchrotron by means of another telephone line. The computer then calculated, according to a previously written stored program, the absolute value of the differential scattering cross-section, its mean square deviation, and other process parameters. The on-line computer also allowed automatically for the background due to non-elastic scattering, chance coincidences etc.

The evaluation of the immense amount of information delivered by the hodoscope detectors would not be possible without the continuous data reduction by means of the real-time on-line reconstruction of the kinematics, which enabled the event to be classified according to a few parameters. Typically, about 10^8 incident particles were handled per hour, corresponding to about 50 000 trigger-selected scattered events.

Finally, there is the possibility of using on-line computers for the analysis of events visualized in bubble and spark chambers [7.042, 7.045].

8. Appendix

8.1. Laplace Transform Calculus

The operational calculus represents probably the most powerful mathematical instrument of pulse techniques. There are numerous monographs devoted to this subject ([8.101] to [8.108]). Moreover, in almost every textbook on pulse techniques the operational calculus or the Laplace transform calculus are dealt with in an appendix ([8.109] to 8.111]). This chapter shall not merely continue the tradition of these appendices, but rather point out the simple approach to the Laplace transformation via an intuitive "naive" operational calculus, and the fundamental differences between the two. More details concerning this subject can be found in the very interesting book by HENNYEY [8.102] and in the lecture texts by STIEFEL [8.108].

8.1.1. Networks

The networks are used for processing the electrical signals (voltages, currents). If the network properties do not depend on the signal amplitude, the network is called linear. Networks consisting of passive components (resistors, condensers, coils) and ideal voltage and current generators are a priori linear, networks with active elements are linear only in the amplitude range where the particular equivalent circuit with passive components applies. In what follows we consider only linear networks.

Terminating points of the connections of two or more components or generators are called NODES, the network parts between two adjacent nodes are called BRANCHES, and a closed signal path through two or more branches is called a MESH. The network problem can be considered as solved, if all branch currents and voltages are known. However, in practice, often only the dependence of a given branch signal ($\rightarrow$ output signal) on the voltage or current of a given generator ($\rightarrow$ input signal) is looked for.

In linear networks the *superposition principle* applies: if the network solutions belonging to two arbitrary input signals are known, the solution belonging to any linear combinations of these input signals can

Appendix

be calculated as the same linear combination of the corresponding partial solutions. Due to the superposition principle, a complex input signal can be expressed as a linear combination of simple elementary functions for which the network solutions are known, and the output signal can be represented by a corresponding linear combination of the elementary solutions.

The most simple functions which are of interest in pulse techniques are the step function $H(t-t_0)$ and the delta function $\delta(t-t_0)$:

$$H(t-t_0) = \begin{cases} 1 & \text{for } t > t_0 \\ 0 & \text{for } t < t_0 \end{cases} \quad (8.101)$$

$$\delta(t-t_0) = \begin{cases} \infty & \text{for } t = t_0 \\ 0 & \text{for } t \neq t_0 \end{cases} \quad \text{with} \quad \int_{-\infty}^{+\infty} \delta(t-t_0)\,dt = 1. \quad (8.102)$$

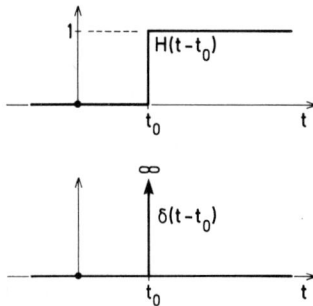

Fig. 8.11 The step function $H(t)$ and the δ-function $\delta(t)$

The step function, for example, can be used to describe a rectangular pulse $V_\sqcap(t)$ having the amplitude V_0, the duration δ, and starting at $t=0$:

$$V_\sqcap(t) = V_0[H(t) - H(t-\delta)]. \quad (8.103)$$

The delta function is suitable for the description of very short processes, e. g. of a short current pulse $I(t)$ with the total charge Q_0 at the photomultiplier output at $t=0$:

$$I(t) = Q_0 \delta(t). \quad (8.104)$$

It must be pointed out that — due to the relation (8.102) — the delta function (t) exhibits the dimension (time)$^{-1}$.

The relations between the particular branch currents of a node or the particular branch voltages of a mesh are described by the so-called Kirchhoff's laws:

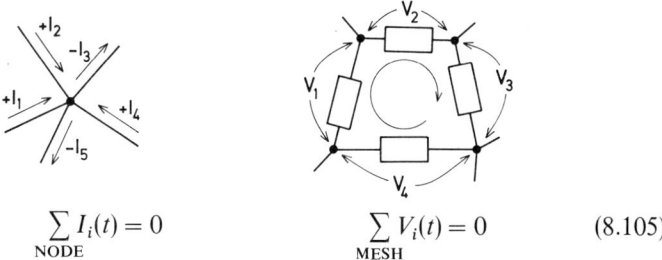

$$\sum_{\text{NODE}} I_i(t) = 0 \qquad \sum_{\text{MESH}} V_i(t) = 0 \qquad (8.105)$$

The particular currents or voltages must be regarded as positive or negative quantities according to their respective polarities and directions (the polarity and direction convention is an arbitrary one). Depending upon the branch component, the following relations hold for the particular branch currents and voltages:

RESISTOR   CONDENSER   COIL

$$V(t) = R \cdot I(t) \qquad V(t) = \frac{1}{C} \int_{-\infty}^{t} I(t')dt' \qquad V(t) = L \cdot \frac{d}{dt} I(t) \qquad (8.106)$$

As can be shown, just as many independent equations of the type (8.105) and (8.106) can be set up for a given network, as there are unknown quantities in that network. For networks consisting only of resistors R as passive components all equations are algebraical and the solution of the equation system is easy. However, condensers and coils yields a differential equation system, the solution of which is much more time consuming.

If only two signals are considered — the known input signal $E(t)$ and the onknown output signal $A(t)$ — the solution of the equation system (8.105), (8.106) finally gives the relationship

$$K\left(R, C, L, \frac{d}{dt}\right) * A(t) = E(t), \qquad (8.107)$$

where the differential operator K is a significant expression of the respective component values and $\frac{d}{dt}$ or $\int dt'$ (the asterisk $*$ denotes a more general operation, contrary to the simple multiplication). The relation (8.107) has a very simple symbolical solution

$$A(t) = K^{-1}\left(R, C, L, \frac{d}{dt}\right) * E(t), \qquad (8.108)$$

Appendix

which however is of no practical importance, since the inverse operator K^{-1} is normally senseless.

As an example, the analysis of the simple network in Fig. 8.12 yields a differential equation

$$I(t) = \frac{V(t)}{R} + C \cdot \frac{dV(t)}{dt} = \left(\frac{1}{R} + C \cdot \frac{d}{dt}\right) * V(t), \qquad (8.109)$$

with a symbolical inversion

$$V(t) = \frac{R}{1 + RC \frac{d}{dt}} * I(t) \qquad (8.110)$$

having no sense.

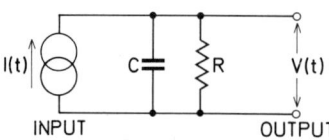

Fig. 8.12. Example of a simple network

Hence, if the advantages and the elegance of the direct solution (8.108) are to be fully utilized, it is necessary either

1. to understand differential operators K^{-1} of the type shown in the relation (8.110), or
2. to transform all time functions into functions of another variable, for which $\frac{d}{dt}$ corresponds to an algebraical operation, so that the expression K^{-1} becomes a pure algebraical one.

8.1.2. Naive Operational Calculus

The first possibility was investigated by OLIVER HEAVISIDE in the 19th century. The corresponding trains of thought are attractively simple, however the calculus has no correct mathematical fundamentals.

For the sake of simpler notation the differential operator $\frac{d}{dt}$ is denoted by the letter p:

$$\frac{d}{dt} * = p *. \qquad (8.111)$$

Obviously

$$\frac{d^2}{dt^2} * = p * (p * \ldots) = p^2 *, \quad \text{and in general} \quad \frac{d^n}{dt^n} * = p^n *. \qquad (8.112)$$

Naive Operational Calculus

For $\frac{1}{p}*(p*\ldots)=\left(\frac{1}{p}\cdot p\right)*=1*$ (identity operator), the operator $\frac{1}{p}*$ must just cancel the operator $p*$. This property is exhibited by the integral operator

$$\int_0^t dt'\,* = \frac{1}{p}*. \qquad (8.113)$$

The lower integration limit is arbitrary chosen to be $t=0$. Hence only signals which are zero for $t<0$ must be considered, i.e. signals of the type $f(t)\cdot H(t)$.[14] In analogy to (8.112) the operator $\frac{1}{p^n}*$ denotes the n-fold integration.

For the interpretation of the operator expressions $F(p)*$ advantageously only their action to some distinct standard functions $f_{st}(t)$ is considered, by which other time functions $F(p)*f_{st}(t)$ are produced (so-called "operational production" [8.102]). Due to the superposition principle as simple an elementary function as possible is chosen for $f_{st}(t)$ — we shall use $f_{st}(t)=\delta(t)$ in what follows. For the sake of clarity the corresponding pairs $F(p)$ and $f(t)=F(p)*\delta(t)$ for some important expression $F(p)$ are summarized in the following table.

Using this small function table, rational functions of p (i.e. fractions of polynomials in p) which arise from the solution of the equation systems (8.105), (8.016) can be interpreted by expanding $F(p)$ in partial fractions $\frac{A}{p+\alpha}$.

Using the notation convention (8.111) the component relations (8.106) read as follows

RESISTOR	CONDENSER	COIL
R	C	L
$V(t)=R\cdot I(t)$	$V(t)=\dfrac{1}{pC}*I(t)$	$V(t)=pL*I(t)$

(8.114)

---

[14] Exactly speaking, $f(t)=0$ for $t\leqslant 0$ is required, since

$$\frac{1}{p}*[p*f(t)] = 1*f(t) = \int_0^t dt'\,\frac{df(t')}{dt'} = f(t)-f(0).$$

However, the calculus holds also for functions which are not defined at $t=0$ (e.g. $\delta(t)$, $H(t)$, etc.).

Appendix

HEAVISIDE initially chose $f_{st} = H(t)$, giving another function table than our choice of $f_{st} = \delta(t)$. The reasons for choosing $f_{st} = \delta(t)$ are discussed by HENNYEY [8.102].

$F(p)$	$f(t)$	Proof:
1	$\delta(t)$	obviously $1 * \delta(t) = \delta(t)$
$\dfrac{1}{p}$	$H(t)$	$\dfrac{1}{p} * \delta(t) = \displaystyle\int_0^t dt'\, \delta(t') = H(t)$
$\dfrac{1}{p^2}$	$t \cdot H(t)$	$\dfrac{1}{p^2} * \delta(t) = \dfrac{1}{p} * H(t) = \displaystyle\int_0^t dt' \cdot H(t') = t \cdot H(t)$
$\dfrac{1}{p^{n+1}}$	$\dfrac{t^n}{n!} \cdot H(t)$	continued integration
$\dfrac{1}{p-\alpha}$	$e^{\alpha t} \cdot H(t)$	general solution technique: apply the series expansion to transcendental functions: $$e^{\alpha t} \cdot H(t) = \left[1 + \alpha t + \alpha^2 \frac{t^2}{2} + \cdots + \alpha^n \frac{t^n}{n!} + \cdots\right] \cdot H(t)$$ $$= \left[\frac{1}{p} + \alpha \frac{1}{p^2} + \alpha^2 \frac{1}{p^3} + \cdots\right] * \delta(t)$$ $$= \frac{1}{p}\left[1 + \frac{\alpha}{p} + \left(\frac{\alpha}{p}\right)^2 + \cdots\right] * \delta(t)$$ (the polynomial in α/p, of course, must converge, otherwise the whole calculus would be meaningless, hence:) $$= \frac{1}{p}\left[\frac{1}{1 - \dfrac{\alpha}{p}}\right] * \delta(p) = \frac{1}{p-\alpha} * \delta(t)$$
$\dfrac{1}{p+\alpha}$	$e^{-\alpha t} \cdot H(t)$	$\alpha \to -\alpha$
$\dfrac{\alpha}{p(p+\alpha)}$	$(1-e^{-\alpha t}) \cdot H(t)$	$\dfrac{\alpha}{p(p+\alpha)} * \delta(t) = \dfrac{\alpha}{p} * \left[\dfrac{1}{p+\alpha} * \delta(t)\right] = \alpha \displaystyle\int_0^t dt' \cdot e^{-\alpha t'} \cdot H(t')$
$\dfrac{p}{p^2+\omega^2}$	$\cos\omega t \cdot H(t)$	$\cos\omega t \cdot H(t) = \dfrac{1}{2}(e^{i\omega t} + e^{-i\omega t}) H(t) = \dfrac{1}{2}\left(\dfrac{1}{p-i\omega} + \dfrac{1}{p+i\omega}\right) * \delta(t)$
$\dfrac{\omega}{p^2+\omega^2}$	$\sin\omega t \cdot H(t)$	$\sin\omega t \cdot H(t) = \omega \displaystyle\int_0^t dt' \cdot \cos\omega t' \cdot H(t') = \dfrac{\omega}{p} * \left[\dfrac{p}{p^2+\omega^2} * \delta(t)\right]$
etc.		

The set of differential equations must not be written explicitly, the operators $\dfrac{1}{pC}*$ and $pL*$ can be treated as algebraical quantities.

Example

For the integrator circuit in Fig. 8.13 we get the simple voltage divider expression

$$V_{out}(t) = \frac{\frac{1}{pC}}{R + \frac{1}{pC}} * V_{in}(t). \tag{8.115}$$

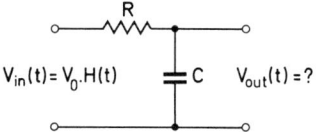

Fig. 8.13. The integrator

Since

$$V_{in}(t) = V_0 \cdot H(t) = V_0 \cdot \frac{1}{p} * \delta(t)$$

we obtain using the function table

$$V_{out}(t) = \frac{\frac{1}{pC}}{R + \frac{1}{pC}} * \left[V_0 \cdot \frac{1}{p} * \delta(t) \right] = V_0 \frac{\frac{1}{RC}}{p\left(p + \frac{1}{RC}\right)} * \delta(t)$$

$$= V_0(1 - \epsilon^{-t/RC}) \cdot H(t) \tag{8.116}$$

Interpretation of the Shifting Operator $\epsilon^{-pT} *$

Because the exponential function can be expanded into an infinite polynomial operator, $\epsilon^{-pT} *$ can be written as

$$\epsilon^{-pT} * f(t) = \left(1 - Tp + \frac{T^2}{2!}p^2 - \frac{T^3}{3!}p^3 + - \cdots \right) * f(t)$$

$$= f(t) - T \cdot \frac{df}{dt} + \frac{T^2}{2!} \cdot \frac{d^2f}{dt^2} - \frac{T^3}{3!} \cdot \frac{d^3f}{dt^3} + - \cdots.$$

This is the Taylor series expansion of the function $f(t-T)$ at $T=0$, which is convergent for any value of t and T. Hence

$$\epsilon^{-pT} * f(t) = f(t-T). \tag{8.117}$$

Since $f(t)$ is an arbitrarily chosen function,

$$\epsilon^{-pT} * \delta(t) = \delta(t-T) \tag{8.118}$$

also holds.

Appendix

The Operational Production of a General Time Function $f(t)$

The differential operators are linear ones, and hence the following relation holds: If

then
$$f(t) = \sum_i c_i f_i(t), \quad \text{with} \quad f_i(t) = F_i(p) * \delta(t),$$
(8.119)
$$f(t) = F(p) * \delta(t), \quad \text{with} \quad F(p) * = \sum_i c_i F_i(p) *$$

(We have already used this trivial relation a few times.)

c_i are constants with respect to time. Using a limit approach the discrete sum of (8.119) becomes an integral.

On the other hand, any general function $f(t)$ can be expressed as a convolution integral with the delta function $\delta(t-T)$

$$f(t) = \int_0^\infty f(T) \cdot \delta(t-T) \cdot dT.$$
(8.120)

where $f(T)$ denotes the respective values of the time function $f(t)$ at $t=T$, which are constants with regard to time t. Using (8.118) and (8.119) the operator $F(p)*$ corresponding to $f(t)$ may be written as

$$F(p) * = \int_0^\infty f(T) \cdot \epsilon^{-pT} \cdot dT *,$$

or — since the integration variable T can be denoted by any other symbol, i.e. also by t — finally as

$$\boxed{F(p) * = \int_0^\infty \epsilon^{-pt} \cdot f(t) \cdot dt *.}$$
(8.121)

Of course the relation (8.121) has merely a symbolical meaning, since p is not a number, but only an operator, and therefore the question of convergency of the integral cannot be answered. Despite this problem, (8.121) can be used to calculate a much more complete function table $f(t) = F(p) * \delta(t)$ as that given above. The general relation for the calculation of $f(t)$ from a given $F(p)*$, i.e. the inversion of (8.121), cannot be derived by similar simple means as in (8.121). Obviously, this "naive" operational calculus without consideration of convergency, has a very labile mathematical foundation.

8.1.3. Laplace Transformation

The second possibility quoted in Chapter 8.1.1 — namely to transform all time functions into functions of a new variable for which $\dfrac{d}{dt}$ becomes an algebraical operation — is the better one from the mathematical point of view. Even if we cannot prove the convergency of all the following expressions due to the restricted scope of this appendix, we know at least that these convergency considerations can be done and have been done. Considering p as a (complex) *variable*, the relation (8.121) yields a transformation with just the desired properties. The function $F(p)$ (which is now a function defined in a part of the complex plane and *not* an operator) is called the *LAPLACE transform* of the original time function $f(t)$, the correspondence between $F(p)$ and $f(t)$ being denoted by £, £$^{-1}$ or by $\triangleq$

$$F(p) = \int_0^\infty \epsilon^{-pt} \cdot f(t) \cdot dt, \qquad (8.122)$$

$$F(p) = \pounds\{f(t)\}, \quad f(t) = \pounds^{-1}\{F(p)\} \quad \text{or simply} \quad F(p) \triangleq f(t).$$

The respective symbols are self-explanatory. In case capitals and minuscules cannot be used for the distinction between the original time function and its Laplace transform, the transform will be denoted by a "roof" sign:

$$\hat{V}(p) = \pounds\{V(t)\}, \quad \hat{I}(p) \triangleq I(t), \quad \text{etc.}$$

Since the relations (8.122) and (8.121) are formally identical, the small function table in Chapter 8.1.2 also correctly describes the Laplace transforms $F(p) \triangleq f(t)$. A much more comprehensive table of Laplace transforms can be found e.g. in the monograph by DOETSCH [8.107], or elsewhere.

8.1.3.1. Rules of the Laplace Transformation

The Laplace transformation is a linear integral transformation, for which the following relation holds

$$f_i(t) \triangleq F_i(p); \quad c_i = \text{const.} \Rightarrow \sum_i c_i f_i(t) \triangleq \sum_i c_i F_i(p). \qquad (8.123)$$

Of course, this is none other than the mathematical expression of the superposition principle.

As desired, the differentation of $f(t)$ corresponds to an arithmetical operation on $F(p)$

$$f(t) \triangleq F(p) \Rightarrow \frac{df}{dt} \triangleq p \cdot F(p) - f(0). \qquad (8.124)$$

Appendix

Exceptionally this relation shall be proved in extenso:

$$\pounds\left\{\frac{df}{dt}\right\} = \int_0^\infty \epsilon^{-pt} \frac{df}{dt} \cdot dt = \left[\epsilon^{-pt} \cdot f(t)\right]_0^\infty - \int_0^\infty \frac{d\epsilon^{-pt}}{dt} \cdot f(t) \cdot dt$$

$$= -f(0) + p\int_0^\infty \epsilon^{-pt} f(t) dt = -f(0) + p \cdot \pounds\{f(t)\}.$$

By repeated differentiation we obtain, in general,

$$f(t) \triangleq F(p) \implies \frac{d^n f}{dt^n} \triangleq p^n \cdot F(p) - p^{n-1} \cdot f(0) - p^{n-2} \cdot f'(0) - \cdots$$
$$- p \cdot f^{(n-2)}(0) - f^{(n-1)}(0). \tag{8.125}$$

Contrary to the simple operational calculus, where the n-th differentiation of $f(t)$ manifested itself only by the appearance of the n-th power of p, in the Laplace transformation additional polynomials of the $(n-1)$-th order in p are found, with values of $f(t)$ and its derivatives at $t=0$. These polynomials obviously allow for the starting conditions of the problem: the simple relation $\frac{df}{dt} \triangleq p \cdot F(p)$ holds only if the considered system (network) was completely at rest at $t=0$.

By inversion of (8.124) we get

$$f(t) \triangleq F(p) \implies \int_0^t dt' \cdot f(t') \triangleq \frac{1}{p} \cdot F(p) \tag{8.126}$$

and by repeated integration finally

$$f(t) \triangleq F(p) \implies \int_0^t dt^{(n)} \int_0^{t^{(n)}} dt^{(n-1)} + \cdots + \int_0^{t''} dt' \cdot f(t') \triangleq \frac{1}{p^n} \cdot F(p). \tag{8.127}$$

When multiplying the variables by a constant, t and p behave inversely

$$f(t) \triangleq F(p); \quad a = \text{const.} > 0 \implies f(at) \triangleq \frac{1}{a} \cdot F\left(\frac{p}{a}\right). \tag{8.128}$$

Obviously by the transformation (8.122) the function $f(t)$ for high values of t is mapped on to regions with small $|p|$ and vice versa. Hence in approximations t corresponds to $\frac{1}{p}$ (cf. Chapter 8.1.3.5).

As in the operational calculus, shifting of the time variable $t \to t - T$ leads to an exponential factor ϵ^{-pT}:

$$f(t) \triangleq F(p); \quad T = \text{const.} > 0 \implies f(t-T) \triangleq \epsilon^{-pT} \cdot F(p), \tag{8.129}$$

as can easily be proved using (8.122).

On the other hand, the substitution $p \to p+\gamma$ yields a multiplication of the time function by the "damping factor" $\epsilon^{-\gamma t}$

$$f(t) \triangleq F(p); \quad \gamma = \text{const.} > 0 \;\Rightarrow\; F(p+\gamma) \triangleq \epsilon^{-\gamma t} \cdot f(t). \quad (8.130)$$

Finally it must be pointed out, that the multiplication of two Laplace transforms corresponds to the convolution of the respective time functions

$$f_1(t) \triangleq F_1(p); \quad f_2(t) \triangleq F_2(p) \;\Rightarrow\; F_1(p) \cdot F_2(p) \triangleq \int_0^t f_1(t') \cdot f_2(t-t') \cdot dt'. \quad (8.131)$$

Of course, the sequence of the two functions f_1 and f_2 in the integral has no influence on the result.

8.1.3.2. Application of the Laplace Transformation in the Network Analysis

Primarily the rules (8.123) to (8.131) are suitable for solving differential equations. However, in network analysis the construction of the set of differential equations can be avoided. For this purpose all time functions are first transformed: $V_i(t) \to \hat{V}_i(p); I_i(t) \to \hat{I}_i(p)$. Due to (8.123) the Kirchhoff's laws (8.105) remain valid also for the signal transforms $\hat{V}_i(p)$ and $\hat{I}_i(p)$.

The component relations

RESISTOR	CONDENSER	COIL
—⟋⟍⟋⟍— R	—∥— C	—⟲⟲⟲— L

$$V_R(t) = R \cdot I_R(t) \qquad V_C(t) = V_C(0) + \frac{1}{C} \int_0^t I_C(t') \cdot dt' \qquad V_L(t) = L \cdot \frac{d}{dt} I_L(t)$$

become

$$\hat{V}_R(p) = R \cdot \hat{I}_R(p) \qquad \hat{V}_C(p) = \frac{1}{pC} \cdot \hat{I}_C(p) \qquad \hat{V}_L(p) = pL \cdot \hat{I}_L(p) \quad (8.132)$$

if the network was completely at rest at $t=0$, i.e. if in all branches $V_C(0)=0$ and $I_L(0)=0$.

Now the network is analyzed using the purely algebraic relations (8.105) and (8.132), and the found solutions are transformed into time functions with the aid of a Laplace transforms table.

Appendix

Example

The differentiator in Fig. 8.14 is solved as a simple voltage divider to

$$\hat{V}_{out} = \frac{R}{R+1/pC} \cdot \hat{V}_{in} = \frac{p}{p+1/RC} \cdot \hat{V}_{in}. \qquad (8.133)$$

With $t \cdot H(t) \triangleq \frac{1}{p^2}$ from the function table we obtain

$$\hat{V}_{in} = \frac{V_0}{t_0} \cdot \frac{1}{p^2} \quad \text{and} \quad \hat{V}_{out} = \frac{V_0}{t_0} \cdot \frac{1}{p(p+1/RC)},$$

leading to (again function table!)

$$V_{out}(t) = V_0 \cdot \frac{RC}{t_0}(1-\epsilon^{-t/RC}) \cdot H(t). \qquad (8.134)$$

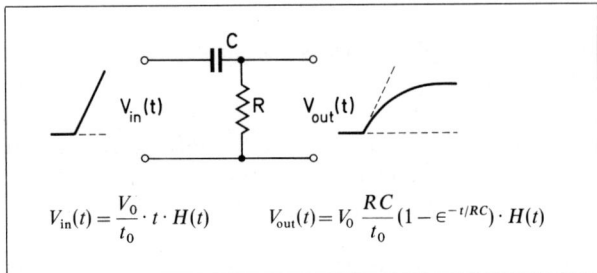

$V_{in}(t) = \frac{V_0}{t_0} \cdot t \cdot H(t) \qquad V_{out}(t) = V_0 \frac{RC}{t_0}(1-\epsilon^{-t/RC}) \cdot H(t)$

Fig. 8.14. The differentiator with a linear ramp input voltage

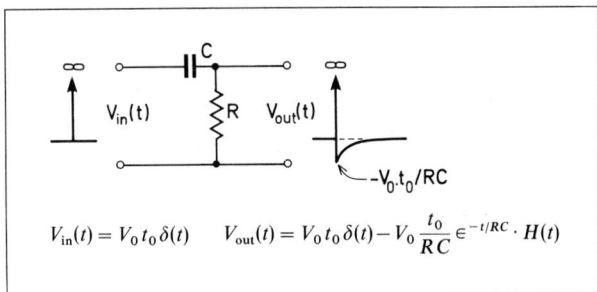

$V_{in}(t) = V_0 t_0 \delta(t) \qquad V_{out}(t) = V_0 t_0 \delta(t) - V_0 \frac{t_0}{RC} \epsilon^{-t/RC} \cdot H(t)$

Fig. 8.15. Transmission of a δ-pulse by the differentiator

Of course, the limiting conditions $V_C(0)=0$ and $I_L(0)=0$ apply only for the intrinsic behaviour of the network and not for the input functions. Since the input signals may be such "physical" functions as $\delta(t)$ or $H(t)$, which often yield a network response with $V_C(0) \neq 0$ or $I_L(0) \neq 0$, the relations $V_C(0)=0$ and $I_L(0)=0$ often do not correctly describe the network behaviour at $t=0$.

The following example shows, how far purely formal calculations can be performed. A very short ($t_0 \ll RC$) voltage pulse having the amplitude V_0 and the duration t_0 is approximately represented by a δ-function, and the response of a differentiator to the δ-function input signal is investigated (Fig. 8.15). Due to $\delta(t) \hat{=} 1$, $\hat{V}_{in} = V_0 \cdot t_0$ holds, and $\hat{V}_{out}$ is

$$\hat{V}_{out} = V_0 t_0 \frac{p}{p+1/RC} = V_0 t_0 \left(1 - \frac{1/RC}{p+1/RC}\right), \quad (8.135)$$

yielding

$$V_{out}(t) = V_0 t_0 \delta(t) - V_0 \frac{t_0}{RC} \cdot \varepsilon^{-t/RC} \cdot H(t).$$

This expression describes correctly not only the distortion-free transmission of the original δ-pulse, but also the resulting undershoot with the amplitude $-V_0 t_0/RC$.

8.1.3.3. Inverse Transformation of Rational Functions $F(p)$

Due to (8.105) and (8.132), the solutions $F(p)$ of networks consisting of lumped components R, C, L are always rational functions

$$F(p) = \frac{P(p)}{Q(p)} = \frac{a_m p^m + a_{m-1} p^{m-1} + \cdots + a_0}{b_n p^n + b_{n-1} p^{n-1} + \cdots + b_0}, \quad (8.136)$$

if the transforms of the input signals themselves are rational functions (this condition being valid in general). The coefficients a_k and b_k are real numbers. Except for expressions yielding δ-functions (cf. e. g. (8.135) with $n=m$), $n>m$ always holds.

For the sake of performing the inverse transformation $F(p)$ is expanded into partial fractions, the expansion being especially simple in the case of simple zeros $p_1, p_2, \ldots, p_n$ of the polynomial $Q(p)$

$$F(p) = \frac{A_1}{p-p_1} + \frac{A_2}{p-p_2} + \cdots + \frac{A_n}{p-p_n}. \quad (8.137)$$

The zeros p_k of $Q(p)$ are the poles of $F(p)$, the corresponding residues A_k can either be calculated by comparison of coefficients or according to the known formula

$$A_k = \frac{P(p_k)}{Q'(p_k)}, \quad \text{with} \quad Q'(p) = \frac{dQ}{dp} \quad (k=1,2,\ldots,n). \quad (8.138)$$

Using the function table the particular partial fractions in (8.137) are easily transformed and in general give exponential functions $\varepsilon^{p_k t}$:

$$\mathcal{L}^{-1}\{F(p)\} = \sum_{k=1}^{n} A_k \cdot \varepsilon^{p_k t} \cdot H(t). \quad (8.139)$$

Due to real coefficients b_k of $Q(p)$ the poles p_k are either real or complex conjugated: $p_{k,k+1} = x_k \pm i\omega_k$. In the second case the sum of the two

Appendix

corresponding exponential functions is a harmonic function

$$A_k e^{p_k t} \cdot H(t) + A_{k+1} e^{p_{k+1} t} \cdot H(t) \approx e^{x_k t} \cdot \sin \omega_k t \cdot H(t), \quad (8.140)$$

the amplitude and phase of which is still determined by the residues A_k and A_{k+1}.

Hence:

> *The inverse transform of a rational function $F(p)$ with simple poles is a sum of damped, limited or unlimited harmonic functions and of exponential and step functions.*

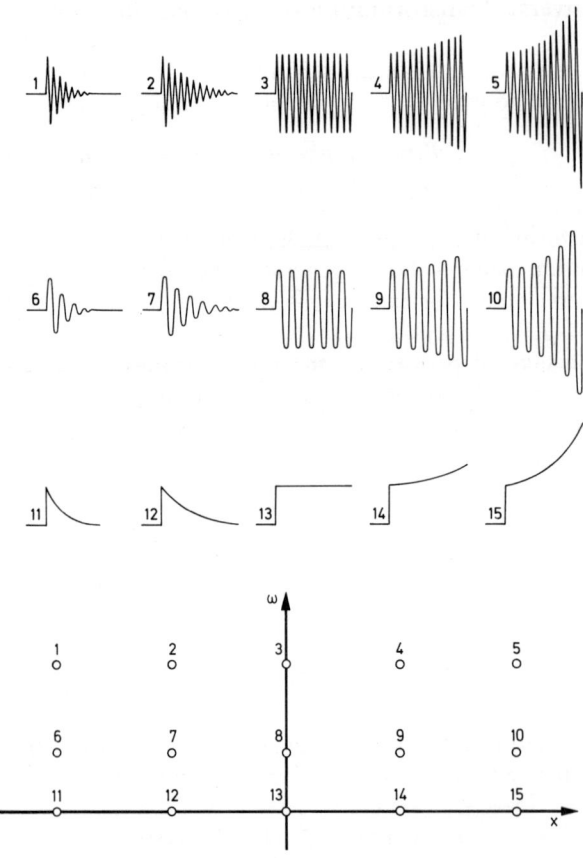

Fig. 8.16. Time functions (inverse transforms) corresponding to various poles of $F(p)$. Only the poles with positive imaginary part are indicated out of the pairs of complex conjugated poles

Fig. 8.16 schematically shows the time functions corresponding to respective poles, depending on the pole position in the complex plane.

For higher order poles of $F(p)$, expressions of the type

$$\frac{1}{(p-p_k)^\rho} \quad (\rho \geqslant 2)$$

appear in the partial fraction expansion of $F(p)$, which can easily be transformed using the relation (8.130):

$$\frac{1}{(p-p_k)^\rho} \triangleq \frac{t^{\rho-1}}{(\rho-1)} \in^{p_k t} \cdot H(t). \qquad (8.141)$$

8.1.3.4. Stability Considerations

In networks with one input and one output signal the network solution is always as follows

$$\hat{A}(p) = G(p) \cdot \hat{E}(p), \qquad (8.142)$$

where $\hat{A}(p)$ and $\hat{E}(p)$ denote the Laplace transforms of the output and input signals, respectively (e.g. $\hat{V}_{out}$ and $\hat{V}_{in}$ in (8.133)). $G(p)$ is denoted as the *transfer function* (or response function).

Since $\delta(t) \triangleq 1$, the inverse transform of $G(p)$, $g(t) = \pounds^{-1}\{G(p)\}$, represents the output signal in the case of a δ-function at the input, i.e. the eigen-oscillations of the network excited by a very short input stimulus.

A network is considered as a stable one, if its eigen-oscillations caused by small distortions at the input, which are always present (noise!), are damped and disappear rapidly. Hence, according to Fig. 8.16 a network is stable if all poles of its transfer function $G(p)$ lie in the left half-plane, i.e. if $\text{Re}\{p_k'\} < 0$. All networks consisting solely of passive components are stable.

8.1.3.5. Approximations

In Chapter 8.1.3.3 we have proved that the solution $f(t)$ of a network is in general a sum of harmonic oscillations $\in^{p_k t}$, which are either damped, or limited, or unlimited, according to the value and sign of the real part of the respective pole p_k. For large values of t, that sum term becomes most significant, the real part of the pole ($\text{Re}\{p_k'\}$) of which is the greatest, i.e. the pole of which lies at the most extreme right of the complex plane (so called *dominant pole*). The dominant pole defines the asymptotical behaviour of $f(t)$ at $t \to \infty$.

Appendix

Using these considerations, for example the value of $V_{out}(t)$ of a passive network with the transfer function $G(p)$ for large values of t can be determined in case of a step input function $V_{in}(t) = V_0 \cdot H(t)$. We have

$$\hat{V}_{out}(p) = G(p) \cdot \hat{V}_{in}(p); \quad \hat{V}_{in}(p) = \frac{V_0}{p};$$

hence

$$\hat{V}_{out}(p) = V_0 \cdot \frac{G(p)}{p}. \quad (8.143)$$

In the partial fraction expansion of $G(p)/p$ besides the fractions corresponding to the poles p_k of $G(p)$, a fraction $G(0)/(p-0)$ appears, corresponding to the new pole $p=0$. However, the pole $p=0$ is dominant, since all poles p_k of $G(p)$ of a passive network lie in the left half-plane. $\pounds^{-1}\left\{\frac{G(0)}{p}\right\} = G(0) \cdot H(t)$ hence describes the asymptotical behaviour of the output signal for large t:

$$\lim_{t \to \infty} V_{out}(t) = V_0 \cdot G(0) \cdot H(t). \quad (8.144)$$

For example, a differentiator (Fig. 8.14) exhibits $G(p) = p/(p + 1/RC)$. For step input signals always $V_{out}(\infty) = 0$, in agreement with $G(0) = 0$.

This behaviour of the inverse transform can be stated in a more general way: *The value of $p \cdot F(p)$ for $p = 0$ yields the value of the inverse transform $f(t) = \pounds^{-1}\{F(p)\}$ for $t \to \infty$.*

In the discussion of the relation (8.128) we have pointed out the reciprocal behaviour of the variables t and p. In fact, this can often be used for simplification of the *transfer function* $G(p)$, if the behaviour of the output time function is of interest but in a restricted range of t. In the corresponding relations used for neglecting particular expression term, only t must be substituted by $1/p$. Advantageously the technique is demonstrated by some examples concerning the differentiator (Fig. 8.14):

In full detail

$$\hat{V}_{out} = G(p) \cdot \hat{V}_{in} = \frac{p}{p + 1/RC} \hat{V}_{in} = \frac{pRC}{pRC + 1} \hat{V}_{in} = \frac{1}{1 + 1/pRC} \hat{V}_{in}. \quad (8.145)$$

The different expressions for $G(p)$ will be used for the derivation of different approximations.

a) *Approximation for very short time intervals, $t \to 0$*

$t \to 0$ yields $\frac{1}{p} \to 0$, or $p \to \infty$, and hence $G(p) \approx 1$. Therefore, $\hat{V}_{\text{out}} \approx \hat{V}_{\text{in}}$; $V_{\text{out}}(t) \approx V_{\text{in}}(t)$: very short signals are transmitted by the differentiator without any distortion.

b) *Approximation for short time intervals $t \ll RC$*

$t \ll RC$ yields $\frac{1}{p} \ll RC$, or $\frac{1}{pRC} \ll 1$, and hence

$$G(p) = \frac{1}{1 + 1/pRC} \approx 1 - \frac{1}{pRC}; \text{ from } \hat{V}_{\text{out}} \approx \left(1 - \frac{1}{pRC}\right) \cdot \hat{V}_{\text{in}}$$

we get

$$V_{\text{out}}(t) \approx V_{\text{in}}(t) - \frac{1}{RC} \int_0^t dt' \cdot V_{\text{in}}(t').$$

In case of $V_{\text{in}}(t) = V_0 \cdot H(t)$ we obtain as expected

$$V_{\text{out}}(t) \approx V_0 \cdot (1 - t/RC) \cdot H(t).$$

c) *Approximation for large time intervals $t \gg RC$*

$t \gg RC$ yields $\frac{1}{p} \gg RC$, or $pRC \ll 1$. Therefore

$$\hat{V}_{\text{out}} = \frac{pRC}{pRC + 1} \hat{V}_{\text{in}} \approx pRC \cdot \hat{V}_{\text{in}} \approx pRC(1 - pRC) \cdot \hat{V}_{\text{in}} \approx pRC \cdot \epsilon^{-pRC} \cdot \hat{V}_{\text{in}}.$$

All three approximation expressions are useful, the last one makes use of the well known approximation $\epsilon^x \approx 1 + x$ for $x \ll 1$. Hence either

$$\hat{V}_{\text{out}} \approx pRC \cdot \hat{V}_{\text{in}} \Rightarrow V_{\text{out}}(t) \approx RC \cdot \frac{d}{dt} V_{\text{in}}(t), \quad \text{(therefore } \textit{differentiator circuit}\text{!)}$$

or

$$\hat{V}_{\text{out}} \approx pRC \cdot \epsilon^{-pRC} \cdot \hat{V}_{\text{in}} \Rightarrow V_{\text{out}}(t) \approx RC \cdot \frac{d}{dt} V_{\text{in}}(t - RC),$$

the last expression also allowing for the small delay of the order of RC introduced by a differentiator.

Appendix

8.2. Noise

8.2.1. General Considerations, Concept of Equivalent Noise Charge

Let us consider a radiation detector connected to an amplifier. Even without any input stimuli (no radiation, no background pulses) there is some voltage $V_{out}(t)$ at the amplifier output, fluctuating in a random manner about its average value (which shall be assumed to be zero in what follows). This phenomenon is called *noise*.

The noise voltage is characterized advantageously by the average of its square, the so-called variance

$$\langle V_{out}^2 \rangle = \lim_{T \to \infty} \frac{1}{T} \int_0^T V_{out}^2(t) \cdot dt, \qquad (8.201)$$

or by the so-called root mean square (rms) deviation σ:

$$\sigma = \sqrt{\langle V_{out}^2 \rangle}. \qquad (8.202)$$

Since the noise voltage is superimposed on any signal voltage, it determines the smallest detectable amplitude, and also causes the amplitudes of an originally monoenergetic signal to fluctuate statistically about its mean. Hence, it is not the absolute value of $\langle V_{out}^2 \rangle$ that is of interest, but its ratio to the signal amplitude under consideration.

The primary signal delivered by the detector is a short charge pulse, which is converted into a voltage pulse first by the input circuit and the shaping stages of the amplifier. Therefore, the comparison between signal and noise can be facilitated, if the latter is expressed as an equivalent charge fluctuation at the amplifier input. The *equivalent noise charge* Q_N is defined as follows:

> A current pulse $I(t) = Q_N \cdot \delta(t)$ applied to the amplifier input would cause an output voltage pulse, the amplitude of which is equal to the rms deviation σ of the noise voltage at the amplifier output.

The equivalent noise charge Q_N can be calculated as follows: From the amplifier transfer function $G(p)$ and the Laplace transform of the hypothetical input signal $I(t) = Q_N \cdot \delta(t)$ the resulting output voltage pulse $V_{out}(t) = \mathcal{L}^{-1}\{Q_N \cdot G(p)\}$ is calculated, and the maximum value of $V_{out}(t)$ is set equal to σ. Solving this equation yields Q_N.

Expressed as equivalent charge Q_N (in coulombs or better in electron charges e), the noise can immediately be compared with the signal. However, it must be considered that the concept of the equivalent noise charge refers to a hypothetical, very short current input pulse. The comparison of Q_N with a real detector signal pulse is thus correct only if the detector pulse is also short compared with the shaping time constants of the amplifier, and if it can also be considered as a δ-function. If not, the actual detector pulse shape $I(t)=Q\cdot a(t)$ must be taken into account by replacing the amplifier transfer function $G(p)$ by $\hat{a}(p)\cdot G(p)$ (with $\hat{a}(p)=\pounds\{a(t)\}$) in the calculation of Q_N, or by stating the amplitude defect of a real signal $(Q\cdot a(t))$ as compared with the ideal one $(Q\cdot\delta(t))$.

8.2.2. Noise Sources

Due to the thermal movement of the charge carriers (electrons), the potential difference across an ohmic resistance fluctuates statistically. This so-called *thermal noise* of a resistor R can be represented by means of a noise voltage generator V_R in series with R (Fig. 8.21), the frequency distribution $d\langle V_R^2\rangle/df$ of the mean square noise voltage being

$$\frac{d\langle V_R^2\rangle}{df} = 4\cdot kT\cdot R. \qquad (8.203)$$

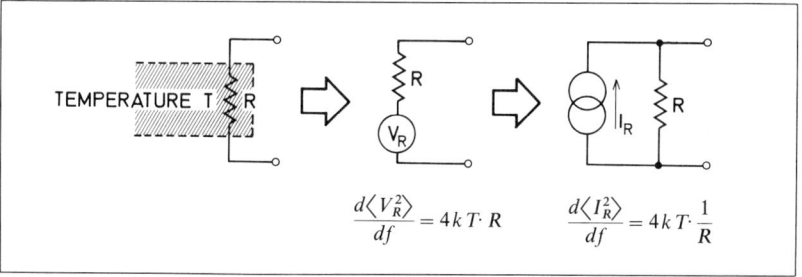

Fig. 8.21. Equivalent circuits for thermal noise of a resistor

Another common representation using a noise current generator I_R in parallel with R, is

$$\frac{d\langle I_R^2\rangle}{df} = 4\cdot kT\cdot\frac{1}{R}. \qquad (8.204)$$

Due to the quantum character ("graininess") of the electrical current consisting of a random flow of charge carriers, any current I exhibits fluctuations with a frequency distribution

$$\frac{d\langle I^2\rangle}{df} = 2\cdot e\cdot I, \qquad (8.205)$$

365

Appendix

irrespective of whether the current is flowing in metal or semiconductor, through a metal-semiconductor junction, from a hot cathode, to a vacuum tube grid or anode[15], etc. This phenomenon is known as *shot-noise*.

Both these noise components are physically well understood. Both of them have a frequency independent spectrum (8.204), (8.205), and are therefore occasionally denoted as "white" noise. In addition, real circuit components (transistors, tubes) exhibit an excess noise $\langle V_f^2 \rangle$, which is not completely understood, and the frequency distribution of which is approximately proportional to $1/f$:

$$\frac{d\langle V_f^2 \rangle}{df} = \frac{A_f}{f}. \tag{8.206}$$

This so-called *flicker noise* is caused by structural changes in the connector materials and by various surface effects.

The influence of the $1/f$ noise on the nuclear spectrometry resolution has been discussed by HATCH [8.216].

8.2.3. The Noise of an Amplifier with the Transfer Function $G(p)$

Fig. 8.22 shows a typical detector-amplifier system. The amplifier has a transfer function $A \cdot G(p)$. However, since both signal and noise are equally amplified, the absolute value of the gain A is of no importance, and for simplicity we can assume $A = 1$.

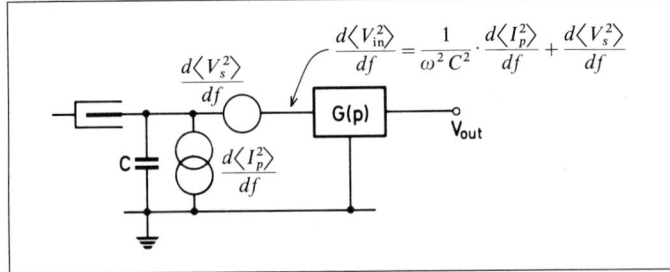

Fig. 8.22. Noise generators in the equivalent circuit of an amplifier

---

[15] Of course, the current fluctuations may be influenced by secondary effects, such as the statistical gain fluctuation in a photomultiplier, space charge effects in a vacuum tube, etc., which must be considered in any special case.

The Noise of an Amplifier with the Transfer Function $G(p)$

Only the first amplifier delivers a significant noise contribution, since the noise generated in the following stages is not fully amplified. (The noise contribution of the second amplifier stage has been discussed by ARBEL [8.215].) Two distinct noise components may be distinguished: a) noise which immediately adds to the signal current, and which must therefore be represented as a current generator I_p in parallel with the input capacity C, and b) noise generated inside the amplifier stage independent of C, which must be represented as a voltage generator V_s in series with C. The *parallel noise* I_p consists of e. g. detector leakage current, first stage input current (grid, base or gate current), and the thermal noise of the detector bias resistor (which is not shown in the simplified diagram of Fig. 8.22). The *series noise* is caused by the random fluctuations of the component current (shot-noise of the anode, collector or channel current), and by the frequency dependent flicker noise. Hence

$$\frac{d\langle I_p^2\rangle}{df} = a; \quad \frac{d\langle V_s^2\rangle}{df} = b + \frac{A_f}{f}, \qquad (8.207)$$

where a and b are system dependent constants, describing the white noise, and A_f is taken from (8.206).

Assuming a normal distribution of the noise amplitudes, the variance of the total noise is simply the arithmetical sum of the variances of the particular noise components. On the other hand, the transfer function $G(p)$ becomes the frequency characteristic of the amplifier, when replacing p by $i\omega$[16]. Hence

$$\frac{d\langle V_{out}^2\rangle}{df} = \frac{d\langle V_{in}^2\rangle}{df} \cdot |G(i\omega)|^2, \qquad (8.208)$$

and the variance of the noise voltage at the amplifier output is

$$\langle V_{out}^2\rangle = \frac{1}{2\pi}\int_0^\infty \left\{\frac{d\langle V_s^2\rangle}{df} + \frac{1}{\omega^2 C^2}\cdot\frac{d\langle I_p^2\rangle}{df}\right\}|G(i\omega)|^2\, d\omega. \qquad (8.209)$$

In what follows only the case of an amplifier with simple differentiator and integrator will be considered. As can be shown [8.201], the signal-

---

[16] This can easily be proved considering the transmission of a harmonic oscillation $\cos\omega t \cdot H(t) \triangleq p/(p^2+\omega^2)$ by a passive network with the transfer function $G(p)$ for $t\to\infty$. i. e. after the transient effects have decayed. The dominant poles of $G(p)\cdot p/(p^2+\omega^2)$ are $\pm i\omega$; hence in the approximation of $t\to\infty$ the output signal is $|G(i\omega)|\cdot\cos(\omega t+\varphi)$, with φ denoting the argument (polar angle) of $G(i\omega)$.

Appendix

to-noise ratio is maximum for $\tau_{\text{diff}} = \tau_{\text{int}} = \tau$. Hence the transfer function $G(p)$ is

$$G(p) = \frac{p\tau}{(1+p\tau)^2}. \tag{8.210}$$

The situation in amplifiers with $G(p)$ other than (8.210) has been discussed by many authors (e. g. [8.201, 8.202]).TSUKUDA [8.203] presents the results of numerical calculations for amplifiers with single or double RC or cable differentiation and single or double integration, for various ratios of $\tau_{\text{int}}/\tau_{\text{diff}}$. The simple case (8.210) which is regarded here is representative of most applications. Multiple integration descreases somewhat the equivalent noise charge Q_N, double RC differentiation or cable clipping increases Q_N. In addition, in these more complicated cases the minimum of Q_N is often reached at $\tau_{\text{diff}} \neq \tau_{\text{int}}$. However, the possible improvement over (8.210) amounts only to some tens per cent (cf. also Chapter 3.13, Table 3.38).

In an accurate timing is desired together with high energy resolution (i. e. low noise), the signal is advantageously treated in two channels with different transfer functions $G(p)$ [8.204].

Introducing (8.207) and $G(p)$ from (8.210) into (8.209), $\langle V_{\text{out}}^2 \rangle$ can be calculated as

$$\langle V_{\text{out}}^2 \rangle = \frac{a}{C^2} \cdot \frac{\tau}{8} + b \cdot \frac{1}{8\tau} + \frac{A_f}{2}, \tag{8.211}$$

where the first term describes the parallel noise, the second term the series noise, and the third term the flicker noise. A hypothetical signal $I(t) = Q_N \cdot \delta(t)$ yields an amplifier output signal (the actual polarity of which is irrelevant)

$$V_{\text{out}}(t) = \frac{Q_N}{C} \cdot \frac{t}{\tau} \epsilon^{-t/\tau} \cdot H(t), \tag{8.212}$$

with the pulse height

$$V_{\text{max}} = \frac{Q_N}{C} \cdot \frac{1}{\epsilon}. \tag{8.213}$$

Making $\langle V_{\text{out}}^2 \rangle = V_{\text{max}}^2$, (8.211) and (8.213) yield

$$Q_N = \frac{1}{e} \sqrt{a\tau + \frac{b}{\tau} C^2 + 4 A_f C^2}, \tag{8.214}$$

where $\epsilon^2 \approx 8$ has been used. Due to the factor $1/e$, the equivalent noise charge Q_N is expressed as the number of electrons e.

8.2.4. Noise in a Charge Sensitive Amplifier

Most low noise amplifiers are connected as charge sensitive stages. Fig. 8.23 shows the equivalent circuit diagram of such a stage. The

Noise in a Charge Senitive Amplifier

output voltage, which corresponds to a given detector signal $I_{sig}(t)$, amounts to (neglecting the polarity)

$$\hat{V}_{out}(p) = \hat{I}_{sig}(p) \frac{1}{p[C_f + (C+C_f)/A_0]} \cdot G(p). \tag{8.215}$$

With $G(p)$ from (8.210) and $I_{sig}(t) = Q_N \cdot \delta(t)$ the amplitude of $V_{out}(t)$ is

$$V_{max} = \frac{Q_N}{C_f + (C+C_f)/A_0} \cdot \frac{1}{e}. \tag{8.216}$$

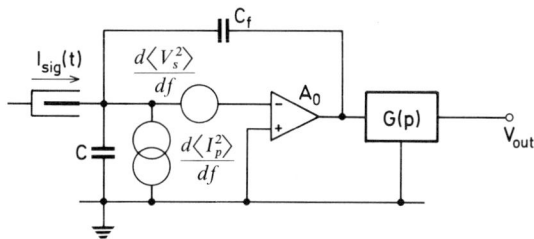

Fig. 8.23. Noise generators in the equivalent circuit of an charge sensitive amplifier

A_0 denotes the difference amplifier gain, C_f is the feedback capacity. A voltage signal V_s injected at the place of the series noise generator causes an output voltage (again, neglecting the polarity)

$$\hat{V}_{out}(p) = \hat{V}_s \cdot \frac{C+C_f}{C_f + (C+C_f)/A_0} \cdot G(p). \tag{8.217}$$

Hence the noise voltage at the amplifier output becomes

$$\langle V_{out}^2 \rangle$$

$$= \frac{1}{2\pi} \int_0^\infty \left\{ \frac{d\langle V_s^2 \rangle}{df} \cdot \frac{(C+C_f)^2}{[C_f + (C+C_f)/A_0]^2} + \frac{d\langle I_p^2 \rangle}{df} \cdot \frac{1}{\omega^2 [C_f + (C+C_f)/A_0]^2} \right\}$$

$$\cdot |G(i\omega)|^2 \cdot d\omega. \tag{8.218}$$

From the comparison of (8.218) and (8.209) as well as (8.216) and (8.213) the equivalent noise charge Q_N is derived to

$$Q_N = \frac{1}{e}\sqrt{a\tau + \frac{b}{\tau}(C+C_f)^2 + 4 \cdot A_f \cdot (C+C_f)^2}. \tag{8.219}$$

369

Appendix

Of course, this result is not surprising, since the signal-to-noise ratio is known to be unaffected by any feedback [8.205]. Nevertheless, the small calculation performed proves that the capacity in (8.219) as in (8.214) represents merely the external "cold" circuit capacities (i.e. C in Fig. 8.22, and $C+C_f$ in Fig. 8.23), and not the dynamical input capacity $C+C_f(1+A_0)$!

8.2.5. Properties of Input Stages with Vacuum Tubes, Bipolar Transistors and FET

The last of the three terms in (8.214) or (8.219) — the flicker noise component $4 \cdot A_f \cdot C^2$ — is independent of the pulse shaping time constant τ. The parallel noise $a\tau$ increases and the series noise $\frac{b}{\tau} \cdot C^2$ decreases with increasing τ. Hence Q_N exhibits a minimum at a certain time constant $\tau = \tau_0$, which can be calculated as

$$\tau_0 = C \cdot \sqrt{\frac{b}{a}}. \qquad (8.220).$$

Normally, depending on the actual situation, τ_0 lies between 0.1 and 100 μsec.

Since very long τ can not be accepted when working at high counting rates, often the optimum condition (8.220) must be renounced. In general, $\tau \approx 1$ μsec is chosen. If $\tau \neq \tau_0$ must be used, the equivalent noise charge Q_N can be reduced by means of a more complex shaping filter $G(p)$ [8.206, 8.207, 8.208].

The parallel noise a is composed of the following components

$$a = 4 \cdot kT/R + 2 \cdot e \cdot I_D + 2 \cdot e \cdot I_{in}. \qquad (8.221)$$

The first term of (8.221), the contribution of the detector bias resistor R, can be neglected, since commonly a very high R (some MΩ) is chosen. The second term describes the fluctuations of the detector leakage current I_D. In ionization chambers $I_D \approx 0$, and its contribution to Q_N can be neglected. In semiconductor detectors the reverse pn-junction current I_D is 10^{-11} to 10^{-7} A, according to the detector type and dimensions. With $I_D = 10^{-7}$ A for example the second term $2 \cdot e \cdot I_D$ may be the most significant contribution to Q_N.

The third term $2 \cdot e \cdot I_{in}$ depends on the input current I_{in} of the used circuit component (grid current in vacuum tubes, base current in bipolar transistors, gate current in FET). In electrometer tubes, and in general in all tubes with low anode voltage and low transconductance, $I_{in} \approx 10^{-12} \ldots 10^{-10}$ A, also in FET ($I_{in} \approx 10^{-11}$ A), and its contribution to Q_N can be neglected. If vacuum tubes with high transconductance are to be used together with detectors with small I_D, the grid current I_{in} may cause difficulties. The high base current ($\gtrsim 10^{-6}$ A) of bipolar transistors is commonly prohibitive for their use in low-noise pre-amplifiers.

The series noise — consisting mostly of the shot noise of the input circuit component — can be described as the thermal noise of an equivalent resistor R_{eq} [17]

$$b = 4 \cdot kT \cdot R_{eq}, \qquad (8.222)$$

with

$$R_{eq} = \frac{2.5}{g_m} \quad \text{for vacuum tubes (e. g. [8.201, 8.207])}, \qquad (8.223)$$

$$R_{eq} = \frac{0.5}{g_m} \quad \text{for bipolar transistors [8.206]}, \qquad (8.224)$$

$$R_{eq} = \frac{0.7}{g_m} \quad \text{for FET [8.209, 8.210, 8.211]}, \qquad (8.225)$$

where g_m denotes the mutual transconductance of the component under consideration, and $T = 290\,°K$. For bipolar transistors $g_m = \frac{e}{kT} \cdot I_C$, hence R_{eq} is dependent on the collector current I_C. For tubes and bipolar transistors a typical value is $g_m \approx 10\,\text{mA/V}$, for FET $g_m \approx 1\,\text{mA/V}$.

The total noise of field-effect transistors can be strongly reduced by cooling them to about 100 to 130 °K [8.212], the corresponding physical effects being not yet completely understood. When cooling FET below 100 °K the noise again starts to increase, due to the decrease of the transconductance at lower temperatures, and to the rise of the fluctuations in the number of activated carriers.

The flicker noise constant A_f for the vacuum tubes is about $10^{-13}\,\text{V}^2$, for FET of about $10^{-12}\,\text{V}^2$. In bipolar transistors the $1/f$ noise can be completely neglected with respect to the other noise components.

The capacity C in (8.214) is composed of two parts, namely the internal capacity C_{int} of the amplifier input, and the external capacity C_{ext} (= detector capacity + capacity of the connection between detector and amplifier): $C = C_{int} + C_{ext}$. Hence the equivalent noise charge Q_N (8.214) decays in two components, the one dependent on C_{ext}, the other independent of it. For practical purposes the equivalent noise charge Q_N can be expounded in power series of C_{ext}, yielding the linear approximation

$$Q_N = Q_{N0} + \frac{dQ_N}{dC} \cdot C_{ext}. \qquad (8.226)$$

---

[17] The discussion of secondary effects, like e. g. the partition noise in pentodes etc., shall be omitted here.

Appendix

Commonly the term Q_{NO} is dominated by a[18], and dQ_N/dC by b and A_f. The factors Q_{NO} and dQ_N/dC sufficiently characterize the amplifier. For example, Fig. 8.24 shows the experimental plot of noise versus external capacity for a preamplifier described by HAHN and MEYER [8.214].

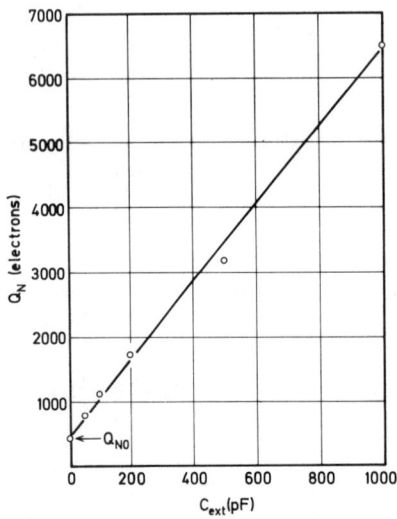

Fig. 8.24. The dependence of the equivalent noise charge Q_N on the external capacity C_{ext} (according to HAHN and MAYER [8.214])

Fig. 8.25 (RADEKA [8.213]) shows the noise performance of amplifiers with vacuum tubes, bipolar and field effect transistors. The shaded areas of the $Q_{NO} - dQ_N/dC$ diagram indicate typical values. Due to very high base current the equivalent noise charge Q_{NO} of bipolar transistors is higher than Q_{NO} of tubes and FET by about one order of magnitude. Because of the low transconductance (high b in (8.214)) FET have higher dQ_N/dC, than vacuum tubes. However, in the present state of the art, FET cooled to about 100°K yield the best overall noise performance.

For completeness, the MOS transistors should be mentioned, the noise performance of which appeared promising. However — at least at present — they are limited by very high $1/f$ noise.

---

[18] Since Q_{NO} serves as a quality parameter of the amplifier, the contribution $2 \cdot e \cdot I_D$ (8.221) of the detector leakage current need not be considered in a.

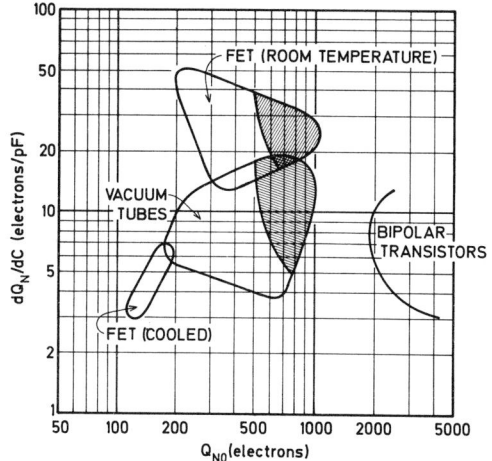

Fig. 8.25. Noise performance of amplifiers with different circuit components expressed in terms of Q_{NO} and dQ_N/dC (according to RADEKA [8.213])

8.2.6. Noise and Resolution

In radiation spectrometry the pulse height is commonly measured in energy units (e. g. keV). The equivalent noise charge Q_N directly determines the rms deviation σ of the signal pulse height, expressed in energy units, when multiplied by W:

$$\sigma = W \cdot Q_N. \tag{8.227}$$

As in Chapter 2, W denotes the energy required to generate one electron charge at the preamplifier input (hence e. g. 35 eV in an air filled ionization chamber, $30/\overline{M}$ eV in a methane filled proportional counter with the gas multiplication factor $\overline{M}$, 3.5 eV in a silicon semiconductor detector, etc.).

Due to the noise Q_N (or σ (8.227)), an infinitely sharp energy line $\eta(E) = N_0 \cdot \delta(E - E_0)$ at $E = E_0$ is broadened to $\tilde{\eta}(E)$:

$$\tilde{\eta}(E) = N_0 \frac{1}{\sigma\sqrt{2\pi}} \epsilon^{-\frac{(E-E_0)^2}{2\delta^2}} \quad \text{(normal distribution)}. \tag{8.228}$$

This so-called normal line shape is shown in Fig. 8.26 If the original enery line $\eta(E)$ exhibits a more general shape — like e. g. in Fig. 2.02 —

Appendix

the noise broadened line $\tilde{\eta}(E)$ is obtained as a convolution of $\eta(E)$ and the normal distribution (8.228):

$$\tilde{\eta}(E) = \int_{-\infty}^{+\infty} \eta(\varepsilon) \cdot \frac{1}{\sigma\sqrt{2\pi}} \cdot e^{-\frac{(E-\varepsilon)^2}{2\delta^2}} \cdot d\varepsilon. \qquad (8.229)$$

If $\eta(E)$ itself is a normal line with rms deviation σ_L, the relation (8.229) yields for $\tilde{\eta}(E)$ a normal line with

$$\sigma_{\text{total}} = \sqrt{\sigma_L^2 + \sigma^2} \qquad (8.230)$$

as standard deviation.

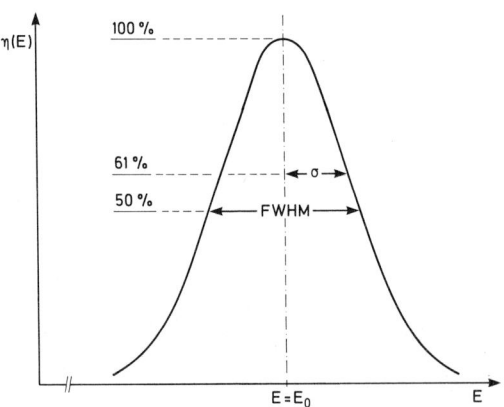

Fig. 8.26. Normal shape of a spectrum line

Normally the resolution is given in terms of full line width at half maximum (FWHM). From Fig. 8.26

$$\text{FWHM} = 2.35 \cdot \sigma_{\text{total}}. \qquad (8.231)$$

As an example, the resolution of the 122 keV Co-57 photo peak will be calculated in the case of a germanium semiconductor detector ($W = 2.6$ eV, $I_D = 1$ nA, $C_{\text{ext}} = 10$ pF) and a vacuum tube preamplifier followed by the main amplifier with single RC differentiator and integrator ($\tau_{\text{diff}} = \tau_{\text{int}} = \tau = 1$ µsec, $Q_{NO} = 200$ electrons, $dQ_N/dC = 7$ electrons/pF). From (8.226) $Q_N = 270$ electrons, to which the contribution $2 \cdot e \cdot I_D \cdot \tau/e^2$ from (8.214) and (8.221) must be added:

$Q_N = \sqrt{2 \cdot I_D \cdot \tau/e + (270)^2} = 292$ electrons. Hence $\sigma = 0.76$ keV. The signal consists of $\Delta E/W = 122\,000/2.6 = 47\,000$ electrons. Assuming a Fano factor $f = 0.5$ the rms deviation of the signal is $\sqrt{47\,000/2} = 153$ electrons, or $\sigma_L = 0.40$ keV. From (8.230) $\sigma_{total} = 0.86$ keV, and from (8.231) the resolution is FWHM ≈ 2 keV.

9. References

1.001 DE WAARD, H., and D. LAZARUS: Modern Electronics. Reading, Massachusetts: Addison-Wesley 1966.
1.002 SAMUELI, J. J., J. PIGUERET, et A. SARAZIN: Instrumentation électronique en physique nucléaire (Mesures de temps et d'énergie). Paris: Masson et Cie., Editeurs 1968.
2.001 FRANZEN, W., and L. W. COCHRAN: Pulse Ionization Chambers and Proportional Counters. In: Nuclear Instruments and their Uses (ed. A. H. SNELL), p. 3. New York: John Wiley & Sons 1962.
2.002 GRAY, L. H.: Proc. Cambridge Phil. Soc. **40**, 72 (1944).
2.003 NEUERT, H.: Kernphysikalische Meßverfahren. Karlsruhe: G. Braun 1966.
2.004 LANDAU, L. D.: J. Phys. USSR **8**, 72 (1944).
2.005 BIRKHOFF, R. D.: In: Handbuch der Physik, Vol. XXXIV. Berlin-Göttingen-Heidelberg: Springer 1958.
2.006 WEGNER, H. E.: Nucl. Electronics IAEA **1**, 427 (1962).
2.007 FANO, U.: Phys. Rev. **72**, 26 (1947).
2.008 WILKINSON, D. H.: Ionisation Chambers and Counters. Cambridge University Press 1950.
2.009 BUNEMAN, O., T. E. CRANSHAW, and J. A. HARVEY: Canad. J. Research **A-27**, 191 (1949).
2.010 FRISCH, O.: Brit. Atomic Energy Com. Report **BR-49** (1944).
2.011 MILLER, H. L.: Rev. Sci. Instr. **27**, 330 (1956).
2.012 TSUKUDA, M.: Nucl. Instr. Methods **14**, 241 (1961).
2.013 OGAWA, I., T. DOKE, and M. TSUKUDA: Nucl. Instr. Methods **13**, 164 (1961).
2.014 BENOIT, R., G. BERTOLINI, et G. B. RESTELLI: Electronique Nucleaire. Paris: SFER 1963, p. 485.
2.015 BOCHAGOV, B. A., A. A. VOROB'EV, and A. P. KOMAR: Izv. Akad. Nauk SSSR, Ser. Fiz. **20**, 1455 (1956).
2.016 ——— Zh. Tekhn. Fiz. **27**, 1575 (1957).
2.017 GILLESPIE, A. B.: Signal, Noise and Resolution in Nuclear Counter Amplifiers. New York: Pergamon Press 1953.
2.018 BALDINGER, E., and W. FRANZEN: Advances in Electronics and Electron Physics **8**, 255 (1956). New York: Academic Press.
2.019 ENGELKEMEIR, D. W., and L. B. MAGNUSSON: Rev. Sci. Instr. **26**, 295 (1955).
2.020 COTTINI, C., E. GATTI, G. GIANNELLI, e G. ROZZI: Nuovo Cimento **3**, 473 (1956).
2.021 HANNA, G. C., D. H. W. KIRKWOOD, and B. PONTECORVO: Phys. Rev. **75**, 985 (1949).
2.022 CURRAN, S. C., and H. W. WILSON: Proportional Counters and Pulse Ionization Chambers. In: Alpha-, Beta- and Gamma-Ray Spectroscopy (ed. K. SIEGBAHN), p. 303. Amsterdam 1965.
2.023 SNYDER, H. S.: Phys. Rev. **72**, 181 (1947).
2.024 CURRAN, S. C., A. L. COCKCROFT, and J. ANGUS: Phil. Mag. **40**, 929 (1949).
2.025 SAUTER, F.: Zt. Naturf. **4a**, 682 (1949).
2.026 HOYT, R. C.: Rev. Sci. Instr. **20**, 178 (1949).
2.027 BISI, A., e L. ZAPPA: Nuovo Cimento **2**, 988 (1955).

References

2.028 WEST, D.: Proc. Phys. Soc. (London) **A-66**, 306 (1953).
2.029 SCHLUMBOHM, H.: Zur Statistik der Elektronenlavinen im Proportionalzählrohr. Diplomarbeit, Hamburg 1958 (quoted in: 2.003).
2.030 BYRNE, J.: Proc. Roy. Soc. Edinburgh **66**, 33 (1962).
2.031 LAUSIART, A., et J. P. MORUCCI: J. Phys. et Rad. **23**, 102-A (1962).
2.032 PRESCOTT, J. R., and P. S. TAKHAR: IRE Trans. Nucl. Sci. **NS-9**/3, 36 (1962)
2.033 — Nucl. Instr. Methods **22**, 256 (1963).
2.034 — Nucl. Instr. Methods **39**, 173 (1965).
2.035 MCCUTCHEN, C. W.: Rev. Sci. Instr. **27**, 106 (1956).
2.036 NAGATANI, T., and Y. SAKAKI: Rev. Sci. Instr. **33**, 556 (1962).
2.037 MAY, F., and F. SEMTURS: Nucl. Instr. Methods **34**, 121 (1965).
2.038 GEIGER, H., and W. MÜLLER: Phys. Zt. **29**, 839 (1928).
2.039 ALDER, F., E. BALDINGER, P. HUBER, and F. METZGER: Helv. Phys. Acta **20**, 73 (1947).
2.040 VAN DUUREN, K., A. J. M. JASPERS, and J. HERMSEN: Nucleonics **17**/6, 86 (1959).
2.041 GEBAUER, H.: A.T.M. J 0 **76**/9 (1961).
2.042 STEVENSON, A.: Rev. Sci. Instr. **23**, 93 (1952).
2.043 — Rev. Sci. Instr. **23**, 93 (1952).
2.044 RAMSEY, W. E.: Rev. Sci. Instr. **23**, 95 (1952).
2.045 PORTER, W. C., and W. E. RAMSEY: Rev. Sci. Instr. **23**, 95 (1952).
2.046 MANDEVILLE, C. E., and M. V. SCHERB: Nucleonics **3**, 2 (1948).
2.047 KELLEY, G. G., W. H. JORDAN, and P. R. BELL: Rev. Sci. Instr. **21**, 330 (1950).
2.048 MANDEVILLE, C. E.: Rev. Sci. Instr. **23**, 94 (1952).
2.049 LOVELESS, F. C., and J. GROSSART: Nucl. Electronics IAEA **2**, 317 (1962).
2.050 ELLIOT, H.: Proc. Phys. Soc. (London) **A-62**, 369 (1949).
2.051 NEHER, H. V., and W. W. HARPER: Phys. Rev. **49**, 940 (1936).
2.052 —, and W. H. PICKERING: Phys. Rev. **53**, 316 (1938).
2.053 GETTING, I. A.: Phys. Rev. **53**, 103 (1938).
2.054 MAIER-LEIBNITZ, H. W.: Rev. Sci. Instr. **19**, 500 (1948).
2.055 CRANE, H. R.: Rev. Sci. Instr. **32**, 953 (1961).
2.056 CROWELL, A. D., and P. R. LOW: Rev. Sci. Instr. **29**, 245 (1958).
2.057 KRAMERS, J. H.: Nucl. Instr. Methods **32**, 37 (1965).
2.058 SIMPSON, I. A.: Phys. Rev. **66**, 39 (1944).
2.059 HODSON, A. L.: J. Sci. Instr. **25**, 11 (1948).
2.060 PORTER, W. C.: Nucleonics **11**/3, 32 (1953).
2.061 GIBSON, W. M., G. L. MILLER, and P. F. DONOVAN: Semiconductor Particle Spectrometers. In: Alpha-, Beta- and Gamma-Ray Spectroscopy (ed. K. SIEGBAHN), p. 345. Amsterdam 1965.
2.062 SEILER, K.: Physik und Technik der Halbleiter. Stuttgart 1964.
2.063 SALOW, H., H. BENEKING, H. KRÖMER, und W. V. MÜNCH: Der Transistor. Physikalische und technische Grundlagen. Berlin 1963.
2.064 DEARNALEY, G., and D. C. NORTHROP: Semiconductor Counters for Nuclear Radiations. London 1964.
2.065 GOULDING, F. S.: Nucl. Instr. Methods **43**, 1 (1966).
2.066 HUTH, G. C., H. E. BERGESON, and J. B. TRICE: Rev. Sci. Instr. **34**, 1283 (1963).
2.067 HANSEN, W. L., and F. S. GOULDING: Nucl. Instr. Methods **29**, 345 (1964).
2.068 BUCK, T. M., G. H. WHEATLEY, and J. W. RODGERS: IEEE Trans. Nucl. Sci. **NS-11**/3, 294 (1964).
2.069 CAPPELLANI, F., and G. RESTELLI: Nucl. Instr. Methods **25**, 230 (1964).
2.070 SIFFERT, P., and A. COCHE: IEEE Trans. Nucl. Sci. **NS-12**/1, 284 (1965).
2.071 KLEMA, E. D.: IEEE Trans. Nucl. Sci. **NS-12**/1, 288 (1965).
2.072 BUSSOLATI, C.: Nucl. Instr. Methods **33**, 293 (1965).
2.073 ANDERSSON-LINDSTRÖM, G., and B. ZAUSIG: Nucl. Instr. Methods **40**, 277 (1966).
2.074 PELL, E. M.: N. R. C. Publ. **871**, 136 (1961).

2.075 — J. Appl. Phys. **31**, 291 (1960).
2.076 MAYER, J. W.: J. Appl. Phys. **33**, 2894 (1962).
2.077 AMMERLAAN, C. A. J., and K. MULDER: Nucl. Instr. Methods **21**, 97 (1963).
2.078 TAVENDALE, A. J.: IEEE Trans. Nucl. Sci. **NS-13**/3, 315 (1966).
2.079 FIEDLER, J. H., L. B. HUGHES, T. J. KENNETT, W. V. PRESTWICH, and B. J. WALL: Nucl. Instr. Methods **40**, 229 (1966).
2.080 SAH, C. T., R. N. NOYCE, and W. SHOCKLEY: Proc. IRE **45**, 1228 (1957).
2.081 GOULDING, F. S., and W. L. HANSEN: Nucl. Instr. Methods **12**, 249 (1961).
2.082 SHOCKLEY, W.: Czech. J. Phys. **B11**, 81 (1961).
2.083 BALDINGER, E., W. CZAJA, and J. GUTMAN: Helv. Phys. Acta **35**, 559 (1962).
2.084 MAYER, J. W.: J. Appl. Phys. **30**, 1937 (1959).
2.085 BUSSOLATTI, C., A. FIORENTINI, and G. FABRI: Phys. Rev. **136A**, 1756 (1964).
2.086 MAYER, J. W.: Nucl. Instr. Methods **43**, 55 (1966).
2.087 FABRI, G., E. GATTI, and V. SVELTO: Phys. Rev. **131**, 134 (1963).
2.088 BRYTSYN, K. I., and A. A. SMIRNOW: Solid State Phys. (USSR) **8**/1, 163 (1966).
2.089 VAN ROOSBROECK, W.: Phys. Rev. **139-A**, 1703 (1965).
2.090 BILGER, H. R.: Nucl. Instr. Methods **40**, 54 (1966).
2.091 PRIOR, A. C.: J. Chem. Phys. Solids **12**, 175 (1960).
2.092 TOVE, P. A., and K. FALK: Nucl. Instr. Methods **12**, 278 (1961).
2.093 CAVALLERI, G., G. FABRI, E. GATTI, and V. SVELTO: Nucl. Instr. Methods **21**, 177 (1963).
2.094 RAMO, S.: Proc. IRE **27**, 584 (1939).
2.095 JEN, C. K.: Proc. IRE **29**, 345 (1941).
2.096 — Proc. IRE **29**, 464 (1941).
2.097 TOVE, P. A., and K. FALK: Nucl. Instr. Methods **29**, 66 (1964).
2.098 ALBERIGI-QUARANTA, A., M. MARTINI, G. OTTAVIANI, and G. ZANARINI: Nucl. Instr. Methods **29**, 173 (1964).
2.099 COTTINI, C., E. GATTI, and V. SVELTO: NAS-NRC Publ. **1184**, 53 (1964).
2.100 MEYER, H.: IEEE Trans. Nucl. Sci. **NS-13**/3, 180 (1966).
2.101 MILLER, G. L., and W. M. GIBSON: Nucl. Electronics IAEA **1**, 477 (1961).
2.102 AXTMANN, R. C., and D. KEDEM: Nucl. Instr. Methods **32**, 70 (1965).
2.103 KUHN, A.: Czech. J. Phys. **16**, 697 (1966).
2.104 FABRI, G., E. GATTI, and V. SVELTO: NAS-NRC Publ. **1184**, 49 (1964).
2.105 POENARU, D. N., and N. VÎLCOV: Nucl. Instr. Methods **36**, 52 (1965).
2.106 DUBRAU, H. J.: Nucl. Instr. Methods **15**, 77 (1962).
2.107 BLANKENSHIP, J. L.: IEEE Trans. Nucl. Sci. **NS-11**/3, 373 (1964).
2.108 —, and S. F. PINASCO: IEEE Trans. Nucl. Sci. **NS-12**/6, 77 (1965).
2.109 LEVENTHAL, E. A.: Nucl. Instr. Methods **35**, 325 (1965).
2.110 CHASE, R. L., W. A. HIGINBOTHAM, and G. L. MILLER: IRE Trans. Nucl. Sci. **NS-8**/1, 147 (1961).
2.111 HEYWOOD, D. R., and B. L. WHITE: Rev. Sci. Instr. **34**, 1050 (1963).
2.112 HAHN, J., and R. O. MEYER: IRE Trans. Nucl. Sci. **NS-9**/4, 20 (1962).
2.113 — — Nucl. Instr. Methods **29**, 277 (1964).
2.114 LANGMANN, H. J., and O. MEYER: Nucl. Instr. Methods **30**, 135 (1964).
2.115 KANDIAH, K.: NAS-NRC Publ. **1184**, 65 (1964).
2.116 TAKEDA, S.: Nucl. Instr. Methods **27**, 269 (1964).
2.117 WAHL, H.: Nucl. Instr. Methods **25**, 247 (1964).
2.118 EMMER, T. L.: IRE Trans. Nucl. Sci. **NS-8**/1, 140 (1961).
2.119 — IRE Trans. Nucl. Sci. **NS-9**/3, 305 (1962).
2.120 SPLICHAL, W. F.: Nucl. Instr. Methods **41**, 156 (1966).
2.121 JONASSON, L. G.: Nucl. Instr. Methods **26**, 104 (1964).
2.122 CHASE, R. L.: In: Semiconductor Nuclear Particle Detectors. NAS-NRC Publ. **871**, 221 (1961).

References

2.123 RADEKA, V., and R. L. CHASE: IEEE Trans. Nucl. Sci. **NS-13**/3, 477 (1966).
2.124 — NAS-NRC Publ. **1184**, 70 (1964).
2.125 BLALOCK, T. V.: IEEE Trans. Nucl. Sci. **NS-11**/3, 365 (1964).
2.126 — IEEE Trans. Nucl. Sci. **NS-13**/3, 457 (1966).
2.127 SMITH, K. F., and J. E. CLINE: IEEE Trans. Nucl. Sci. **NS-13**/3, 468 (1966).
2.128 ELAD, E.: Nucl. Instr. Methods **37**, 327 (1965).
2.129 NYBAKKEN, T. W., and V. VALI: Nucl. Instr. Methods **33**, 164 (1965).
2.130 MEYER, O.: Nucl. Instr. Methods **33**, 164 (1965).
2.131 RADEKA, V.: Nucleonics **23**/7, 52 (1965).
2.132 FAIRSTEIN, E.: In: Semiconductor Nuclear Particle Detectors. NAS-NRC Publ. **871** (1961).
2.133 — IRE Trans. Nucl. Sci. **NS-8**/1, 129 (1961).
2.134 ALBERIGI-QUARANTA, A., M. MARTINI, G. OTTAVIANI, and G. ZANARINI: Nucl. Instr. Methods **32**, 352 (1965).
2.135 MOTT, W. E., and R. B. SUTTON: Scintillation and Čerenkov Counters. In: Handbuch der Physik (ed. S. FLÜGGE), Vol. LXV, p. 86. Berlin-Göttingen-Heidelberg: Springer 1958.
2.136 MURRAY, R. B.: Scintillation Counters. In: Nuclear Instruments and their Uses (ed. A. H. SNELL), Vol. 1, p. 82. New York: John Wiley & Sons 1962.
2.137 SCHRAM, E.: Organic Scintillation Detectors. Amsterdam: Elsevier Publ. Comp. 1963.
2.138 NEILER, J. H., and P. R. BELL: The Scintillation Method. In: Alpha-, Beta- and Gamma-Ray Spectroscopy (ed. K. SIEGBAHN), p. 245. Amsterdam 1965.
2.139 BIRKS, J. B.: The Theory and Practice of Scintillation Counting. Oxford: Pergamon Press 1964.
2.140 GWIN, R., and R. B. MURRAY: Phys. Rev. **131**, 508 (1963).
2.141 HREHUSS, G.: Nucl. Instr. Methods **8**, 344 (1960).
2.142 KOWALSKI, E., R. ANLIKER, and K. SCHMID: Int. J. Appl. Rad. Isotopes **18**, 307 (1967).
2.143 BALDINGER, E., and W. FRANZEN: Amplitude and Time Measurement in Nuclear Physics. In: Advances in Electronics and Electron Physics (ed. L. MARTON) **8**, 255 (1956).
2.144 GOL'DANSKIJ, V. I., A. V. KUCENKO, and M. I. PODGORECKIJ: Statistika otstschetov pri registracii jadernych tschastic. Moscow 1959.
2.145 LEWIS, I. A. D., and F. H. WELLS: Millimicrosecond Pulse Techniques. London: Pergamon Press 1959.
2.146 TANASESCU, T.: IRE Trans. Nucl. Sci. **NS-7**/2—3, 39 (1960).
2.147 PFEFFER, G., H. LAMI, G. LAUSTRIAT, and A. COCHE: Nucl. Instr. Methods **23**, 74 (1963).
2.148 BONITZ, M., W. MEILING, and F. STARY: Nucl. Instr. Methods **29**, 309 (1964).
2.149 D'ALESSIO, J. T., P. K. LUDWIG, and M. BURTON: Rev. Sci. Instr. **35**, 1015 (1964).
2.150 — — IEEE Trans. Nucl. Sci. **NS-12**/1, 351 (1965).
2.151 KUCKUCK, R. W., and J. CHONG LEE: IEEE Trans. Nucl. Sci. **NS-12**/1, 356 (1965).
2.152 DEMARIA, A. J., D. A. STETSER, and H. HEYMAN: Appl. Phys. Letters **8**, 174 (1966).
2.153 DIDOMENICO, JR., M., J. E. GEUSIC, H. M. MARCOS, and R. G. SMITH: Appl. Phys. Letters **8**, 180 (1966).
2.154 CROWELL, M. H.: IEEE J. Quant. Electr. **QE-1**/1, 12 (1965).
2.155 WEBER, H. P., E. MATHIEU, and K. P. MEYER: J. Appl. Phys. **37**, 3584 (1966).
2.156 FRANKEN, P. A., and J. F. WARD: Rev. Mod. Phys. **35**, 23 (1963).
2.157 NEWTON, T. D.: Phys. Rev. **78**, 490 (1950).
2.158 BAY, Z.: Phys. Rev. **77**, 419 (1950).
2.159 —, R. R. MEIJER, and G. PAPP: Phys. Rev. **82**, 754 (1951).
2.160 SJÖLIN, P. G.: Nucl. Instr. Methods **37**, 45 (1965).
2.161 MCGUIRE, R. L., E. C. YATES, D. G. CRANDALL, and C. R. HATCHER: IEEE Trans. Nucl. Sci. **NS-12**/1, 24 (1965).

2.162 YATES, E. C., and D. G. CRANDALL: IEEE Trans. Nucl. Sci. **NS-13**/3, 153 (1966).
2.163 Philips Photomultiplier Tubes, Philips Bulletin 23/007/D/E-3-'63.
2.164 BREITENBERGER, E.: Scintillation Spectrometer Statistics. Progr. Nucl. Phys. (ed. O. R. FRISCH) **4**, 56 (1955). London: Pergamon Press.
2.165 BAICKER, J. A.: IRE Trans. Nucl. Sci. **NS-7**/2—3, 74 (1960).
2.166 ST. JOHN, R. M.: Rev. Sci. Instr. **32**, 370 (1961).
2.167 HARMAN, G. G.: Rev. Sci. Instr. **30**, 743 (1959).
2.168 FRANKLIN, A. R., W. W. HOLLOWAY JR., and D. H. MCMAHON: Rev. Sci. Instr. **36**, 232 (1965).
2.169 MORTON, G. A., and K. W. ROBINSON: Nucleonics **4**/2, 25 (1949).
2.170 PACKARD, L. E.: Instrumentation for Internal Sample Liquid Scintillation Counting. In: Liquid Scintillation Counting (ed. C. G. BELL, F. N. HAYES), Vol. 50. Oxford: Pergamon Press 1958.
2.171 NISHIWAKI, Y., H. KAWAI, Y. OSHIMA, and M. KOYAMA: Japan. J. Appl. Phys. **1**/4, 237 (1962).
2.172 SWANK, R. K.: Limits of Sensitivity of Liquid Scintillation Counters. In: Liquid Scintillation Counting (ed. C. G. BELL, F. N. HAYES), Vol. 23. Oxford: Pergamon Press 1958.
2.173 FORTE, M., and A. ANZANI: Metrology of Radionuclides. IAEA Wien, 269 (1960).
2.174 LANDIS, D., and F. S. GOULDING: NAS-NRC Publ. **1184**, 143 (1964).
2.175 — — Nucl. Instr. Methods **33**, 303 (1965).
2.176 DAMERELL, C. J. S.: Nucl. Instr. Methods **15**, 171 (1962).
2.177 BIRK, M., Q. A. KERNS, and R. F. TUSTING: IEEE Trans. Nucl. Sci. **NS-11**, 129 (1964).
2.178 KRALL, H. R.: IEEE Trans. Nucl. Sci. **NS-12**/1, 39 (1965).
2.179 Discussion contribution by D. A. MACK, NAS-NRC Publ. **1184**, 88 (1964).
2.180 FARINELLI, U., and R. MALVANO: Rev. Sci. Instr. **29**, 699 (1958).
2.181 ROOSE, U. J.: Nucl. Instr. Methods **36**, 333 (1965).
2.182 GÖRLICH, P., H. J. POHL, S. V. MUCHIN, and J. N. SEMENYUSCHKIN: IEEE Trans. Nucl. Sci. **NS-13**/3, 112 (1966).
2.183 MUCHIN, S. V., H. J. POHL, S. V. RICHNIZKY, J. N. SEMENYUSCHKIN, and J. FOLTIN: Nucl. Instr. Methods **33**, 339 (1965).
2.184 JOVANOVIC, D. T., B. M. STOJANOVIC, and R. P. ILIC: Nucl. Instr. Methods **22**, 113 (1963).
2.185 IZUMI, J., and M. KOKUBU: Nucl. Instr. Methods **28**, 349 (1964).
2.186 SHEEN, E. M., and C. A. RATCLIFFE: Nucl. Instr. Methods **31**, 339 (1964).
2.187 FREVERT, L., and W. D. KREISEL: Nucl. Instr. Methods **34**, 69 (1965).
2.188 WAGNER, S. W.: Stromversorgung elektronischer Schaltungen und Geräte. Hamburg: R. v. Decker's Verlag 1964.
2.189 PAGANO, R., C. J. S. DAMERELL, and R. D. CHERRY: Rev. Sci. Instr. **33**, 955 (1962).
2.190 WALTON, P. W.: Rev. Sci. Instr. **35**, 518 (1964).
2.191 BELLETTINI, G., C. BEMPORAD, C. CERRI, and L. FOÀ: Nucl. Instr. Methods **21**, 106 (1963).
2.192 — — — — Nucl. Instr. Methods **27**, 38 (1964).
2.193 GIBSON, W. A.: Rev. Sci. Instr. **37**, 631 (1966).
2.194 DAVIS, H. F., and H. A. SPORE: IEEE Trans. Nucl. Sci. **NS-12**/4, 245 (1965).
2.195 NESS, S., and R. J. SMITH-SAVILLE: Nucl. Instr. Methods **40**, 175 (1966).
2.196 BARNA, A.: Nucl. Instr. Methods **24**, 247 (1963).
2.197 MURRAY, R. B., and J. J. MANNING: IRE Trans. Nucl. Sci. **NS-7**/2—3, 80 (1960).
2.198 CAMERON, J. F., C. G. CLAYTON, and R. A. SPACKMAN: Nucl. Electronics IAEA **1**, 95 (1962).
2.199 RHODE, R. E.: IEEE Trans. Nucl. Sci. **NS-12**/1, 16 (1965).
2.200 JUNG, H., PH. PANUSSI, and J. JÄNECKE: Nucl. Instr. Methods **9**, 121 (1960).

References

2.201 CHÉRY, R.: J. Phys. Radium **21,** 679 (1960).
2.202 MICHAELIS, W., H. SCHMIDT, and C. WEITKAMP: Nucl. Instr. Methods **21,** 65 (1963).
2.203 SCHNEIDER, H., and C. WEINGARDT: Nucl. Instr. Methods **40,** 305 (1966).
2.204 COVELL, D. F.: Nucl. Instr. Methods **36,** 229 (1965).
2.205 DE WAARD, H.: Nucleonics **13**/7, 36 (1955).
2.206 HAUN, S., and D. KAMKE: Nucl. Instr. Methods **8,** 331 (1960).
2.207 SCHERBATSKOY, S. A.: Rev. Sci. Instr. **32,** 599 (1961).
2.208 VALCKX, F. P. G.: Nucl. Instr. Methods **10,** 234 (1961).
2.209 RIJKS, H. J.: Nucl. Instr. Methods **14,** 76 (1961).
2.210 MARLOW, K. W.: Nucl. Instr. Methods **15,** 188 (1962).
2.211 DEMUYNCK, J. L., and O. J. SEGAERT: Nucl. Instr. Methods **16,** 358 (1962).
2.212 DIXON, J.: Nucl. Instr. Methods **25,** 26 (1963).
2.213 DUDLEY, R. A., and R. SCARPATETTI: Nucl. Instr. Methods **25,** 297 (1964).
2.214 AGENO, M., and C. FELICI: Rev. Sci. Instr. **34,** 997 (1963).
2.215 STENMAN, F.: Nucl. Instr. Methods **29,** 107 (1964).
2.216 PATWARDHAN, P. K.: Nucl. Instr. Methods **31,** 169 (1964).
2.217 BLACK, J. L., and E. VALENTINE: Nucl. Instr. Methods **31,** 325 (1964).
2.218 KERNS, Q. A., and R. F. TUSTING: NAS-NRC Publ. **1184,** 220 (1964).
2.219 HINRICHSEN, P. F.: IEEE Trans. Nucl. Sci. **NS-11**/3, 420 (1964).
2.220 COMUNETTI, A.: Nucl. Instr. Methods **37,** 125 (1965).
2.221 BORBAS, R., D. J. DOYLE, J. H. AITKEN, and L. JONES: Nucl. Instr. Methods **37,** 183 (1965).
2.222 WILLIAMS, D., G. F. SNELLING, and J. PICKUP: Nucl. Instr. Methods **39,** 141 (1966).
2.223 PAKKANEN, A., and F. STENMAN: Nucl. Instr. Methods **44,** 321 (1966).
2.224 TAMM, U.: Nucl. Instr. Methods **40,** 355 (1966).
2.225 BRIMHALL, J. E., and L. A. PAGE: Nucl. Instr. Methods **35,** 328 (1965).
2.226 MOYER, B. J.: A Survey of Čerenkov Counter Techniques. In: Nuclear Instruments and their Uses (ed. A. H. SNELL), p. 166. New York 1962.
2.227 BIRK, M., Q. A. KERNS, and R. F. TUSTING: IEEE Trans. Nucl. Sci. **NS-11**/3, 129 (1964).
2.228 BOLLINGER, L. M., and G. E. THOMAS: Rev. Sci. Instr. **32,** 1044 (1961).
2.229 RÖDEL, W.: Nucl. Instr. Methods **41,** 169 (1966).
2.230 MOON, L. L.: Nucl. Instr. Methods **48,** 162 (1967).
2.231 CARVER, J. H., and P. MITCHELL: Nucl. Instr. Methods **52,** 130 (1967).
2.232 CHARLES, M. W., and B. A. COOKE: Nucl. Instr. Methods **61,** 31 (1968).
2.233 BENNETT, E. F.: Nucl. Instr. Methods **48,** 170 (1967).
2.234 GLÄSER, M.: Nucl. Instr. Methods **61,** 217 (1968).
2.235 KLEIN, C. A.: IEEE Trans. Nucl. Sci. **NS-15**/3, 214 (1968).
2.236 BALLAND, J. C., J. PIGUERET, and J. J. SAMUELI: Nucl. Instr. Methods **52,** 351 (1967).
2.237 MOSZYŃSKI, M., W. KURCEWICZ, and W. PRZYBORSKI: Nucl. Instr. Methods **61,** 173 (1968).
2.238 —, and W. PRZYBORSKI: Nucl. Instr. Methods **64,** 244 (1968).
2.239 LIBS, G., G. DE ROSNY: Nucl. Instr. Methods **44,** 39 (1966).
2.240 GOLDSWORTHY, W. W.: Nucl. Instr. Methods **52,** 343 (1967).
2.241 ELAD, E., and M. NAKAMURA: Nucl. Instr. Methods **42,** 315 (1966).
2.242 PINASCO, S. F.: Nucl. Instr. Methods **46,** 355 (1967).
2.243 HARRIS, JR., R. J., and W. B. SHULER: Nucl. Instr. Methods **51,** 341 (1967).
2.244 FERRARI, A. M. R.: Nucl. Instr. Methods **52,** 179 (1967).
2.245 ELAD, E., and M. NAKAMURA: Nucl. Instr. Methods **54,** 308 (1967).
2.246 — — IEEE Trans. Nucl. Sci. **NS-15**/1, 283 (1968).
2.247 — — IEEE Trans. Nucl. Sci. **NS-15**/3, 477 (1968).
2.248 GOLDSWORTHY, W. W.: Nucl. Instr. Methods **54,** 301 (1967).
2.249 WILCOX, G. E.: IEEE Trans. Nucl. Sci. **NS-14**/1, 460 (1967).

2.250 CHEVALIER, P.: Nucl. Instr. Methods **50**, 346 (1967).
2.251 TAWARA, H.: Nucl. Instr. Methods **42**, 318 (1966).
2.252 FRANKE, H. G., and H. SCHMEING: Nucl. Instr. Methods **52**, 171 (1967).
2.253 JOHNSON, J., and D. PORAT: Rev. Sci. Instr. **38**, 1796 (1967).
2.254 MATHÉ, GY.: Nucl. Instr. Methods **63**, 117 (1968).
2.255 EBERHARDT, E. H.: IEEE Trans. Nucl. Sci. **NS-14**/2, 7 (1967).
2.256 LODGE, J. A., P. MUFF, R. B. OWEN, and D. SMOUT: IEEE Trans. Nucl. Sci. **NS-15**/1, 491 (1968).
2.257 KREHBIEL, H.: Nucl. Instr. Methods **54**, 302 (1967).
2.258 WEITKAMP, C., G. G. SLAUGHTER, W. MICHAELIS, and H. SCHMIDT: Nucl. Instr. Methods **61**, 122 (1968).
2.259 BACCI, C., V. BIDOLI, and R. BALDINI-CELIO: Nucl. Instr. Methods **57**, 100 (1967).
3.001 KOVARIK, A. F.: Phys. Rev. **13**, 153 (1919).
3.002 GREINACHER, H.: Z. f. Physik **23**, 361 (1924).
3.003 — Z. f. Physik **36**, 364 (1926).
3.004 WYNN-WALLIAMS, C. E., and F. A. B. WARD: Proc. Roy. Soc. London **A 131**, 391 (1931).
3.005 FAIRSTEIN, E., and J. HAHN: Part I., Nucleonics **23**/7, 56 (1965).
3.006 — — Part II. Nucleonics **23**/9, 81 (1965).
3.007 — — Part III. Nucleonics **23**/11, 50 (1965).
3.008 — — Part IV. Nucleonics **24**/1, 54 (1966).
3.009 — — Part V. and Appendix, Nucleonics **24**/3, 68 (1966).
3.010 SHEA, R. F.: Transistortechnik (German translation, 2nd edition). Stuttgart 1962.
3.011 — Amplifier Handbook. New York: McGraw-Hill 1966.
3.012 LITTAUER, R.: Pulse Electronics. New York: McGraw-Hill 1965.
3.013 FRÄNZ, K., and H. PAUCKSCH: Nucl. Instr. Methods **27**, 125 (1964).
3.014 THOMASON, J. G.: Linear Feedback Analysis. London: Pergamon Press 1955.
3.015 ARBEL, A. F.: Nucl. Instr. Methods **32**, 341 (1965).
3.016 —, and J. BAR-DAVID: Nucl. Instr. Methods **7**, 153 (1960); **9**, 244 (1960).
3.017 KUHLENKAMP, A.: Der Regler. Stuttgart: DVA 1963.
3.018 FAIRSTEIN, E.: Rev. Sci. Instr. **25**, 1134 (1954).
3.019 SCHLEGEL, H. R., u. A. NOWAK: Impulstechnik, p. 237 ff. Prien/Chiemsee: C. F. Winter'sche Verlagsbuchhandlung 1961.
3.020 MADELUNG, E.: Die mathematischen Hilfsmittel des Physikers, 7th ed. Berlin-Göttingen-Heidelberg: Springer 1964.
3.021 FAIRSTEIN, E.: Rev. Sci. Instr. **27**, 483 (1956).
3.022 — Electrometers and Amplifiers. In: Nuclear Instruments and their Uses (ed. A. H. SNELL), p. 194. New York: John Wiley & Sons 1962.
3.023 STODDARD, H. F.: NAS-NRC Publ. **467**, 26 (1957).
3.024 CHASE, R. L., and V. SVELTO: IRE Trans. Nucl. Sci. **NS-8**/3, 45 (1961).
3.025 HAHN, J., and V. GUIRAGOSSIAN: IEEE Trans. Nucl. Sci. **NS-10**/3, 44 (1963).
3.026 FAIRSTEIN, E.: Rev. Sci. Instr. **27**, 475 (1956).
3.027 — NAS-NRC Publ. **467**, 23 (1957).
3.028 NOWLIN, C. H., and J. L. BLANKENSHIP: Rev. Sci. Instr. **36**, 1830 (1965).
3.029 — — to be published!
3.030 BLANKENSHIP, J. L., and C. H. NOWLIN: IEEE Trans. Nucl. Sci. **NS-13**/3, 495 (1966).
3.031 COTTINI, C., E. GATTI, and E. ZAGLI: Energia Nucleare **6**, 588 (1959).
3.032 GOULDING, F. S., R. W. NICHOLSON, and J. B. WAUGH: Nucl. Instr. Methods **8**, 272 (1960).
3.033 PATRONIS, E. T.: Nucl. Instr. Methods **22**, 83 (1963).
3.034 BLALOCK, T. V.: Rev. Sci. Instr. **36**, 1448 (1965).
3.035 GOLDSWORTHY, W. W.: Rev. Sci. Instr. **36**, 1643 (1965).
3.036 DEN HARTOG, H., and F. A. MULLER: Physica **13**, 571 (1947).

References

3.037 WILSON, R.: Phil. Mag. **41**, 66 (1950).
3.038 NOWLIN, C. H., J. L. BLANKENSHIP, and T. V. BLALOCK: Rev. Sci. Instr. **36**, 1063 (1965).
3.039 GILLESPIE, A. B.: Signal, Noise and Resolution in Nuclear Counter Amplifiers. London: Pergamon Press 1953.
3.040 BLANKENSHIP, J. L.: IEEE Trans. Nucl. Sci. **NS-11**/3, 373 (1964).
3.041 FAIRSTEIN, E.: IRE Trans. Nucl. Sci. **NS-8**/1, 129 (1961).
3.042 CHASE, R. L.: Nuclear Pulse Spectrometry. New York: McGraw-Hill 1961.
3.043 ARBEL, A. F.: NAS-NRC Publ. **1184**, 79 (1964).
3.044 CAMPBELL, N. R., and V. J. FRANCIS: J. Instr. Elec. Eng. (London) **3**, 93, 45 (1946).
3.045 DE LOTTO, I., and D. DOTTI: Nucl. Instr. Methods **39**, 281 (1965).
3.046 SOUČEK, B.: Nucl. Instr. Methods **28**, 306 (1964).
3.047 — Rev. Sci. Instr. **36**, 1582 (1965).
3.048 DE LOTTO, I., D. DOTTI, and D. MARIOTTI: Nucl. Instr. Methods **40**, 169 (1966).
3.049 ROZEN, S.: Nucl. Instr. Methods **11**, 316 (1961).
3.050 SCHWARZSCHILD, A.: Nucl. Instr. Methods **21**, 1 (1963).
3.051 WEISBERG, H.: Nucl. Instr. Methods **32**, 138 (1965).
3.052 GUPTA, S. K., K. V. K. JYENGAR, and P. J. BHALERAO: Nucl. Instr. Methods **44**, 123 (1966).
3.053 MONIER, L. F., and G. E. TRIPARD: Rev. Sci. Instr. **37**, 316 (1966).
3.054 CHASE, R. L., W. A. HIGINBOTHAM: Rev. Sci. Instr. **23**, 34 (1952).
3.055 COLLINGE, B., C. WEST, and G. H. LLOYD: Nucl. Instr. Methods **35**, 313 (1965).
3.056 LARSEN, R. N.: Nucl. Instr. Methods **32**, 147 (1965).
3.057 LASCARIS, C., and L. PAPADOPOULOS: Nucl. Instr. Methods **31**, 250 (1964).
3.058 GOLDSWORTHY, W. W.: IEEE Trans. Nucl. Sci. **NS-10**/1, 61 (1963).
3.059 GOULDING, F. S., and D. LANDIS: NAS-NRC Publ. **1184**, 124 (1964).
3.060 MARLOW, K. W.: Nucl. Instr. Methods **15**, 188 (1962).
3.061 PAKKANEN, A., and F. STENMAN: Nucl. Instr. Methods **44**, 321 (1966).
3.062 ARQUE-ALMARAZ, H.: Nucl. Instr. Methods **32**, 283 (1965).
3.063 PATWARDHAN, P. K.: Nucl. Instr. Methods **31**, 169 (1964).
3.064 STRAUSS, M. G., and R. BRENNER: Rev. Sci. Instr. **36**, 1857 (1965).
3.065 SIKORSKY, E.: IEEE Trans. Nucl. Sci. **NS-10**/1, 42 (1963).
3.066 SHOCKLEY, W.: Electrons and Holes in Semiconductors. New York: Van Nostrand 1950/1956.
3.067 HIRAMOTO, T.: Nucl. Instr. Methods **32**, 141 (1965).
3.068 KAISER, R. C.: NAS-NRC Publ. **1184**, 140 (1964).
3.069 KAHN, H. L.: Rev. Sci. Instr. **33**, 235 (1962); **35**, 135 (1964).
3.070 SAH, C. T.: IRE Trans. Electron Devices **ED-9**, 94 (1962).
3.071 GIANNELLI, G., and L. STANCHI: Nucl. Instr. Methods **8**, 79 (1960).
3.072 PATERSON, W. L.: Rev. Sci. Instr. **34**, 1311 (1963).
3.073 COOKE-YARBOROUGH, E. H.: NAS-NRC Publ. **1184**, 136 (1964).
3.074 LUNSFORD, J. S.: Rev. Sci. Instr. **36**, 461 (1965).
3.075 WÅHLIN, L.: Nucl. Instr. Methods **14**, 281 (1961).
3.076 VINCENT, C. H., and D. KAINE: IEEE Trans. Nucl. Sci. **NS-9**/3, 327 (1962).
3.077 GOLDSWORTHY, W. W.: IEEE Trans. Nucl. Sci. **NS-12**/1, 336 (1965).
3.078 AITKEN, J. H.: Nucl. Instr. Methods **14**, 343 (1961).
3.079 GRIFFITHS, R. J., K. M. KNIGHT, C. J. CANDY, and A. J. COLE: Nucl. Instr. Methods **15**, 309 (1962).
3.080 BAYER, R.: Nuclear Electronics, IAEA **2**, 337 (1962).
3.081 TSUKUDA, M.: Nucl. Instr. Methods **25**, 265 (1963).
3.082 KUHLMANN, W. R., and B. SCHIMMER: Nucl. Instr. Methods **40**, 113 (1966).
3.083 KONRAD, M.: Nuclear Electronics, IAEA **2**, 405 (1962).
3.084 MILLER, G. L., and V. RADEKA: NAS-NCR Publ. **1184**, 104 (1964).

3.085 GRUNBERG, J., V. SOLD, and V. GALIL: Nucl. Instr. Methods **34**, 311 (1965).
3.086 GERE, E. A., and G. L. MILLER: IEEE Trans. Nucl. Sci. **NS-11**/3, 382 (1964).
3.087 GRÜTER, H.: Nucl. Instr. Methods **33**, 159 (1965).
3.088 BRISCOE, W. L.: Rev. Sci. Instr. **29**, 401 (1958).
3.089 COFFEY, W. L.: Rev. Sci. Instr. **36**, 1580 (1965).
3.090 HORN, L. S., and B. I. KHASANOV: Nucl. Instr. Methods **40**, 267 (1966).
3.091 EMMER, T. L.: IRE Trans. Nucl. Sci. **NS-9**/3, 305 (1962).
3.092 KANDIAH, K.: NAS-NRC Publ. **1184**, 119 (1964).
3.093 FABRI, G., E. GATTI, and V. SVELTO: Nucl. Instr. Methods **15**, 237 (1962).
3.094 BERTOLACCINI, M., C. BUSSOLATI, and S. COVA: Nucl. Instr. Methods **32**, 31 (1965).
3.095 MILLMAN, J., and H. TAUB: Pulse and Digital Circuits. New York: McGraw Hill 1956.
3.096 GOULDING, F. S.: NAS-NRC Publ. **1184**, 121 (1964).
3.097 BARNA, A., and J. H. MARSHALL: Rev. Sci. Instr. **35**, 881 (1964).
3.098 FELDMAN, M.: Rev. Sci. Instr. **36**, 241 (1965).
3.099 HILLMAN, J., and T. H. PUCKETT: Proc. IRE **43**, 27 (1955).
3.100 COLI, M., and S. LUPINI: Nucl. Instr. Methods **34**, 235 (1965).
3.101 KELLER, K. B.: Rev. Sci. Instr. **35**, 1360 (1964).
3.102 CHAGNON, P. R.: Rev. Sci. Instr. **32**, 68 (1961).
3.103 GINGELL, C. E. L.: IEEE Trans. Nucl. Sci. **NS-10**/3, 32 (1963).
3.104 CHAPLIN, G. B. B., and A. J. COLE: Nucl. Instr. Methods **7**, 45 (1960).
3.105 VALCKX, F. P. G., and A. DYMANUS: Nucl. Instr. Methods **7**, 197 (1960).
3.106 SASAKI, A., and M. TSUKUDA: Nucl. Instr. Methods **33**, 252 (1965).
3.107 LIU, F. F., and F. J. LOEFFLER: Nucl. Instr. Methods **12**, 124 (1961).
3.108 SEILER, K.: Physik und Technik der Halbleiter. Stuttgart: WVG 1964.
3.109 KO, W. H.: IRE Trans. Electron Devices **ED-8**/2, 123 (1961).
3.110 KELLY, G.: Nucleonics **10**/4, 34 (1952).
3.111 MEYER, M. A.: Nucl. Instr. Methods **1**, 62 (1957).
3.112 SARAZIN, A.: Nucl. Instr. Methods **8**, 70 (1960).
3.113 CUMMIUS, W. F., and D. R. BRANUM: Rev. Sci. Instr. **31**, 1247 (1960).
3.114 WEDDIGEN, C., and E. L. HAASE: Nucl. Instr. Methods **33**, 157 (1965).
3.115 CRAIB, J. F.: Electronics **24**/6, 129 (1951).
3.116 GERSHO, A.: Proc. IEEE **54**, 1574 (1966).
3.117 ESPLEY, D. S.: J. IEE **93**, 314 (1946).
3.118 GINZTON, E. L., W. R. HEWLETT, J. H. JASBERG, and J. D. NOE: Proc. IRE **36**, 956 (1948).
3.119 HOSTON, W. H., J. H. JASBERG, and J. D. NOE: Proc. IRE **38**, 748 (1950).
3.120 YU, Y. P., H. E. KALLMAN, and P. S. CHRISTALDI: Electronics **24**/7, 106 (1951).
3.121 PERCIVAL, W. S.: J. Televis. Soc. **7**, 445 (1955).
3.122 BÉNÉTEAU, P. J., and L. BLASER: A 175 Mc Distributed Amplifier Using Silicon Mesa Transistors. Fairchild Application Data APP-14/2.
3.123 SALOW, H., H. BENEKING, H. KRÖMER, u. W. V. MÜNCH: Der Transistor. Berlin-Göttingen-Heidelberg: Springer 1963.
3.124 ELMORE, W. C.: J. Appl. Phys. **19**, 55 (1948).
3.125 WILLIAMS, C. W., and J. H. NEILER: IRE Trans. Nucl. Sci. **NS-9**/5, 1 (1962).
3.126 ALBERIGI-QUARANTA, A., and M. MARTINI: Nucl. Instr. Methods **25**, 125 (1963).
3.127 LIUVILL, J. G., et G. DANON: Electronique Nucléaire, p. 829. Paris: SFER 1963.
3.128 RUSH, CH. J.: Rev. Sci. Instr. **35**, 149 (1964).
3.129 BALDINGER, E., and A. SIMMEN: ZAMP **15**, 71 (1964).
3.130 REDDI, V. G. K.: Transistor Pulse Amplifiers. Fairchild Application Data APP-32/2.
3.131 COLI, M., S. LUPINI, V. SILVESTRINI, and G. PENSO: Nucl. Instr. Methods **33**, 298 (1965).
3.132 LUNSFORD, J. S.: Rev. Sci. Instr. **35**, 1483 (1964).

References

3.133 VERWEIJ, H.: Nucl. Instr. Methods **20**, 323 (1963).
3.134 — Nucl. Instr. Methods **24**, 39 (1963).
3.135 ALBERIGI-QUARANTA, A., and M. MARTINI: NAS-NRC Publ. **1184**, 89 (1964).
3.136 — Nucl. Instr. Methods **23**, 169 (1963).
3.137 EPSTEIN, R. J.: Nucl. Instr. Methods **24**, 333 (1963).
3.138 LAVAILLE, A.: Electronique Nucléaire, p. 761. Paris: SFER 1963.
3.139 SCHAPPER, M. A.: Nucl. Instr. Methods **27**, 172 (1964).
3.140 JACKSON, H. G.: Nucl. Instr. Methods **33**, 161 (1965).
3.141 AGENO, M., and C. FELICI: Nucl. Instr. Methods **16**, 59 (1962).
3.142 DUBROVSKIJ, I. A.: Exptl. Instr. Tech. (UdSSR, 1966) 2/5-12 "Wideband and Pulse Transistorized Amplifiers for the Nanosecond Range" (Review).
3.143 KANDIAH, K.: Nuclear Electronics IAEA **2**, 11 (1962).
3.144 FRÄNZ, K.: Nucl. Instr. Methods **47**, 217 (1967).
3.145 STRAUSS, M. G., I. S. SHERMAN, R. BRENNER, S. J. RUDNICK, R. N. LARSEN, and H. M. MANN: Rev. Sci. Instr. **38**, 725 (1967).
3.146 FERRARI, A. M. R., and E. FAIRSTEIN: Nucl. Instr. Methods **63**, 218 (1968).
3.147 GORNI, S.: Nucl. Instr. Methods **47**, 74 (1967).
3.148 RADEKA, V.: Rev. Sci. Instr. **38**, 1397 (1967).
3.149 BERTOLACCINI, M., C. BUSSOLATI, and E. GATTI: Nucl. Instr. Methods **42**, 286 (1966).
3.150 NYGAARD, K.: Nucl. Instr. Methods **54**, 98 (1967).
3.151 PINASCO, S. F.: Nucl. Instr. Methods **47**, 71 (1967).
3.152 BERTOLACCINI, M., C. BUSSOLATI, and E. GATTI: Nucl. Instr. Methods **41**, 173 (1966).
3.153 KONRAD, M.: IEEE Trans. Nucl. Sci. **NS-15**/1, 268 (1968).
3.154 BERTOLACCINI, M., C. BUSSOLATI, S. COVA, I. DE LOTTO, and E. GATTI: Nucl. Instr. Methods **61**, 84 (1968).
3.155 — — — — — Nucl. Instr. Methods **62**, 221 (1968).
3.156 GOLDSWORTHY, W.: Nucl. Instr. Methods **62**, 93 (1968).
3.157 WEISE, K., Nucl. Instr. Methods **61**, 241 (1968).
3.158 SCHUSTER, H.-J.: Nucl. Instr. Methods **63**, 342 (1968).
3.159 RADEKA, V., and N. KARLOVAC: Nucl. Instr. Methods **52**, 86 (1967).
3.160 — IEEE Trans. Nucl. Sci. **NS-15**/3, 455 (1968).
3.161 DE WIT, P., and A. C. WOLFF: Nucl. Instr. Methods **61**, 237 (1968).
3.162 WHITE, G.: Nucl. Instr. Methods **45**, 270 (1966).
3.163 GOLDSWORTHY, W.: IEEE Trans. Nucl. Sci. **NS-14**/1, 70 (1967).
3.164 CONNELLY, J. A., and J. F. PIERCE: Nucl. Instr. Methods **64**, 7 (1968).
3.165 AMSEL, G., R. BOSSHARD, and C. ZAJDE: IEEE Trans. Nucl. Sci. **NS-14**/1, 1 (1967).
3.166 FUSCHINI, E., C. MARONI, and P. VERONESI: Nucl. Instr. Methods **41**, 153 (1966).
3.167 GRACOVETSKY, S., and J.-F. LOUDE: Nucl. Instr. Methods **63**, 349 (1968).
3.168 MOSZYŃSKI, M., J. JASTRZEBSKI, and B. BENGSTON: Nucl. Instr. Methods **47**, 61 (1967).
3.169 WILLIAMS, C. W.: IEEE Trans. Nucl. Sci. **NS-15**/1, 297 (1968).
3.170 REMIGOLSKY, B., and L. TEPPER: Nucl. Instr. Methods **53**, 29 (1967).
3.171 BYRD, J. S.: Nucl. Instr. Methods **48**, 296 (1967).
3.172 TUROS, A., and A. ZIEMIŃSKI: Nucl. Instr. Methods **44**, 119 (1966).
3.173 FISHER, P. S., and D. K. SCOTT: Nucl. Instr. Methods **49**, 301 (1967).
3.174 CHAMINADE, R., J. C. FAIVRE, and J. PAIN: Nucl. Instr. Methods **49**, 217 (1967).
3.175 MANFREDI, P. F., P. MARANESI, and A. RIMINI: Rev. Sci. Instr. **38**, 1253 (1967).
3.176 MILLS, A. P.: Nucl. Instr. Methods **50**, 132 (1967).
3.177 VISENTIN, R.: Nucl. Instr. Methods **64**, 21 (1968).
3.178 SMITH, B.: Nucl. Instr. Methods **55**, 138 (1967).
3.179 SCHUSTER, H. J.: Nucl. Instr. Methods **58**, 179 (1968).
3.180 GOYOT, M., J. PIGUERET, J. REMILLIEUX, J.-J. SAMUELI, and A. SARAZIN: Nucl. Instr. Methods **53**, 87 (1967).

References

3.181 LOOTEN, A., E. BALDINGER, and A. SIMMEN: Nucl. Instr. Methods **53**, 128 (1967).
3.182 OWENS, A. R., and G. WHITE: Nucl. Instr. Methods **49**, 291 (1967).
3.183 MANFREDI, P. F., and A. RIMINI: Nucl. Instr. Methods **49**, 71 (1967).
3.184 NYBAKKEN, T. W.: Nucl. Instr. Methods **53**, 331 (1967).
3.185 TOJO, A.: Nucl. Instr. Methods **50**, 45 (1967).
3.186 MILLARD, J. K.: Rev. Sci. Instr. **38**, 169 (1967).
3.187 GOYOT, M., J.-J. SAMUELI, and A. SARAZIN: Nucl. Instr. Methods **46**, 149 (1967).
4.001 LITTAUER, R.: Pulse Electronics. New York: McGraw-Hill 1965.
4.002 SCHMITT, O. H.: J. Sci. Instr. **15**, 24 (1938).
4.003 MILLMAN, J., and H. TAUB: Pulse and Digital Circuits. New York: McGraw-Hill 1956.
4.004 ROBINSON, L. B.: Rev. Sci. Instr. **32**, 1057 (1961).
4.005 GOULDING, F. S.: NAS-NRC Publ. **1184**, 121 (1964).
4.006 —, and R. A. MCNAUGHT: Nucl. Instr. Methods **8**, 282 (1960).
4.007 CHASE, R. L.: Nuclear Pulse Spectrometry. New York: McGraw-Hill 1961.
4.008 VERWEIJ, H.: Nucl. Instr. Methods **10**, 308 (1961).
4.009 LECOMTE, J. L., and R. ALLEMAND: Nuclear Electronics, IAEA **2**, 399 (1962).
4.010 KANDIAH, K.: Nuclear Electronics, IAEA **2**, 239 (1962).
4.011 LARSEN, R. N.: Nucl. Instr. Methods **32**, 147 (1965).
4.012 KANDIAH, K.: Proc. IEE, II **101**, 239 (1954).
4.013 BARABASCHI, S., C. COTTINI, et E. GATTI: Nuovo Cimento, Ser. 10, **2**, 1042 (1955).
4.014 BINARD, L., and E. GOMSKI: Nuclear Electronics, IAEA **3**, 167 (1962).
4.015 GIANNELLI, G., and L. STANCHI: Nucl. Instr. Methods **8**, 79 (1960); **9**, 244 (1960).
4.016 STRAUSS, M. G., and R. BRENNER: Rev. Sci. Instr. **36**, 1857 (1965).
4.017 VAN RENNES, A. B.: Nucleonics 10/7, 20 (1952); 10/8, 22 (1952); 10/9, 32 (1952); 10/10, 50 (1952).
4.018 SOLD, U., and S. BROJDO: Nucl. Instr. Methods **26**, 147 (1964).
4.019 POLLY, P.: Nucl. Instr. Methods **16**, 214 (1962).
4.020 BRAFMAN, H.: Nucl. Instr. Methods **32**, 321 (1965).
4.021 AUDRIEUX, H., J. GUITTON, and L. HUGOLIN: Nuclear Electronics, IAEA **2**, 329 (1962).
4.022 WELTER, L. M.: Rev. Sci. Instr. **36**, 487 (1965).
4.023 GATTI, E., e F. PIVA: Nuovo Cimento, Ser. 9, **10**, 984 (1953).
4.024 — Nuovo Cimento, Ser. 9, **11**, 153 (1954).
4.025 COLOMBO, S., C. COTTINI, e E. GATTI: Nuovo Cimento, Ser. 10, **5**, 748 (1957).
4.026 MORI, G. M.: Nucl. Instr. Methods **27**, 348 (1964).
4.027 MOODY, N. F., W. J. BATTELI, and W. D. HOWELL: Rev. Sci. Instr. **22**^{c} 551 (1959).
4.028 BRANDT, B., and U. CAPPELLER: Nuclear Electronics, IAEA **2**, 167 (1962).
4.029 BONITZ, M., and J. BERLOWITSCH: Nucl. Instr. Methods **4**, 133 (1959).
4.030 WEINZIERL, P.: Rev. Sci. Instr. **27**, 226 (1956).
4.031 JOHANSSON, B.: Nucl. Instr. Methods **1**, 274 (1957).
4.032 FAIRSTEIN, E.: A Pulse Crossover Pickoff Gate for Use with a Medium Speed Coincidence Circuit, ORNL Instrument and Controls Div. Ame. Report (1. July 1957).
4.033 GRUHLE, W.: Nucl. Instr. Methods **4**, 112 (1959).
4.034 CHASE, R. L.: Rev. Sci. Instr. **31**, 945 (1960).
4.035 GATTI, E., F. VAGHI, and E. ZAGLIO: Nuclear Electronics, IAEA **3**, 105 (1962).
4.036 EMMER, T. L.: The use of time discrimination in pulse height analysis. NAS-NRC Publ. **1184**, 112 (1964).
4.037 STRAUSS, M. G.: Rev. Sci. Instr. **34**, 1248 (1963).
4.038 BELL, R. E.: Nucl. Instr. Methods **42**, 211 (1966).
4.039 —, and M. H. JØRGENSEN: Canad. J. Phys. **38**, 652 (1960).
4.040 ESAKI, L.: Phys. Rev. **109**, 603 (1958).

References

4.041 DE BLUST, E., V. MANDL, et I. DE LOTTO: Electronique Nucléaire, p. 785. Paris: SFER 1963.
4.042 LACOUR, J.: Nuclear Electronics, IAEA **3**, 179 (1962).
4.043 WINTER, J.: Nucl. Instr. Methods **28**, 229 (1964).
4.044 VAN ZURK, R.: Nucl. Instr. Methods **16**, 157 (1962).
4.045 BANNER, M., and J. TEIGER: Nucl. Instr. Methods **31**, 205 (1964).
4.046 RIGHINI, B.: Nucl. Instr. Methods **29**, 89 (1964).
4.047 WARD, C. B., and C. M. YORK: Nucl. Instr. Methods **23**, 213 (1963).
4.048 COLI, M.: Nucl. Instr. Methods **39**, 297 (1966).
4.049 PAPADOPOULOS, L.: J. Sci. Instr. **43**, 202 (1966).
4.050 HVAM, T., and M. SMEDSDAL: Nucl. Instr. Methods **24**, 55 (1963).
4.051 PANDARESE, F., and F. VILLA: Nucl. Instr. Methods **20**, 319 (1963).
4.052 ORMAN, P. R.: Nucl. Instr. Methods **21**, 121 (1963).
4.053 KANDIAH, K.: NAS-NRC Publ. **1184**, 117 (1964).
4.054 ALSTON, W. J., and J. E. DRAPER: Nucl. Instr. Methods **35**, 155 (1965).
4.055 WIEGAND, C.: Nucl. Instr. Methods **20**, 313 (1963).
4.056 GARVEY, J.: Nucl. Instr. Methods **29**, 137 (1964).
4.057 SUSSKIND, A.: Notes on Analog-Digital Conversion Techniques. New York: John Wiley & Sons 1957.
4.058 BORUCKI, L., u. J. DITTMANN: Digitale Messtechnik. Berlin-Heidelberg-New York: Springer 1966.
4.059 WILKINSON, D. H.: Proc. Cambridge Phil. Soc. **46**, 508 (1950).
4.060 FRANK, S. G. F., O. R. FRISCH, and G. G. SCARROTT: Phil. Mag. **42**, 603 (1951).
4.061 GUILLON, H., J. Phys. Radium **14**, 128 (1953).
4.062 KANDIAH, K.: Nucl. Instr. Methods **2**, 112 (1958).
4.063 — Nuclear Electronics, IAEA **2**, 11 (1962).
4.064 GUILLON, H.: Nucl. Instr. Methods **43**, 240 (1966).
4.065 MANFREDI, F. F., and A. RIMINI: NAS-NRC Publ. **1184**, 186 (1964).
4.066 SZAVITS, O.: Nucl. Instr. Methods **39**, 293 (1966).
4.067 COOKE-YARBOROUGH, E. H.: NAS-NRC Publ. **1184**, 169 (1964).
4.068 GERE, E. A., and G. L. MILLER: IEEE Trans. Nucl. Sci. **NS-13**/3, 508 (1966).
4.069 COMISKEY, G. F., R. A. KARLIN, and R. O. CARLSON: IEEE Trans. **NS-12**/1, 325 (1965).
4.070 EMMER, T. L.: IEEE Trans. **NS-12**/1, 329 (1965).
4.071 ARBEL, A. F.: Nuclear Electronics, IAEA **2**, 3 (1962).
4.072 GOURSKI, V., et H. GOUILLON: Electronique Nucléaire, p. 313. Paris: SFER 1963.
4.073 BONSIGNORI, C., D. MALOSTI, and U. PELLEGRINI: Nucl. Instr. Methods **20**, 362 (1963).
4.074 STRAUSS, M. G.: Rev. Sci. Instr. **34**, 335 (1963).
4.075 KANDIAH, K.: NAS-NRC Publ. **1184**, 177 (1964).
4.076 CATZ, PH., and J. MAJÉROWICZ: Nucl. Instr. Methods **45**, 59 (1966).
4.077 CHASE, R. L.: IRE Trans. Nucl. Sci. **NS-9**/3, 275 (1962).
4.078 STANFORD, G. S.: Nucl. Instr. Methods **34**, 1 (1965).
4.079 DUDLEY, R. A., and R. SCARPATETTI: Nucl. Instr. Methods **25**, 297 (1964).
4.080 CHASE, R. L.: IRE Trans. Nucl. Sci. **NS-9**/1, 119 (1962).
4.081 BYINGTON, P., and C. JOHNSTONE: IRE National Convention Record, Part **10**, 204 (1955).
4.082 ARQUE-ALMARAZ, H.: Nucl. Instr. Methods **33**, 61 (1965).
4.083 COSTRELL, L., and R. E. BRUECKMAN: Nuclear Electronics, IAEA **2**, 29 (1962).
4.084 RUMPHORST, R. F., C. DAUM, and L. A. CH. KOERTS: Nuclear Electronics, IAEA **3**, 195 (1962).
4.085 DRAPER, J. E., and W. J. ALSTON: Rev. Sci. Instr. **30**, 805 (1959).
4.086 FRANZ, K., and J. SCHULZ: NAS-NRC Publ. **1184**, 172 (1964).

4.087 LENG, J., and P. K. PATWARDHAN: NAS-NRC Publ. **1184**, 180 (1964).
4.088 COTTINI, C., E. GATTI, and V. SVELTO: Nucl. Instr. Methods **24**, 241 (1963).
4.089 — — — Electronique Nucléaire, p. 309. Paris: SFER 1963.
4.090 GÅSSTROM, R. V.: Nuclear Electronics, IAEA 317 (1959).
4.091 MACMAHON, J. P.: Nuclear Electronics, IAEA 291 (1959).
4.092 PIZER, H. I.: Nucl. Instr. Methods **20**, 358 (1963).
4.093 ALBERIGI-QUARANTA, A., and B. RIGHINI: Nucl. Instr. Methods **20**, 355 (1963).
4.094 OVEN, R. B.: IRE Trans. Nucl. Sci. **NS-9**/3, 285 (1962).
4.095 AMMERLAAN, C. A. J., R. F. RUMPHORST, and L. A. CH. KOERTS: Nucl. Instr. Methods **22**, 189 (1963).
4.096 SCHEER, J. A.: Nucl. Instr. Methods **22**, 45 (1963).
4.097 ALEXANDER, T. K., J. D. PEARSON, A. E. LITHERLAND, and C. BRONDE: Phys. Rev. Letters **13**, 86 (1964).
4.098 MATHIESON, E., et P. W. SANFORD: Electronique Nucléaire, p. 65. Paris: SFER 1963.
4.099 SAYRES, A., and M. COPPOLA: Rev. Sci. Instr. **35**, 431 (1964).
4.100 MENDELL, R. B., and S. A. KORFF: Rev. Sci. Instr. **34**, 1356 (1963).
4.101 CRAWFORD, R. L., and A. ERTEZA: Nucl. Instr. Methods **30**, 303 (1964).
4.102 DVORAK, R. F., and R. W. FERGUS: IEEE Trans. Ncul. Sci. **NS-11**/3, 415 (1964).
4.103 LEGG, J. C.: Nucl. Instr. Methods **36**, 343 (1965).
4.104 BROOKS, F. D.: Nucl. Instr. Methods **4**, 151 (1959).
4.105 SUHAMI, A., and D. OPHIR: Nucl. Instr. Methods **30**, 141 (1964).
4.106 ALEXANDER, T. K., and F. S. GOULDING: Nucl. Instr. Methods **13**, 244 (1961).
4.107 LANDIS, D., and F. S. GOULDING: NAS-NRC Publ. **1184**, 143 (1964).
4.108 PEELE, R. W., and T. A. LOVE: NAS-NRC Publ. **1184**, 146 (1964).
4.109 ROUSH, M. L., M. A. WILSON, and W. F. HORNYAK: Nucl. Instr. Methods **31**, 112 (1964).
4.110 MATHÉ, GY., and B. SCHLENK: Nucl. Instr. Methods **27**, 10 (1964).
4.111 FÜLLE, R., GY. MATHÉ, and D. NETZBAND: Nucl. Instr. Methods **35**, 250 (1965).
4.112 MATHÉ, GY.: Nucl. Instr. Methods **39**, 356 (1966).
4.113 BASS, R., W. KESSEL, and G. MAJONI: Nucl. Instr. Methods **30**, 237 (1964).
4.114 — — Nucl. Instr. Methods **34**, 169 (1965).
4.115 NADAV, E., and B. KAUFMAN: Nucl. Instr. Methods **33**, 289 (1965).
4.116 SCHWEIMER, W.: Nucl. Instr. Methods **39**, 393 (1966).
4.117 VARGA, L.: Nucl. Instr. Methods **14**, 24 (1961).
4.118 GATTI, E., and F. DE MARTINI: Nuclear Electronics, IAEA **2**, 265 (1962).
4.119 TAKAMI, Y., and M. HOSOE: Nucl. Instr. Methods **31**, 347 (1964).
4.120 NIZAN, A. I., and E. ELAD: Nucl. Instr. Methods **47**, 210 (1967).
4.212 — — Nucl. Instr. Methods **51**, 270 (1967).
4.122 CHASE, R. L., and L. R. POULO: IEEE Trans. Nucl. Sci. **NS-14**/1, 83 (1967).
4.123 PATZELT, R.: Nucl. Instr. Methods **59**, 283 (1968).
4.124 WILLIAMS, C. W.: IEEE Trans. Nucl. Sci. **NS-15**/1, 297 (1968).
4.125 GERE, E. A., and G. L. MILLER: IEEE Trans. Nucl. Sci. **NS-14**/1, 89 (1967).
4.126 KLEIN, S. S., L. HULSTMAN, and J. BLOK: Nucl. Instr. Methods, **60**, 88 (1968).
4.127 BERNARD, P., J. CHAMBON, J. MEY, and R. VAN ZURK: Nucl. Instr. Methods **60**, 213 (1968).
4.128 SATTLER, E.: Nucl. Instr. Methods **64**, 221 (1968).
4.129 LEWYN, L. L.: IEEE Trans. Nucl. Sci. **NS-14**/1, 126 (1967).
4.130 WAUGH, J. B. S.: Nucl. Instr. Methods **61**, 121 (1968).
4.131 GEDCKE, D. A., and W. J. MCDONALD: Nucl. Instr. Methods **56**, 148 (1967).
4.132 GRIEDER, P. K. F.: Nucl. Instr. Methods **56**, 229 (1967).
4.133 GEDCKE, D. A., and W. J. MCDONALD: Nucl. Instr. Methods **58**, 253 (1968).
4.134 COMPTON, JR., P. D., and W. A. JOHNSON: IEEE Trans. Nucl. Sci. **NS-14**/1, 116 (1967).

References

4.135 ABBATISTA, N., M. COLI, and V. L. PLANTAMURA: Nucl. Instr. Methods **44**, 29 (1966).
4.136 — — — Nucl. Instr. Methods **43**, 383 (1966).
4.137 NUTT, R.: IEEE Trans. Nucl. Sci. **NS-14**/1, 110 (1967).
4.138 TOVE, P. A., E. PETRUSSON, and Z. H. CHO: Nucl. Instr. Methods **47**, 249 (1967).
4.139 MANTAKAS, CH.: Nucl. Instr. Methods **48**, 179 (1967); Erratum: Nucl. Instr. Methods **51**, 357 (1967).
4.140 GORODETZKY, J., P.-L. WENDEL, CH. RING, and R. ARMBRUSTER: Nucl. Instr. Methods **45**, 72 (1966).
4.141 TURKO, B.: Nucl. Instr. Methods **56**, 261 (1967).
4.142 LYCKLAMA, H., and T. J. KENNETT: Nucl. Instr. Methods **59**, 56 (1968).
4.143 GARDNER, F. M., and R. MCKEETHEN: Nucl. Instr. Methods **46**, 121 (1967).
4.144 STRAUSS, M. G., L. L. SIFTER, F. R. LENKSZUS, and R. BRENNER: IEEE Trans. Nucl. Sci. **NS-15**/3, 518 (1968).
4.145 GREENBLATT, J., K. S. KUCHELA, and N. K. SHERMAN: Nucl. Instr. Methods **49**, 86 (1967).
4.146 SCHUSTER, H.-J.: Nucl. Instr. Methods **63**, 182 (1968).
4.147 HRISOHO, A., Nucl. Instr. Methods **55**, 344 (1967).
4.148 ABBATTISTA, N., M. COLI, and V. L. PLANTAMURA: Nucl. Instr. Methods **59**, 163 (1968).
4.149 ROBINSON, L. B., F. GIN, and F. S. GOULDING: Nucl. Instr. Methods **62**, 237 (1968).
4.150 COLOMBO, G., and L. STANCHI: IEEE Trans. Nucl. Sci. **NS-15**/1, 291 (1968).
4.151 ALBERIGI-QUARANTA, A., B. RIGHINI, and R. VOLTA: Nucl. Instr. Methods **54**, 199 (1967).
4.152 JOHNSON, F. A.: Nucl. Instr. Methods **58**, 134 (1968).
4.153 TAMM, U., W. MICHAELIS, and P. COUSSIEU: Nucl. Instr. Methods **48**, 301 (1967).
4.154 TOJO, A.: Nucl. Instr. Methods **50**, 38 (1967).
4.155 SOUČEK, B., and R. L. CHASE: Nucl. Instr. Methods **50**, 71 (1967).
4.156 SABBAH, B., and A. SUHAMI: Nucl. Instr. Methods **58**, 102 (1968).
4.157 ABE, K., N. KAWAMURA, and N. MUTSURO: Nucl. Instr. Methods **63**, 105 (1968).
4.158 JONES, D. W.: IEEE Trans. Nucl. Sci. **NS-15**/3, 491 (1968).
5.001 BELL, R. E.: Coincidence Techniques and the Measurement of Short Mean Lives. In: Alpha-, Beta- and Gamma-Ray Spectroscopy (ed. K. SIEGBAHN), Vol. 2, p. 905. Amsterdam: North-Holland Publishing Company 1965.
5.002 DE BENEDETTI, S., and R. W. FINDLEY: Handbuch der Physik (ed. S. FLÜGGE), Vol. LXV, p. 222. Berlin-Göttingen-Heidelberg: Springer 1958.
5.003 MIEHE, J. A., E. OSTERTAG and A. COCHE:IEEE Trans. Nucl. Sci. **NS-13**/3, 127 (1966).
5.004 BELL, R. E., H. E. PETCH: Phys. Rev. **76**, 1409 (1949).
5.005 —, R. L. GRAHAM, and H. E. PETCH: Canad. J. Phys. **30**, 35 (1952).
5.006 ASPELUND, O.: Nucl. Instr. Methods **23**, 1 (1963).
5.007 SUGARMAN, R. M., F. MERRITT, and W. HIGINBOTHAM: Nanosecond counter circuit manual, BNL 711 (1962); cf. e.g. R. M. SUGARMAN und F. MERRITT: Experimental performance of high-speed limiters for fast coincidence circuits, BNL Millimicro-Note No. 3 (1960) and Transistor limiters with overload protection, BNL Millimicro-Note No. 4 (1960).
5.008 VERGEZAC, P., and J. KAHANE: Nucl. Instr. Methods **26**, 317 (1964).
5.009 BARNA, A., J. H. MARSHALL, and M. SANDS: Nucl. Instr. Methods **7**, 124 (1960).
5.010 SIDI, M., and U. SOLD: Nucl. Instr. Methods **21**, 89 (1963).
5.011 WHETSTONE, A., and S. KOUNOSU: Rev. Sci. Instr. **33**, 423 (1962).
5.012 BJERKE, A. E., Q. A. KERNS, and T. A. NUNAMAKER: Nucl. Instr. Methods **15**, 249 (1962).
5.013 WIEGAND, C.: Nucl. Instr. Methods **20**, 313 (1963).
5.014 ORMAN, P. R.: Nucl. Instr. Methods **21**, 121 (1963).

References

5.015 GARVEY, J.: Nucl. Instr. Methods **29**, 137 (1964).
5.016 SCHEER, J. A.: Nucl. Instr. Methods **22**, 45 (1963).
5.017 WILLIAMS, C. W., and J. A. BIGGERSTAFF: Nucl. Instr. Methods **25**, 370 (1964).
5.018 BELL, R. E.: Nucl. Instr. Methods **42**, 211 (1966).
5.019 GATTI, E., F. VAGHI, and E. ZAGLIO: Nuclear Electronics, IAEA **3**, 105 (1962).
5.020 —, and U. SVELTO: Nucl. Instr. Methods **39**, 309 (1966).
5.021 SCHWARZSCHILD, A.: Nucl. Instr. Methods **21**, 1 (1963).
5.022 BENOIT, P., C. AUBRET, et J. C. DUMAS: Electronique Nucléaire, p. 747. Paris: SFER 1963.
5.023 STUCKENBERG, H. J.: In: Kernphysikalische Meßverfahren (editor H. NEUERT), p. 408. Karlsruhe: Verlag G. Braun 1966.
5.024 AGOURIDIS, D. C.: Rev. Sci. Instr. **33**, 1396 (1962).
5.025 RAGSDALE, R. H.: Rev. Sci. Instr. **34**, 450 (1963).
5.026 BLAUGRUND, A. E., and Z. VAGER: Nucl. Instr. Methods **29**, 131 (1964).
5.027 EMMER, T. L.: NAS-NRC Publ. **1184**, 112 (1964).
5.028 BOTHE, W.: Z. f. Phys. **59**, 1 (1930).
5.029 FISCHER, J., and J. MARSHALL: Rev. Sci. Instr. **23**, 417 (1952).
5.030 ROSSI, B.: Nature **125**, 636 (1930).
5.031 deBENEDETTI, S., and H. J. RICHINGS: Rev. Sci. Instr. **23**, 37 (1952).
5.032 DE VRIES, H.: Nuclear Electronics IAEA **3**, 59 (1962).
5.033 BRUNNER, W.: Nucl. Instr. Methods **30**, 109 (1964).
5.034 DUMAS, J. C., C. AUBRET, and P. BENOIT: Nucl. Instr. Methods **21**, 323 (1963).
5.035 GOULDING, F. S., and R. A. MCNAUGHT: Nucl. Instr. Methods **8**, 282 (1960).
5.036 GARWIN, R.: Rev. Sci. Instr. **24**, 618 (1953).
5.037 BAKER, S. C.: Nucl. Instr. Methods **12**, 20 (1961).
5.038 BALDINGER, E., P. HUBER, and K. P. MEYER: Rev. Sci. Instr. **19**, 473 (1948).
5.039 BAY, Z.: Phys. Rev. **83**, 242 (1951); Rev. Sci. Instr. **22**, 398 (1951).
5.040 MINTON, G. H.: J. Res. NBS **57**, 119 (1956).
5.041 MEILING, W., J. SCHINTELMEISTER, and F. STARY: Nucl. Instr. Methods **11**, 355 (1961).
5.042 KULLANDER, S.: Nucl. Instr. Methods **24**, 342 (1963).
5.043 NEDDERMEYER, S. H., E. J. ALTHAUS, W. ALLISON, and E. R. SCHATZ: Rev. Sci. Instr. **18**, 488 (1947).
5.044 LEWIS, I. A. D., and F. H. WELLS: Millimicrosecond Pulse Techniques. London: Pergamon Press 1959.
5.045 CHASE, R. L.: Nuclear Pulse Spectrometry. New York: McGraw-Hill 1961.
5.046 BJERKE, A. E., Q. A. KERNS, and T. A. NUNAMAKER: Nucl. Instr. Methods **12**, 25 (1961).
5.047 GATTI, E., and V. SVELTO: Nucleonics **23**/7, 62 (1965).
5.048 BONITZ, M.: Nucl. Instr. Methods **22**, 238 (1963).
5.049 SCHUMANN, R. W.: Rev. Sci. Instr. **27**, 686 (1956).
5.050 OLLIVIER, B., R. POUSSOT, and J. THÉNARD: Nuclear Electronics, IAEA **2**, 81 (1962).
5.051 LUNDY, R. A.: Rev. Sci. Instr. **34**, 146 (1963).
5.052 DURAND, P., et P. GIRAUD: Electronique Nucléaire, p. 643. Paris: SFER 1963.
5.053 WHITTAKER, J. K., et P. CAVANAGH: Electronique Nucléaire, p. 679. Paris: SFER 1963.
5.054 — Nucl. Instr. Methods **28**, 293 (1964).
5.055 IVANOV, A. A.: Instr. Exptl. Techn. (UdSSR, 1966) 113.
5.056 GIANELLI, G.: Nuclear Electronics, IAEA **2**, 49 (1962).
5.057 THÉNARD, J.: Nuclear Electronics, IAEA **2**, 101 (1962).
5.058 —, et G. VICTOR: Electronique Nucléaire, p. 333. Paris: SFER 1963.
5.059 — — Nucl. Instr. Methods **26**, 45 (1964).
5.060 MEYER, H.: Electronique Nucléaire, p. 633. Paris: SFER 1963.

References

5.061 DURAND, P., P. GIRAUD, G. BRUDERMÜLLER, et B. REUTER: Electronique Nucléaire, p. 651. Paris: SFER 1963.
5.062 BELL, R. E., and M. H. JØRGENSEN: Canad. J. Phys. **38**, 652 (1960).
5.063 RODDA, J. L., J. E. GRIFFIN, and M. G. STEWART: Nucl. Instr. Methods **23**, 137 (1963).
5.064 THIEBERGER, P.: Nucl. Instr. Methods **44**, 349 (1966).
5.065 WEBER, W., C. W. JOHNSTONE, and L. CRANBERG: Rev. Sci. Instr. **27**, 166 (1956).
5.066 MEILING, W., J. SCHINTELMEISTER, and F. Stary: Nucl. Instr. Methods **21**, 275 (1963).
5.067 CULLIGAN, G., and N. LIPMAN: Rev. Sci. Instr. **31**, 1209 (1960).
5.068 OPHIR, D.: Nucl. Instr. Methods **28**, 237 (1964).
5.069 BRUN, J. C., R. BRENIL, and C. VICTOR: Nuclear Electronics, IAEA **3**, 99 (1962).
5.070 WIEBER, D. L.: Nucl. Instr. Methods **24**, 269 (1963).
5.071 RODDICK, R. G., and F. J. LYNCH: IEEE Trans. Nucl. Sci. **NS-11**/3, 399 (1964).
5.072 SCHWEIMER, W.: Nucl. Instr. Methods **32**, 190 (1965).
5.073 MAYDAN, D.: Nucl. Instr. Methods **34**, 229 (1965).
5.074 BRAFMAN, H.: Nucl. Instr. Methods **34**, 239 (1965).
5.075 CHRISTIANSEN, J.: Nuclear Electronics, IAEA **3**, 93 (1962).
5.076 BLOESS, D., and F. MÜNNICH: Nucl. Instr. Methods **28**, 286 (1964).
5.077 KOWALSKI, E.: Nucl. Instr. Methods **52**, 357 (1967).
5.078 HENEBRY, W. M., and A. RASIEL: IEEE Trans Nucl. Sci. **NS-13**/2, 64 (1966).
5.079 BALLINI, R., and E. POMELAS: Nucl. Instr. Methods **11**, 331 (1961).
5.080 WEISBERG, H. L., and S. BERKO: IEEE Trans. Nucl. Sci. **NS-11**/3, 406 (1964).
5.081 WEISBERG, H.: Nucl. Instr. Methods **32**, 133 (1965).
5.082 GREEN, R. E., and R. E. BELL: Nucl. Instr. Methods **3**, 127 (1958).
5.083 TINTA, F.: Nuclear Electronics, IAEA **3**, 71 (1962).
5.084 ROGÉRIEUX, M., and P. VERGEZ: Nuclear Electronics, IAEA **3**, 205 (1962).
5.085 GORODETZKY, S., R. RICHERT, R. MANQUENOUILLE, and A. KNIPPER: Nucl. Instr. Methods **7**, 50 (1960).
5.086 JUNG, H.: Nucl. Instr. Methods **24**, 197 (1963).
5.087 SUNYAR, A. W.: Proc. 2nd Conf. Peacef. Atom. En. **14**, 347 (1958).
5.088 JONES, G.: J. Sci. Instr. **37**, 318 (1960).
5.089 VERGEZAC, P., et J. KAHANE: Electronique Nucléaire, p. 629. Paris: SFER 1963.
5.090 GRIN, G. A., and C. JOSEPH: Nucl. Instr. Methods **24**, 331 (1963).
5.091 BELL, J., S. J. TAO, and J. H. GREEN: Nucl. Instr. Methods **35**, 213 (1965).
5.092 —, J. H. GREEN, and S. J. TAO: Nucl. Instr. Methods **35**, 320 (1965).
5.093 SIMMS, P. C.: Rev. Sci. Instr. **32**, 894 (1961).
5.094 JONES, G., and W. R. FALK: Nucl. Instr. Methods **37**, 22 (1965).
5.095 BONITZ, M., and E. J. BERLOVICH: Nucl. Instr. Methods **9**, 13 (1960).
5.096 COTTINI, C., e E. GATTI: Nuovo Cimento **4**, 1550 (1956).
5.097 AMRAM, Y.: Proc. Conference EANDC, Karlsruhe 1964, p. 268.
5.098 DE LOTTO, I., E. GATTI, and F. VAGHI: Proc. Conference EANDC, Karlsruhe 1964, p. 291.
5.099 — — — Alta Frequenza **33**, 839 (1964).
5.100 LEFEVRE, H. W., and J. T. RUSSELL: Rev. Sci. Instr. **30**, 159 (1959).
5.101 CRESSWELL, J., and P. WILDE: Proc. Conference EANDC, Karlsruhe 1964, p. 300.
5.102 GRAHAM, R. L., J. S. GEIGER, R. E. BELL, and R. BARTON: Nucl. Instr. Methods **15**, 40 (1962).
5.103 WEBER, J.: Nucl. Instr. Methods **25**, 285 (1964).
5.104 FALK, W., G. JONES, and R. ORTH: Nucl. Instr. Methods **33**, 345 (1965).
5.105 SEYFRIED, P., and S. DEIKE: Nucl. Instr. Methods **39**, 313 (1966).
5.106 LANGKAU, R.: Nucl. Instr. Methods **45**, 351 (1966).
5.107 IACI, G., and M. LO SAVIO: Nucl. Instr. Methods **65**, 103 (1968).

5.108 VAN ZURK, R.: Nucl. Instr. Methods **46**, 125 (1967).
5.109 MURN, R.: Nucl. Instr. Methods **63**, 233 (1968).
5.110 BERNAOLA, O. A., A. FILEVICH, and P. THIEBERGER: Nucl. Instr. Methods **50**, 299 (1967).
5.111 ABBATISTA, N., V. L. PLANTAMURA, and M. COLI: Nucl. Instr. Methods **49**, 155 (1967).
5.112 GORNI, S., G. HOCHNER, E. NADAV, and H. ZMORA: Nucl. Instr. Methods **53**, 349 (1967).
5.113 MICHAELIS, W.: Nucl. Instr. Methods **61**, 109 (1968.)
5.114 FOUAN, J. P., and. J. P. PASSERIEUX: Nucl. Instr. Methods **62**, 327 (1968).
5.115 SHERMAN, I. S., R. G. RODDICK, and A. J. METZ: IEEE Trans. Nucl. Sci. **NS-15**/3, 500 (1968).
5.116 WAUGH, J. B. S.: IEEE Trans. Nucl. Sci. **NS-15**/3, 509 (1968).
5.117 VAN ZURK, R.: Nucl. Instr. Methods **53**, 45 (1967).
5.118 VERWEIJ, H.: Nucl. Instr. Methods **41**, 181 (1966).
5.119 ABBATTISTA, N., M. COLI, and V. L. PLANTAMURA: Nucl. Instr. Methods **44**, 153 (1966).
5.120 CONRAD, R.: Nucl. Instr. Methods **48**, 229 (1967).
5.121 BUCHER, W. P., and C. E. HOLLANDSWORTH: Rev. Sci. Instr. **38**, 1259 (1967).
5.122 METZ, A. J.: Rev. Sci. Instr. **38**, 1445 (1967).
5.123 RIEDINGER, M., F. SCHMITT, G. METZGER, G. SUTTER, P. CHEVALLIER, and C. WINTER: Nucl. Instr. Methods **47**, 100 (1967).
5.124 MOSZYŃSKI, M.: Nucl. Instr. Methods **56**, 141 (1967).
5.125 BARNA, A., and B. RICHTER: Nucl. Instr. Methods **59**, 141 (1968).
5.126 GRIFFIN, J. E., and W. W. SOUDER: Nucl. Instr. Methods **64**, 85 (1968).
5.127 LEUNG, C. Y., and D. A. L. PAUL: IEEE Trans. Nucl. Sci. **NS-15**/3, 531 (1968).
5.128 STANCHI, L.: IEEE Trans. Nucl. Sci. **NS-15**/1, 315 (1968).
5.129 ROTA, A., G. BERTOLINI, and M. COCCHI: IEEE Trans. Nucl. Sci. **NS-14**/1, 152 (1967).
5.130 POLLY, P.: Nucl. Instr. Methods **49**, 341 (1967).
5.131 WHITTAKER, J. K.: Nucl. Instr. Methods **45**, 138 (1966).
5.132 AVIDA, R., and S. GORNI: Nucl. Instr. Methods **52**, 125 (1967).
5.133 JOHNSON, F. A.: Nucl. Instr. Methods **59**, 237 (1968).
5.134 OGATA, A., S. J. TAO, and J. H. GREEN: Nucl. Instr. Methods **60**, 141 (1968).
5.135 ECKHAUSE, M., R. T. SIEGEL, and R. E. WELSH: Nucl. Instr. Methods **43**, 365 (1966).
5.136 YOUSEFKHANI, A.: Nucl. Instr. Methods **53**, 91 (1967).
5.137 SEN, P., and A. P. PATRO: Nucl. Instr. Methods **59**, 289 (1968).
5.138 HAUSER, V., G. KNISSEL, J. MORITZ, and V. SCHNEIDER: Rev. Sci. Instr. **38**, 1220 (1967).
5.139 DARDINI, C., G. IACI., M. LO SAVIO, and R. VISENTIN: Nucl. Instr. Methods **47**, 233 (1967).
5.140 WHITE, G.: Nucl. Instr. Methods **55**, 157 (1967).
5.141 SEN, P., and A. P. PATRO: Nucl. Instr. Methods **60**, 335 (1968).
5.142 CHO, Z. H., L. GIDEFELDT, and L. ERIKSSON: Nucl. Instr. Methods **52**, 273 (1967); Erratum: Nucl. Instr. Methods **57**, 357 (1967).
5.143 KIESLER, R., and B. RIGHINI: Nucl. Instr. Methods **56**, 357 (1967).
5.144 FRANKE, H. G., and R. FRITZ: Nucl. Instr. Methods **52**, 163 (1967).
5.145 BAKER, C. A., C. J. BATTY., and L. E. WILLIAMS: Nucl. Instr. Methods **59**, 125 (1968).
5.146 NADAV, E., M. PALMAI, and D. SALZMANN: Nucl. Instr. Methods **59**, 173 (1968).
6.001 MILLMAN, J., and H. TAUB: Pulse and Digital Circuits. New York: McGraw-Hill 1956.

References

6.002 PRESSMAN, A. I.: Design of Transistorized Circuits for Digital Computer. New York: John F. Rider Publisher 1959, cf. also german translation: Digitale Schaltungen mit Transistoren. Stuttgart: Berliner Union 1964.
6.003 STAUTON, W. A.: Pulse Technology. New York: John Wiley & Sons 1964.
6.004 RUMPF, K. H., u. M. PULVERS: Transistor Elektronik. Berlin: VEB Verlag Technik 1964.
6.005 WEBER, S.: Modern digital circuits (109 papers from Electronics, 1961 to 1963). New York: McGraw-Hill 1964.
6.006 DEAN, K. J.: Digital Instruments. London: Chapman and Hall Ltd 1965.
6.007 BORUCKI, L., u. J. DITTMAN: Digitale Meßtechnik. Berlin-Heidelberg-New York: Springer 1966.
6.008 BRILLOUIN, L.: Science and Information Theory. New York: Academic Press 1962.
6.009 WIENER, N.: Kybernetik. Düsseldorf: Econ-Verlag GmbH 1963.
6.010 SHANNON, C. E.: Trans. AIEE **57,** 713 (1938).
6.011 WEYH, U.: Elemente der Schaltungsalgebra. München: R. Oldenbourg 1964.
6.012 SPEISER, A. P.: Digitale Rechenanlagen. Berlin-Heidelberg-New York: Springer 1965.
6.013 KHAMBATA, A. J.: Introduction to integrated semiconductor circuits. New York: John Wiley & Sons 1963.
6.014 FOGLESONG, R. L.: The design of high speed all transistor logic circuits. SGS Fairchild Bollettino Applicazioni Semiconduttori, BAS 46, 1963.
6.015 VERWEIJ, H.: Nucl. Instr. Methods **20,** 323 (1963).
6.016 JACKSON, H. G., L. B. ROBINSON, and D. L. WIEBER: Nucl. Instr. Methods **30,** 261 (1964).
6.017 BERGMAN, R. H., and M. COOPERMAN: RCA Review **23,** 152 (1962).
6.018 COOPERMAN, M.: IEEE Trans. El. Comp. **EC-13,** 18 (1964).
6.019 HAZONI, Y.: Nucl. Instr. Methods **13,** 95 (1961).
6.020 DUCHEMIN, J. P.: Electronique Nucléaire, p. 755. Paris: SFER 1963.
6.021 VERROUST, G., and C. VICTOR: Nuclear Electronics, IAEA **3,** 19 (1962).
6.022 RADEKA, V.: Nucl. Instr. Methods **22,** 153 (1963).
6.023 GOTO, E., K. MURATA, and K. NAKAZAWA, et al.: IRE Trans. El. Comp. **EC-9,** 25 (1960).
6.024 LEBAIL, P.: Electronique Nucléaire, p. 769. Paris: SFER 1963.
6.025 SUGARMAN, R., W. A. HIGINBOTHAM, and A. H. YOUDA: Nuclear Electronics, IAEA **3,** 3 (1962).
6.026 BALDINGER, E.: Nucl. Instr. Methods **20,** 309 (1963).
6.027 WEBER, J.: Nucl. Instr. Methods **26,** 325 (1964).
6.028 SPIEGEL, P.: Rev. Sci. Instr. **31,** 754 (1960).
6.029 RABINOVICI, B.: Rev. Sci. Instr. **33,** 1391 (1962).
6.030 STANCHI, L.: Proc. IEEE **54,** 68 (1966).
6.031 FOOTE, R. S., and D. JOHNSON: Rev. Sci. Instr. **35,** 1126 (1964).
6.032 BALDINGER, E., and A. SIMMEN: Nucl. Instr. Methods **33,** 363 (1965).
6.033 MAXWELL, L., and C. MARAZZI: Proc. IEE **113,** 271 (1966).
6.034 ALEXANDER, T. K., and D. R. HEYWOOD: Nucl. Instr. Methods **13,** 83 (1961).
6.035 KUCHELA, K. S.: Nucl. Instr. Methods **16,** 287 (1962).
6.036 BONDÁR, L.: Nucl. Instr. Methods **24,** 280 (1963).
6.037 COOKE-YARBOROUGH, E. H., E. A. SAYLE, and J. P. KERRY: Nucl. Instr. Methods **30,** 106 (1964).
6.038 TARCZY-HORNOCH, M.: Electronic Design 9/2, 34 (1961).
6.039 ENGELMANN, R.: Electronics 36/46, 34 (1963).
6.040 BIRK, M., and H. BRAFMAN et al.: Nuclear Electronics, IAEA **2,** 429 (1962).
6.041 RADEKA, V.: Nuclear Electronics, IAEA **2,** 437 (1962).
6.042 REKER, H.: Nucl. Instr. Methods **29,** 299 (1964).

6.043 VINCENT, C. H.: Nucl. Instr. Methods **29**, 306 (1964).
6.044 AMRAM, Y., H. GUILLON, and J. THÉNARD: Nuclear Electronics, IAEA **2**, (1961).
6.045 BRUN, J. C., P. ANTOINE, G. CORBÉ, J. G. SCHILLER, and C. VICTOR: Nuclear Electronics, IAEA **3**, 267 (1962).
6.046 MCNAUGHT, R., and A. PEARSON: Proc. Conference EANDC, Karlsruhe 1964, p. 409.
6.047 HORSTMANN, H.: Proc. Conference EANDC, Karlsruhe 1964, p. 413.
6.048 BRINI, D., A. GANDOLFI, and G. L. TABELLINI: Nucl. Instr. Methods **8**, 46 (1960).
6.049 OXLEY, A. J.: Nucl. Instr. Methods **26**, 77 (1964).
6.050 ROWLES, J. B., R. A. W. STEELS, and C. H. VINCENZ: Nucl. Instr. Methods **27**, 129 (1964).
6.051 COULEUR, J. F.: IRE Trans. El. Comp. **EC-7**/4 (1958).
6.052 MEYERHOFF, A. J.: Digital Applications of Magnetic Devices. New York: John Wiley & Sons 1960.
6.053 QUARTLY, C. J.: Square-Loop Ferrite Circuitry. London: Iliffe Books Ltd. 1962.
6.054 STEINBUCH, K.: Taschenbuch der Nachrichtenverarbeitung. Berlin-Göttingen-Heidelberg: Springer 1962.
6.055 ALEXANDRE, B., G. ANTIER, et G. GRUNBERG: Electronique Nucléaire, p. 657. Paris: SFER 1963.
6.056 EMMER, T. L.: IEEE Trans. Nucl. Sci. **NS-12**/1, 329 (1965).
6.057 DAKIN, C. J., and C. E. G. COOKE: Circuits for digital equipment, p. 362. London: Iliffe Books Ltd. 1967.
6.058 ELMORE, W. C., and M. SANDS: Electronics. New York: McGraw-Hill 1949.
6.059 GIANNELLI, G., and V. MANDL: Rev. Sci. Instr. **31**, 623 (1960).
6.060 MARVIN, J. F., W. D. MILLER, and M. K. LOKEN: Rev. Sci. Instr. **31**, 1238 (1960).
6.061 THOMAS, S.: IEEE Trans. Nucl. Sci. **NS-10**/1, 36 (1963).
6.062 COOKE-YARBOROUGH, E. H., and E. W. PULSFORD: Proc. IRE **98**/II, 196, (1951).
6.063 VINCENT, C. H., and J. B. ROWLES: Nucl. Instr. Methods **22**, 201 (1963).
6.064 — —, and R. A. W. STEELS: Nucl. Instr. Methods **26**, 221 (1964).
6.065 WERNER, M.: Nucl. Instr. Methods **34**, 103 (1965).
6.066 KOSTIC, V. N., and B. J. KOVAC: Nuclear Electronics, IAEA **2**, 445 (1962).
6.067 ZADICARIO, J., J. GRÜNBERG, and U. SOLD: Nucl. Instr. Methods **33**, 238 (1965).
6.068 DAKIN, C. J., and C. E. G. COOKE: Circuits for digital equipment. London: Iliffe Books Ltd. 1967.
6.069 TAN, Z. C.: Nucl. Instr. Methods **63**, 333 (1968).
6.070 MURATA, Y.: Nucl. Instr. Methods **64**, 349 (1968).
6.071 BALDINGER, E., and A. SIMMEN: Nucl. Instr. Methods **57**, 141 (1967).
6.072 TAN, Z. C.: Rev. Sci. Instr. **38**, 1415 (1967).
6.073 —, and P. C. MAXWELL: Nucl. Instr. Methods **63**, 230 (1968).
6.074 — — Nucl. Instr. Methods **53**, 133 (1967), Erratum: Nucl. Instr. Methods **57**, 357 (1967).
6.075 FRANZ, H. W., D. FICK, and E. FLÜGEL: Nucl. Instr. Methods **46**, 106 (1967).
6.076 SEVERUS, R. P., T. F. TURNER, and A. R. KOELLE: IEEE Trans. Nucl. Sci. **NS-14**/3, 1078 (1967).
6.077 BUDGE, E. C.: IEEE Trans. Nucl. Sci. **NS-14**/3, 1074 (1967).
6.078 POLYCHRONAKIS, G., and G. PHILOKYPROU: Nucl. Instr. Methods **44**, 90 (1966).
6.079 VINCENT, C. H.: Nucl. Instr. Methods **47**, 157 (1967).
6.080 TOLMIE, R. W., and Q. BRISTOW: IEEE Trans. Nucl. Sci. **NS-14**/1, 158 (1967).
7.001 PECINA, R. J., and W. WEIDEMANN: Nucl. Instr. Methods **27**, 285 (1964).
7.002 ISELIN, F.: Nucl. Instr. Methods **20**, 330 (1963).
7.003 BARNA, A., and D. HORELICK: Nucl. Instr. Methods **35**, 341 (1965).
7.004 JACKSON, H. G., L. B. ROBINSON, and D. L. WIEBER: Nucl. Instr. Methods **30**, 261 (1964).

References

7.005 McGinnis, G. A., R. J. Pecina, and L. L. Prucha: Nucl. Instr. Methods **36**, 255 (1965).
7.006 Hutchinson, G. W., and G. G. Scarrott: Phil. Mag. **42**, 792 (1951).
7.007 Byington, P. W., and C. W. Johnstone: IRE Conv. Record **3**, 204 (1955).
7.008 Schumann, R. W., and J. P. McMahon: Rec. Sci. Instr. **27**, 675 (1956).
7.009 Emmer, T. L.: IEEE Trans. **NS-12**/1, 329 (1965).
7.010 Chase, R. L.: Nuclear Pulse Spectrometry. New York: McGraw-Hill 1961.
7.011 — IRE Trans. **NS-9**/3, 275 (1962).
7.012 Goulding, F. S.: In: Alpha-, Beta- and Gamma-Ray Spectroscopy (ed. K. Siegbahn), Vol. 1, p. 413. Amsterdam: North Holland Publishing Company 1965.
7.013 Standord, G. S.: Nucl. Instr. Methods **34**, 1 (1965).
7.014 Guillon, H.: Nucl. Instr. Methods **43**, 240 (1966).
7.015 Speiser, A. P.: Digitale Rechenanlagen. Berlin-Heidelberg-New York: Springer 1965.
7.016 Lynch, F. J., and J. B. Baumgartner: Proc. 2nd Int. Conf. Mössbauer Effect, p. 54. New York: John Wiley & Sons 1962.
7.017 Chase, R. L.: IRE Natl. Conv. Record **Pt. 9**, 196 (1959).
7.018 Alexander, T. K., and L. B. Robinson: Nucleonics **20**/5, 70 (1962) and Nuclear Electronics, IAEA **2**, 173 (1962).
7.019 Litherland, A. E., and D. A. Bromley: Nucl. Instr. Methods **6**, 176 (1960).
7.020 Strauss, M. G.: Nucl. Instr. Methods **29**, 69 (1964).
7.021 Norbeck, E.: Proc. Conference EANDC, Karlsruhe 1964, p. 366.
7.022 Birk, M., T. Braid, and R. Detenbeck: Rev. Sci. Instr. **29**, 203 (1958).
7.023 Rockwood, C. C., and M. G. Strauss: Rev. Sci. Instr. **32**, 1211 (1961).
7.024 Wells, F. H., I. N. Hooton, and J. G. Page: J. Brit. Instr. Rad. Engrs. **20**, 749 (1960).
7.025 Amram, Y.: Nuclear Electronics, IAEA **2**, 73 (1962).
7.026 Pagès, A.: Nuclear Electronics, IAEA **2**, 185 (1962).
7.027 Zhukov, G. P., G. I. Zabyakin, et V. D. Shibayev: Electronique Nucléaire, p. 575. Paris: SFER 1963.
7.028 Spinrad, R. J.: Data Systems for Multiparameter Analysis. Ann. Rev. Nucl. Sci. **14**, 239 (1964).
7.029 — Digital Systems for Data Handling. In: Progress in Nuclear Techniques and Instrumentations (ed. F. J. M. Farley), Vol. 1, p. 221. Amsterdam: North-Holland Publishing Company 1965.
7.030 Egelstaff, P. A., and E. R. Rae: Proc. Conference EANDC, Karlsruhe 1964, p. 83.
7.031 Hooton, I. N.: Proc. Conference EANDC, Karlsruhe 1964, p. 338.
7.032 Souček, B.: Rev. Sci. Instr. **36**, 750 (1965).
7.033 —, Nucl. Instr. Methods **36**, 181 (1965).
7.034 —, IEEE Trans. Nucl. Sci. **NS-13**/1, 183 (1966).
7.035 Bianchi, G., C. R. Corge, and J. P. Meinadier: Proc. Conference EANDC, Karlsruhe 1964, p. 174.
7.036 Kandiah, K.: NAS-NRC Publ. **1184**, 322 (1964).
7.037 O'Kelley, G. D: Proc. Conference EANDC, Karlsruhe 1964, p. 385.
7.038 Dimmler, G., and G. Krüger: Proc. Conference EANDC, Karlsruhe 1964, p. 393.
7.039 Whalen, J. F.: Proc. Conference EANDC, Karlsruhe 1964, p. 399.
7.040 Mack, D. A.: NAS-NRC Publ. **1184**, 320 (1964).
7.041 Heath, R. L.: NAS-NRC Publ. **1184**, 313 (1964).
7.042 Lindenbaum, S. J.: On-Line Computer Techniques in Nuclear Research. Ann. Rev. Nucl. Sci. **16**, 619 (1966).
7.043 Foley, K. J., S. J. Lindenbaum, W. A. Love, S. Ozaki, J. J. Russell, and L. C. L. Yuan: Phys. Rev. Letters **10**, 376, 543 (1963); **11**, 425, 503 (1963); Nucl. Instr. Methods **30**, 45 (1964).
7.044 Jones, J. A.: Nucleonics **25**/1, 34 (1967).
7.045 Butler, J. W., and M. K. Butler: Nucleonics **25**/2, 44 (1967).

7.046 BROWN, R. M., and E. MUELLER: Nucleonics **25**/3, 48 (1967).
7.047 COLOMBO, G., and L. STANCHI: Nucl. Instr. Methods **42**, 104 (1966).
7.048 LÖVBORG, L.: Nucl. Instr. Methods **54**, 137 (1967).
7.049 SCHIAVUTA, E, and F. SOSO: Nucl. Instr. Methods **60**, 36 (1968).
7.050 TAKABA, S.: Nucl. Instr. Methods **58**, 223 (1968).
7.051 HARMS, J.: Nucl. Instr. Methods **53**, 192 (1967).
7.052 EULER, B.: Nucl. Instr. Methods **61**, 311 (1968).
7.053 FULLWOOD, R. R.: Nucl. Instr. Methods **50**, 261 (1967).
7.054 CASTRICA, F.: Nucl. Instr. Methods **48**, 157 (1967).
7.055 WAUGH, J. B. S.: Nucl. Instr. Methods **58**, 293 (1968).
7.056 STEGER, W. W.: IEEE Trans. Nucl. Sci. **NS-14**/1, 655 (1967).
7.057 GEMMELL, S. D.: Nucl. Instr. Methods **46**, 1 (1967).
7.058 JONES, J. A.: IEEE Trans. Nucl. Sci. **NS-14**/1, 576 (1967).
7.059 LIDOFSKY, L. J.: IEEE Trans. Nucl. Sci. **NS-15**/1, 93 (1968).
8.101 WAGNER, K. W.: Operatorentechnik nebst Anwendungen in Physik und Technik. Leipzig: Joh. Ambr. Barth 1940.
8.102 HENNYEY, Z.: Linear Electric Circuits. Oxford: Pergamon Press 1963.
8.103 CARSLAW, H. S., and J. C. JAEGER: Operational Methods in Applied Mathematics. Oxford: University Press 1948.
8.104 DOETSCH, G.: Handbuch der Laplace-Transformation, Volume I. Basel: Birkhäuser Verlag 1950.
8.105 — Handbuch der Laplace-Transformation, Volume II. Basel: Birkhäuser Verlag 1955.
8.106 — Handbuch der Laplace-Transformation, Volume III. Basel: Birkhäuser Verlag 1966.
8.107 — Anleitung zum praktischen Gebrauch der Laplace-Transformation. München: Oldenbourg 1956.
8.108 STIEFEL, E., u. J. NIEVERGELT: Fourier- und Laplace-Transformationen. Zürich: Verein der Mathematiker und Physiker an der ETH 1961.
8.109 SCHLEGEL, H. R., u. A. NOWAK: Impulstechnik. Prien/Chiemsee: C. F. Wintersche Verlagsbuchhandlung 1961.
8.110 THOMASON, J. G.: Linear Feedback Analysis. London: Pergamon Press 1955.
8.111 LEWIS, I. A. D., and F. H. WELLS: Millimicrosecond Pulse Techniques. London: Pergamon Press 1959.
8.201 GILLESPIE, A. B.: Signal, Noise and Resolution in Nuclear Counter Amplifiers. London: Pergamon Press 1953.
8.202 BALDINGER, E., and W. FRANZEN: Amplitude and Time Measurement in Nuclear Physics. In: Advances in Electronics and Electron Physics **8**, 255 (1956).
8.203 TSUKUDA, M.: Nucl. Instr. Methods **14**, 241 (1961).
8.204 DEARNALEY, G., and A. B. WHITEHEAD: Nucl. Instr. Methods **12**, 205 (1961).
8.205 THOMASON, J. G.: Linear Feedback Analysis. London: Pergamon Press 1955.
8.206 BILGER, H. R.: Nucl. Instr. Methods **40**, 54 (1966).
8.207 BLANKENSHIP, J. L.: IEEE Trans. Nucl. Sci. **NS-11**/3, 373 (1964).
8.208 FAIRSTEIN, E.: IRE Trans. Nucl. Sci. **NS-8**/1, 129 (1961).
8.209 VAN DER ZIEL, A.: Proc. IEEE **51**, 461, (1963); **51**, 1670 (1963); **50**, 1808 (1962).
8.210 RADEKA, V.: Field-Effect Transistors in Charge-Sensitive Preamplifiers. NAS-NRC Publ. **1184**, (1964).
8.211 —, IEEE Trans. Nucl. Sci. **NS-11**/3, 358 (1964).
8.212 BLALOCK, T. V.: IEEE Trans. Nucl. Sci. **NS-11**/3, 365 (1964).
8.213 RADEKA, V.: Nucleonics **23**/7, 52 (1965).
8.214 HAHN, J., and R. MAYER: IRE Trans. Nucl. Sci. **NS-9**/4, 20 (1962).
8.215 ARBEL, A. F.: IEEE Trans. Nucl. Sci. **NS-15**/4, 2 (1968).
8.216 HATCH, K.: IEEE Trans. Nucl. Sci. **NS-15**/1, 303 (1968).

10. Subject Index

Accumulator 312, 343
Active filter amplifiers 95
Added-step technique 175, 178, 203
Adder, operational 120
Addition, digital 310
AIKEN code 275, 313
Amplifiers, active filter pulse shaping 95
—, charge sensitive 11, 14, 38
—, — —, noise in 368
—, current 77
—, distributed 143
—, fast 142
—, linear 75
—, operational 117
—, stability 83
—, variable-gain 73, 115
—, voltage 76
—, window type 128
Amplitude balance 202
Analog circuits 75–150
— computers 117
— representation, definition of 268
— signals, arithmetic operations on 117
— -to-digital converters 151–212
Analyzer, multi-channel 198, 332
—, single-channel 169, 189
AND gate 271
Anticoincidence 170, 231
ANTILOG-converter, operational 120, 122
Arithmetical analog circuits 117
— digital circuits 308
Associative memory 339

Baseline, fluctuation of 107
— restorer 162
BCD codes (binary coded decimal) 275, 322
BELL, GRAHAM and PETCH coincidence circuit 233
Bidirectional linear gates 135
— scaler 302

Binary scaler 300
Bipolar pulse 87, 99
— —, baseline restorer for 163
— —, linear gates for 135
Biquinary code 275, 295, 308
Bit 269
Bitripole 295
Boolean algebra 269
Bootstrap feedback 41, 325
BOTHE coincidence circuit 235
Buffer register 199

Cable differentiator 64, 95, 102
Campbell's theorem 106
Cascode 13, 18, 40, 46
Čerenkov counters 74
Chronotron 240
Clamping diode 162
CML logic 286, 291, 308
Coincidence circuits 227–243
Comparison of digital-encoded numbers 308
Computer 342
Conjunction 271
Count rate meters 324
CULLIGAN's LC-converter 252
Current hogging 280
— mode operation 91
— -to-voltage converter 77, 81
Cusp pulse 103

Dark current 56
Darlington amplifier 18
Data output 320
— transmission 310
DCTL logic 278
Dead time, in GM counters 26
Decimal scaler 300
— -to-dual conversion 316
Delay line differentiator 64, 95, 102, 222
Delayed coincidence measurements, analysis of 53

399

Subject Index

Delaying of digital signals 276
Delta function $\delta(t)$ 348
DE MORGAN theorem 274
Depletion layer 28
Derandomizer 200, 255
Descriptor storage 338
Difference scaler 302
Differential coincidence circuit 240
Differentiation of digital signals 276
Differentiator, operational 120
Diffusion voltage 29
Digital circuits 268–326
— count rate meter 326
— encoding of pulse height 191
— — — time interval 243
— gates 269
— period meter 326
— representation, definition of 268
— -to-analog converters 322
Diode clamp 162
— pump 194, 324
Discriminators, added-step technique 175
—, differential 169, 189
—, integral 158
—, multi-channel 177
—, pulse height 152–191
—, pulse shape 205
—, tunnel diode 183
Disjunction 271
Disjunctive normal form 272
Display of data, optical 306, 340
DTL logic 282
Dual-to-decimal conversion 314
Dual scaler 300
Dynamic scaler 305

ECTL logic 286
Electrometer amplifiers 3
Electronics, definition 1
Encoding matrix 307
Energy loss, Fano factor 5
— —, Landau theory 5
— —, statistics of 5
— resolution 56, 373
— W per one charge-carrier pair 5, 31
— W per one photo cathode electron 50
Equivalent noise charge 364

Fano factor 16, 32
— —, definition 5
Fan-out 221, 226
Fast-slow coincidence systems 68, 216, 224, 247

Feedback 78
—, parallel 80, 117
—, positive 152
—, shunt 80
Ferrite core memories 317
Flicker noise 366
Flip-flop 164, 286–296
— —, tunnel diode type 292
Fluorescence decay time 51, 52
Frequency response 77
Function generators, using diodes 125
FWHM 374

GARWIN coincidence circuit 238, 258
Gated oscillator 198
Gates, digital 269
—, linear 131
Geiger-Müller-Counter 20
— —, halogen 22
— —, non-self-quenching 20
— —, self-quenching 21
Germanium 32
GOTO twin-TD flip-flop 294

Half-adder 311
Heaviside operational calculus 350
High tension power supply 66
Hodoscope 345
Hutchinson-Scarrott multichannel analyzer 333
Hysteresis 153, 184, 318

Indication of the scaler state 306
Information amount, definition of 269
Integrator, operational 120
Inverter, digital 271
—, operational 120
Ionization chamber 4–14
— —, gridded 10, 12, 13

JK flip-flop 289

Kandiah discriminator 168
Kicksorter 191

Laplace transformation 39, 347–363
— —, rules of 354
Leading edge timing 207, 224
Leakage current of a semiconductor detector 30
Light pen 342
— pulse generators, fast 52, 74

400

Limiters   113, 218
Linear gates   131
Linearity   76, 83, 127
Lithium drifted detectors   30
LLL logic   283
LOG-converter, operational   120, 122
Logical digital circuits   308
Long-tailed pair   44

Memories   316
Miller integrator   325
Mixer   226, 229
Mobility of the charge-carriers   6, 34
Moebius-strip scaler   300
Mössbauer experiment   336
Multi-channel analyzer   198, 332
—-— discriminator   177
— -parameter analyzer   336
Multiplication, operational   127
— factor   14, 20
— process, statistics of   16
Multiscaler arrays   329
Multistop time-to-pulse-height converter   255
Multivibrator, bistable   164
—, monostable   165, 185
—, principle   152
—, tunnel diode   183

NAND gate   271
Negation, logical   271
Network analysis   347
— synthesis   98
Noise   43, 103, 364–375
—, measurement of   48
—, parallel   367
—, pick-up of   18, 114
—, series   367
—, suppression of   58, 59
—, thermal   56, 365
Nonlinearity, differential   83, 106, 195, 201, 202
—, integral   83
NOR gate   271
Normal distribution   373

On-line computer   342
Operational calculus   350
OR gate   271
Overlap converter   246, 256
Overload recovery   98, 111

Parallel representation of digital signals   276, 313
— -to-series converter   298
Parametric diode   45
Particle identification   127
Photomultiplier   49
—, electron propagation time of   51
—, gated operation of   64
—, statistics of the multiplication process   17, 55–56
Pile-up effect   61, 92, 106
pin-Counter   29
pn-Counter   28
Pole-zero compensation techniques   97
POLYA distribution   17
Preamplifiers, for ionization chamber   11
—, for proportional counters   17
—, for scintillation counters   60
—, for semiconductor detectors   38
Preset count   328
— time   328
Proportional counter   14–19
Pulse shape, in GM counters   20
— —, — ionization chambers   6
— —, — proportional counters   14
— —, — scintillation counters   51
— —, — semiconductor detectors   34
— — discriminator   38, 53, 205
— shapers   217
— shaping   80, 90–106
— —, active filter   95
— —, delay line differentiator   95, 102, 222
— —, double differentiator   99
— —, pole-zero compensation techniques   97
— —, single differentiator   92
— stretchers   137, 195

Quenching circuits for GM counters   24

Radiation detectors   4–74
Raman frequency   31
Random access memories   316
Rate meters   324
RCTL logic   280
Reactor instrumentation   3
Read amplifier   319
— -out of the scaler state   306, 320, 330, 340
Real-time operation   344
Recovery time, in GM counters   26
Regeneration of digital signals   276

401

Subject Index

Reserve gain   40
Resolution, energy   56, 373
—, time   213, 225
—, —, measurement of   264
Ring scalers   299
Rise time, in multistage amplifiers   89, 144
Root-locus method   83
Rossi coincidence circuit   236
RS flip-flop   287
RTL logic   280, 290, 307

Scale-of-two   287
Scalers   299–308, 328
Scaler-timer assembly   328
— with feedback   300
Schmitt trigger   154
Scintillation counters   48–74
Semiconductor detectors   27–48
Series representation of digital signals   276
Shannon theorem   274
Shift registers   296
Shot noise   366
Silicon   32
Single channel analyzer   169, 189
Sliding scale principle   204
Source strength measurement   214
Stabilizers for scintillation counters   70
Start-stop converter   246, 250
Step function $H(t)$ 348
Successive binary approximation type converter   202
Sum effects   106
— pulses, elimination of   110
Superposition principle   347
Synchronous scaler   304

T flip-flop   287
Thermal noise   56
Time information, conservation in a discriminator   179
— —, evaluation of   213–267
— —, in GM counter   23
— —, — ionization chamber   10
— —, — proportional counter   16
— -of-flight method   213, 243
Timer   328
Time-to-pulse-height converter   246
Timing, leading edge   207, 224
—, zero crossover   180, 190, 207, 221, 224
Transfer function   361, 366
Transient response   77, 79, 84
Transitron   44
Transmission of data   310
Triangle pulse   104
TTL logic   284
Tunnel diode circuits   45, 183, 221, 234, 292
Twisted-ring scaler   300

Variable-gain attenuators   73, 115
Vernier chronotron   263
Vernier principle   261
Virtual ground   118
Voltage mode operation   91
— -to-current converter   77

Walking   180, 217, 247
—, correction networks allowing for   248
Wilkinson converter   192, 262
Window amplifiers   128, 166, 188

Zero crossover timing   170, 190, 207, 221, 224

QC
787
C6
.K65
1970

ASHEVILLE-BUNCOMBE TECHNICAL COLLEGE

3 3312 00010 3150

74-1329

Kowalski, Emil
Nuclear electronics

DISCARDED

JUN 1 8 2025

Asheville-Buncombe Technical Institute
LIBRARY
340 Victoria Road
Asheville, North Carolina 28801